Natural Antioxidants and Aquatic Animal Health

Natural Antioxidants and Aquatic Animal Health

Guest Editor

Rui Jia

MDPI

Basel • Beijing • Wuhan • Barcelona • Belgrade • Novi Sad • Cluj • Manchester

Guest Editor
Rui Jia
Freshwater Fisheries Research Center
Chinese Academy of Fishery Sciences
Wuxi
China

Editorial Office
MDPI AG
Grosspeteranlage 5
4052 Basel, Switzerland

This is a reprint of the Special Issue, published open access by the journal *Antioxidants* (ISSN 2076-3921), freely accessible at: www.mdpi.com/journal/antioxidants/special_issues/2M8Q1M6W60.

For citation purposes, cite each article independently as indicated on the article page online and using the guide below:

Lastname, A.A.; Lastname, B.B. Article Title. *Journal Name* **Year**, *Volume Number*, Page Range.

ISBN 978-3-7258-3314-6 (Hbk)
ISBN 978-3-7258-3313-9 (PDF)
https://doi.org/10.3390/books978-3-7258-3313-9

Contents

About the Editor

Rui Jia

Dr. Rui Jia, Associate Professor at the Freshwater Fisheries Research Center of the Chinese Academy of Fishery Sciences, specializes in fish physiology and healthy aquaculture. With a distinguished academic career, Dr. Jia has authored or co-authored 41 peer-reviewed papers. His research has received substantial support from numerous grants, including those from the National Natural Science Foundation of China and the Jiangsu Provincial Natural Science Foundation.

Preface

As the demand for aquatic products continues to rise, ensuring the health and well-being of aquatic animals has become increasingly critical. In aquaculture systems, farmed animals are inevitably exposed to a variety of environmental stressors, which can lead to oxidative stress. Prolonged or intense oxidative stress poses significant threats, including inhibited growth and development, suppressed immune function, and increased susceptibility to tissue damage and disease outbreaks. To mitigate these adverse effects, natural antioxidants derived from medicinal plants have emerged as powerful and sustainable tools in aquaculture.

Over the past few decades, extensive research has been conducted to explore the effects of natural antioxidants on the health of aquatic animals. Despite significant progress, several key aspects remain to be elucidated. These include the primary molecular mechanisms and pathways underlying the beneficial effects of natural antioxidants, as well as their absorption, metabolism, and interactions with intestinal microbiota in aquatic species. To address these gaps and advance the field, this Special Issue is dedicated to investigating the relationship between natural antioxidants and aquatic animal health.

This Special Issue highlights outstanding research efforts focused on the beneficial effects of antioxidants as dietary additives in aquaculture. It comprises 12 papers covering a range of topics, primarily exploring how antioxidants improve the health of aquatic animals, mitigate oxidative stress induced by external stressors, and elucidate the mechanisms of antioxidant defense systems in helping aquatic species cope with adverse environmental conditions. These studies provide a robust theoretical foundation for the application of natural antioxidants in aquaculture practices.

As the Guest Editor, I extend my deepest gratitude to the authors, reviewers, and editorial team, whose dedication and expertise have been instrumental in the development of this Special Issue. Their collective efforts have ensured the high quality and scientific rigor of the contributions included herein. I would also like to offer special thanks to the Managing Editor, Mr. Terrance Yu, for his steadfast support and guidance throughout the process.

Rui Jia
Guest Editor

antioxidants

Editorial

Natural Antioxidants and Aquatic Animal Health

Rui Jia

Key Laboratory of Integrated Rice-Fish Farming Ecology, Ministry of Agriculture and Rural Affairs, Freshwater Fisheries Research Center, Chinese Academy of Fishery Sciences, Wuxi 214081, China; jiar@ffrc.cn

1. Introduction

In intensive aquaculture, aquatic animals inevitably encounter multiple stressors due to environmental fluctuations, triggering various stress responses, with oxidative stress being the most prevalent [1,2]. Numerous factors within the aquatic environment, including stocking density, temperature, salinity, heavy metals, bacteria, and viruses, can induce the production of reactive oxygen species (ROS), leading to oxidative stress [3–5]. Excessive ROS can trigger lipid peroxidation, protein misfolding, and DNA damage in aquatic animals, leading to tissue damage and impaired physiological functions [6]. Prolonged oxidative stress can decrease growth performance, compromise the immune system, and potentially lead to mortality [7]. Furthermore, the development and progression of various diseases in aquatic animals have been conclusively linked to oxidative stress [8]. To mitigate the adverse effects induced by stressors, promising research results have shown that incorporating medicinal plants and their extracts into diets is an effective and eco-friendly strategy [9]. Many medicinal plants, rich in active compounds such as polysaccharides, alkaloids, tannins, saponins, glycosides, flavonoids, steroids, and essential oils, have shown significant antioxidant properties [10]. However, the relationship between antioxidants and the health of aquatic animals still necessitates extensive research, as the primary molecular mechanisms and pathways underlying these beneficial effects remain largely unclear.

The Special Issue "Natural Antioxidants and Aquatic Animal Health" is designed to elucidate the critical role of natural antioxidants in enhancing the health and well-being of aquatic species. It provides insights into the mechanisms through which aquatic animals respond to oxidative stress and how dietary antioxidants impact their health. Here, we present an overview of the Special Issue, which encompasses 12 original articles.

check for updates

Received: 24 January 2025
Accepted: 27 January 2025
Published: 5 February 2025

Citation: Jia, R. Natural Antioxidants and Aquatic Animal Health. *Antioxidants* **2025**, *14*, 185. https://doi.org/10.3390/antiox14020185

2. Overview of Published Articles

Antioxidants, used as dietary additives in aquatic animals, have attracted considerable attention due to their multifaceted roles, such as promoting growth, enhancing antioxidant status, and improving metabolic functions [11]. Jia et al. [Contribution 1] discovered that dietary proanthocyanidins, a natural antioxidant, enhanced muscle nutrients by increasing antioxidant capacity and the levels of polyunsaturated fatty acids (n-3 and n-6) in *Cyprinus carpio*. Furthermore, their research revealed that proanthocyanidins modified intestinal functions through the upregulation of the sphingolipid catabolic process and the lysosomal pathway. This was coupled with a reduction in intestinal cholesterol absorption and an increase in the diversity of the intestinal microbiota. Interestingly, Cheng et al. [Contribution 2] documented that microbial antioxidants, novel compounds fermented by probiotics, enhanced growth, boosted levels of antioxidant enzymes (T-SOD and Gpx), and, as a result, improved survival rates of *Procambarus clarkii*. The work from Song et al. [Contribution 3] focused on the effects dietary lysophosphatidylcholine (LPC) on uptake of astaxanthin, an efficient antioxidant in shrimp feeds. The results showed that LPC supplementation could facilitate the deposition of dietary astaxanthin into farmed shrimp, thereby increasing its beneficial effects. Also, LPC could independently influence the regulation of body color and cholesterol transport in *Litopenaeus vannamei.*

It is noteworthy that various studies have focused on the use of antioxidants to alleviate adverse stimuli and improve health in aquatic animals. In the Gibel carp (*Carassius auratus gibelio*) [Contribution 4], two prevalent antioxidants, taurine and vitamin C, found extensively in various medicinal plants, have been shown to improve growth, antioxidant capacity, immunity, and hypoxia tolerance. Similarly, berberine, a natural alkaloid prevalent in various medicinal plants and known for its antioxidative properties, has shown protective effects in *Oreochromis niloticus* [Contribution 5]. Here, berberine protected against liver damage induced by a high-fat diet (HFD) through the regulation of lipid metabolism, enhancement of antioxidant status, and modulation of immune responses, likely mediated by the Nrf2, TLR2/MyD88/NF-κB, and PPARα signaling pathways. An in vitro study demonstrated that NAD+ precursors mitigate hepatocyte damage caused by high glucose levels in *Megalobrama amblycephala* [Contribution 6]. This protective effect is facilitated through the activation of Sirt1/3, enhancement of redox defense, inhibition of inflammatory responses and apoptosis, and modulation of glucose metabolism. Jiang et al. [Contribution 7] found that dietary supplementation with phosphatidylserine (PS) mitigated stress responses, redox imbalances, and immunosuppression induced by high stocking density in *M. amblycephala*. They noted that the Nrf2 pathway plays a critical role in the beneficial effects of PS.

It has been reported that the antioxidant defense system plays a crucial role in how aquatic animals respond to adverse stimuli [12]. Jin et al. [Contribution 8] found that alkalinity exposure led to a rising trend in superoxide dismutase (SOD) activity and also impacted the oxidative phosphorylation signaling pathway in *Macrobrachium nipponense*. The study by Xue et al. [Contribution 9] indicated that long-term high-fat diet feeding in *Aplodinotus grunniens* induced oxidative stress, which suppressed antioxidant capacity and increased both apoptosis and autophagy in gut cells. He et al. [Contribution 10] evaluated the response mechanisms of *Eriocheir sinensis* to oxidative stress triggered by hydrogen peroxide (H_2O_2), a prevalent oxidant. Their findings revealed that low concentrations of H_2O_2 activated the antioxidant system, enhancing the organism's ability to cope with adverse stimuli. Conversely, high concentrations of H_2O_2 decreased the antioxidant capacity, suggesting that while mild oxidative stress activates the antioxidant system as a defense mechanism, excessive oxidative stress overwhelms this system and accelerates cellular damage. A study from Liang et al. [Contribution 11] suggested that inadequate nutrition has been linked to decreased levels of key antioxidant enzymes—superoxide SOD, catalase (CAT), and glutathione peroxidase (GPx)—in *Micropterus salmoides*, potentially due to the inhibition of the Nrf2 pathway. Additionally, antioxidative parameters were employed to assess and optimize anesthetic practices for mud crabs (*Scylla paramamosain*), and it was found that clove oil is a safe and optimal anesthetic agent for this species, as it exerts no physiological stress [Contribution 12].

3. Conclusions

These research articles in this Special Issue highlight the beneficial effects of antioxidants as dietary additives and provide a theoretical basis for their application in aquaculture. Additionally, these studies offer new insights into the role of antioxidant defense systems in mitigating adverse stimuli in aquatic animals. However, the relationship between natural antioxidants and aquatic animals is rendered complex due to the diversity of natural antioxidants and variations among experimental animals. This complexity underscores the necessity for continued research in this field. We extend our gratitude to all the contributors for their innovative research. We hope that this Special Issue will inspire further scientific exploration into the effects of natural antioxidants on the health of aquatic animals, advancing our knowledge in this field.

Funding: This research received no external funding.

Conflicts of Interest: The authors declare no conflicts of interest.

List of Contributions

1. Jia, R.; Hou, Y.; Feng, W.; Nomingerel, M.; Li, B.; Zhu, J. Multi-Omics Analysis to Understand the Effects of Dietary Proanthocyanidins on Antioxidant Capacity, Muscle Nutrients, Lipid Metabolism, and Intestinal Microbiota in Cyprinus carpio. *Antioxidants* **2023**, *12*, 2095.
2. Cheng, Z.; Shi, J.; Qian, C.; Li, J.; Wu, X.; Kong, I.; Li, J. The Enhanced Growth Performance and Antioxidant Capacity of Juvenile Procambarus clarkii Fed with Microbial Antioxidants. *Antioxidants* **2025**, *14*, 135.
3. Song, Z.; Liu, Y.; Liu, H.; Ye, Z.; Ma, Q.; Wei, Y.; Xiao, L.; Liang, M.; Xu, H. Dietary Lysophosphatidylcholine Improves the Uptake of Astaxanthin and Modulates Cholesterol Transport in Pacific White Shrimp Litopenaeus vannamei. *Antioxidants* **2024**, *13*, 505.
4. Zhang, L.; Zhang, L.; Liang, H.; Huang, D.; Ren, M. Effects of Taurine and Vitamin C on the Improvement of Antioxidant Capacity, Immunity and Hypoxia Tolerance in Gibel Carp (Carrassius auratus gibeilo). *Antioxidants* **2024**, *13*, 1169.
5. Jia, R.; Hou, Y.; Zhang, L.; Li, B.; Zhu, J. Effects of Berberine on Lipid Metabolism, Antioxidant Status, and Immune Response in Liver of Tilapia (Oreochromis niloticus) under a High-Fat Diet Feeding. *Antioxidants* **2024**, *13*, 548.
6. Dong, Y.; Wang, X.; Wei, L.; Liu, Z.; Chu, X.; Xiong, W.; Liu, W.; Li, X. The Effectiveness of Four Nicotinamide Adenine Dinucleotide (NAD+) Precursors in Alleviating the High-Glucose-Induced Damage to Hepatocytes in Megalobrama amblycephala: Evidence in NAD+ Homeostasis, Sirt1/3 Activation, Redox Defense, Inflammatory Response, Apoptosis, and Glucose Metabolism. *Antioxidants* **2024**, *13*, 385.
7. Jiang, Y.; Liu, Z.; Zhang, L.; Liu, W.; Li, H.; Li, X. Phosphatidylserine Counteracts the High Stocking Density-Induced Stress Response, Redox Imbalance and Immunosuppression in Fish Megalobrama ambylsephala. *Antioxidants* **2024**, *13*, 644.
8. Jin, S.; Xu, M.; Gao, X.; Jiang, S.; Xiong, Y.; Zhang, W.; Qiao, H.; Wu, Y.; Fu, H. Effects of Alkalinity Exposure on Antioxidant Status, Metabolic Function, and Immune Response in the Hepatopancreas of Macrobrachium nipponense. *Antioxidants* **2024**, *13*, 129.
9. Xue, M.; Xu, P.; Wen, H.; Chen, J.; Wang, Q.; He, J.; He, C.; Kong, C.; Li, X.; Li, H. A High-Fat-Diet-Induced Microbiota Imbalance Correlates with Oxidative Stress and the Inflammatory Response in the Gut of Freshwater Drum (Aplodinotus grunniens). *Antioxidants* **2024**, *13*, 363.
10. He, Q.; Feng, W.; Chen, X.; Xu, Y.; Zhou, J.; Li, J.; Xu, P.; Tang, Y. H_2O_2-Induced Oxidative Stress Responses in Eriocheir sinensis: Antioxidant Defense and Immune Gene Expression Dynamics. *Antioxidants* **2024**, *13*, 524.
11. Liang, H.; Ren, M.; Zhang, L.; Mi, H.; Yu, H.; Huang, D.; Gu, J.; Teng, T. Excessive Replacement of Fish Meal by Soy Protein Concentrate Resulted in Inhibition of Growth, Nutrient Metabolism, Antioxidant Capacity, Immune Capacity, and Intestinal Development in Juvenile Largemouth Bass (Micropterus salmoides). *Antioxidants* **2024**, *13*, 809.
12. Zhu, L.; Qi, S.; Shi, C.; Chen, S.; Ye, Y.; Wang, C.; Mu, C.; Li, R.; Wu, Q.; Wang, X. Optimizing Anesthetic Practices for Mud Crab: A Comparative Study of Clove Oil, MS-222, Ethanol, and Magnesium Chloride. *Antioxidants* **2023**, *12*, 2124.

References

1. Lushchak, V.I. Environmentally induced oxidative stress in aquatic animals. *Aquat. Toxicol.* **2011**, *101*, 13–30. [CrossRef] [PubMed]
2. Tort, L. Stress and immune modulation in fish. *Dev. Comp. Immunol.* **2011**, *35*, 1366–1375. [CrossRef] [PubMed]

3. Livingstone, D. Contaminant-stimulated reactive oxygen species production and oxidative damage in aquatic organisms. *Mar. Pollut. Bull.* **2001**, *42*, 656–666. [CrossRef] [PubMed]
4. Wang, Y.; Ni, J.; Nie, Z.; Gao, J.; Sun, Y.; Shao, N.; Li, Q.; Hu, J.; Xu, P.; Xu, G. Effects of stocking density on growth, serum parameters, antioxidant status, liver and intestine histology and gene expression of largemouth bass (Micropterus salmoides) farmed in the in-pond raceway system. *Aquac. Res.* **2020**, *51*, 5228–5240. [CrossRef]
5. De Freitas Souza, C.; Baldissera, M.D.; Verdi, C.M.; Santos, R.C.V.; Da Rocha, M.I.U.M.; da Veiga, M.L.; da Silva, A.S.; Baldisserotto, B. Oxidative stress and antioxidant responses in Nile tilapia Oreochromis niloticus experimentally infected by Providencia rettgeri. *Microb. Pathog.* **2019**, *131*, 164–169. [CrossRef]
6. Ghosh, N.; Das, A.; Chaffee, S.; Roy, S.; Sen, C.K. Chapter 4—Reactive Oxygen Species, Oxidative Damage and Cell Death. In *Immunity and Inflammation in Health and Disease*; Chatterjee, S., Jungraithmayr, W., Bagchi, D., Eds.; Academic Press: Cambridge, MA, USA, 2018; pp. 45–55.
7. Lauridsen, C. From oxidative stress to inflammation: Redox balance and immune system. *Poult. Sci.* **2019**, *98*, 4240–4246. [CrossRef]
8. Slaninova, A.; Smutna, M.; Modra, H.; Svobodova, Z. REVIEWS Oxidative stress in fish induced by pesticides. *Neuroendocrinol. Lett.* **2009**, *30*, 2.
9. Zhu, F. A review on the application of herbal medicines in the disease control of aquatic animals. *Aquaculture* **2020**, *526*, 735422. [CrossRef]
10. Tadese, D.A.; Song, C.; Sun, C.; Liu, B.; Liu, B.; Zhou, Q.; Xu, P.; Ge, X.; Liu, M.; Xu, X.; et al. The role of currently used medicinal plants in aquaculture and their action mechanisms: A review. *Rev. Aquac.* **2022**, *14*, 816–847. [CrossRef]
11. Hu, X.; Ma, W.; Zhang, D.; Tian, Z.; Yang, Y.; Huang, Y.; Hong, Y. Application of Natural Antioxidants as Feed Additives in Aquaculture: A Review. *Biology* **2025**, *14*, 87. [CrossRef] [PubMed]
12. Pamplona, R.; Costantini, D. Molecular and structural antioxidant defenses against oxidative stress in animals. *Am. J. Physiol.-Regul. Integr. Comp. Physiol.* **2011**, *301*, R843–R863. [CrossRef] [PubMed]

antioxidants

MDPI

Article

Multi-Omics Analysis to Understand the Effects of Dietary Proanthocyanidins on Antioxidant Capacity, Muscle Nutrients, Lipid Metabolism, and Intestinal Microbiota in *Cyprinus carpio*

Rui Jia [1,2], Yiran Hou [1,2], Wenrong Feng [1], Munkhjargal Nomingerel [2], Bing Li [1,2,*] and Jian Zhu [1,2,*]

1 Key Laboratory of Integrated Rice-Fish Farming Ecology, Ministry of Agriculture and Rural Affairs, Freshwater Fisheries Research Center, Chinese Academy of Fishery Sciences, Wuxi 214081, China; jiar@ffrc.cn (R.J.); houyr@ffrc.cn (Y.H.); fengwenrong@ffrc.cn (W.F.)

2 Wuxi Fisheries College, Nanjing Agricultural University, Wuxi 214081, China; nomibgerel999@gmail.com

* Correspondence: lib@ffrc.cn (B.L.); zhuj@ffrc.cn (J.Z.)

Abstract: Proanthocyanidins (Pros), a natural polyphenolic compound found in grape seed and other plants, have received significant attention as additives in animal feed. However, the specific mechanism by which Pros affect fish health remains unclear. Therefore, the aim of this study was to investigate the potential effects of dietary Pro on common carp by evaluating biochemical parameters and multi-omics analysis. The results showed that Pro supplementation improved antioxidant capacity and the contents of polyunsaturated fatty acids (n-3 and n-6) and several bioactive compounds. Transcriptomic analysis demonstrated that dietary Pro caused an upregulation of the sphingolipid catabolic process and the lysosome pathway, while simultaneously downregulating intestinal cholesterol absorption and the PPAR signaling pathway in the intestines. Compared to the normal control (NC) group, the Pro group exhibited higher diversity in intestinal microbiota and an increased relative abundance of *Cetobacterium* and *Pirellula*. Furthermore, the Pro group had a lower Firmicutes/Bacteroidetes ratio and a decreased relative abundance of potentially pathogenic bacteria. Collectively, dietary Pro improved antioxidant ability, muscle nutrients, and the diversity and composition of intestinal microbiota. The regulation of lipid metabolism and improvement in muscle nutrients were linked with changes in the intestinal microbiota.

Keywords: proanthocyanidins; intestinal microbiota; multi-omics; muscle nutrients; PPAR signaling pathway; *Cyprinus carpio*

Citation: Jia, R.; Hou, Y.; Feng, W.; Nomingerel, M.; Li, B.; Zhu, J. Multi-Omics Analysis to Understand the Effects of Dietary Proanthocyanidins on Antioxidant Capacity, Muscle Nutrients, Lipid Metabolism, and Intestinal Microbiota in *Cyprinus carpio*. *Antioxidants* **2023**, *12*, 2095. https://doi.org/10.3390/antiox12122095

Academic Editors: Erchao Li and Alessandra Napolitano

Received: 15 November 2023
Revised: 7 December 2023
Accepted: 9 December 2023
Published: 10 December 2023

1. Introduction

Aquaculture plays a vital role in supplying high-quality protein and essential micronutrients for human consumption, contributing to human health and overall well-being. However, in intensive aquaculture production, several factors, such as high stocking density, excessive feeding, and fluctuating conditions, have made it more susceptible to disease outbreaks [1]. To mitigate economic losses, various veterinary drugs, especially antibiotics and chemical agents, are extensively utilized in aquaculture for disease prevention and treatment [2]. Despite their effectiveness, the use of veterinary drugs is increasingly limited due to their adverse effects on the environment and human health [3]. As consumers are becoming more concerned about organic and environmentally friendly food, the utilization of natural products such as plant extracts or probiotics in aquaculture has been proposed as a possible solution [4,5]. In recent years, medicinal plants and their extracts have received considerable attention as eco-friendly and efficient alternatives to chemical agents [6]. They have been found to offer various beneficial effects to aquatic animals, such as stress reduction, growth promotion, appetite stimulation, and immunity improvement [7]. These effects are attributed to the presence of active compounds such as polysaccharides, alkaloids, tannins, saponins, flavonoids, or essential oils [6,8]. Furthermore, medicinal plants and their derivatives are commonly used as diet additives in aquaculture due to their ease of preparation, low cost, and minimal adverse effects on both the fish and the environment [4,9].

Proanthocyanidins (Pros), also known as condensed tannins, are a type of natural polyphenolic compound distributed in various plant parts, including stems, leaves, flowers, seeds, and fruit [10]. They are polymerized by flavan-3-ol subunits (i.e., catechin and epicatechin) and produced by the flavonoid biosynthetic pathway [11]. In recent years, there has been a growing interest in the research into Pros due to their attractive nutritional properties and perceived health benefits. Pros possess a variety of biological activities, such as antioxidative, free radical scavenging, anti-inflammatory, immune-stimulating, anti-viral, and cardio-protective features [12]. An in vitro study showed that Pros have the strongest antioxidant activity among more than 100 types of representative phenolic compounds [13]. They have also been found to improve intestinal permeability and increase microbial diversity in response to diet-induced unfavorable changes in the intestine [14,15]. Moreover, Pros can alleviate cardiovascular diseases [16], ameliorate the pesticide rotenone-induced mitochondrial respiration anomalies [17], and inhibit inflammation by modulating the NF-κB pathway [18]. Therefore, it is recommended to moderately enhance the intake of Pro-enriched food, as this may contribute to the prevention of chronic diseases and improve health conditions in both humans and animals [19].

In aquaculture, Pros have been investigated as dietary supplements, and their positive effects on various fish species have been demonstrated. Dietary Pro was reported to promote growth and improve serum biochemistry parameters related to health status in tilapia (*Oreochromis niloticus*) [20]. Wang et al. [21] showed that the weight gain, feed utilization, and growth performance of juvenile American eel (*Anguilla rostrata*) were increased when Pros were incorporated into their diet. In another study, Pros alleviated cadmium toxicity in pearl gentian grouper (*Epinephelus fuscoguttatus* female × *Epinephelus lanceolatus* male) via enhancing antioxidant ability [22]. Furthermore, Pros have been demonstrated to attenuate growth retardation and low immunoglobulin level induced by histamine in American eels [23], as well as mitigate hepatic lipid accumulation and inflammation caused by a high-fat diet in grass carp (*Ctenopharyngodon idella*) [24]. It is apparent that previous studies in aquatic animals have primarily focused on investigating the impact of Pros on growth and biochemical parameters. However, there is a scarcity of data regarding alterations in muscle nutrients, lipid metabolism, and intestinal microbiota in cultured fish after Pros feeding, particularly the lack of multi-omics analysis data to elucidate the underlying mechanism of Pros' actions.

With the rapid advancement of molecular biology and bioinformatics, multi-omics methodologies are gaining recognition as powerful tools for understanding the biological processes of aquatic organisms and their interactions with environmental factors. These methodologies, including genomics, transcriptomics, proteomics, metabolomics, and microbiomics, can provide comprehensive insights into the intricate workings of biology systems. In response to environmental stressors, multi-omics analysis has revealed new pathways in fish that could be pivotal in understanding how these organisms adapt and survive [25,26]. Moreover, multi-omics analyses have facilitated the elucidation of crucial mechanisms involved in fish diseases, contributing to the treatment options and comprehension of pathogenic processes [27]. For example, Li et al. employed both transcriptomic and proteomic analyses to elucidate the beneficial impact of resveratrol on lipid metabolism disorders induced by a high-fat diet in red tilapia [28]. Extensive research evidence suggests that multi-omics approaches are capable of revealing a wider array of mechanisms in the fields of toxicology, pharmacology, physiology, and pathology of fish.

Common carp (*Cyprinus carpio*), a globally distributed and consumed species, is commonly used as an experimental animal in various research fields including pharmacology, toxicology, pathology, and nutrition. In aquaculture, it is frequently utilized for screening medicinal plant extracts and assessing their pharmacological effects [29,30]. Therefore, in this study, we selected common carp as the experimental animal to evaluate the impact of Pros on muscle nutrients, lipid metabolism, and intestinal microbiota via multi-omics analysis. For this purpose, the fish were fed diets supplemented with different doses of Pro (0, 0.2, 0.4, and 0.8 g/kg diet) for 10 weeks, and the effects on growth performance, antioxidant status, muscle metabolites, and intestinal transcriptome profiling and microbiota were investigated. To the best of our knowledge, this is the first report to evaluate the beneficial effects of Pro on cultured fish using multi-omics analysis.

2. Materials and Methods

2.1. Experimental Design and Sampling

The experimental common carp (average weight 41.6 ± 2.15 g) were obtained from the farm of Freshwater Fisheries Research Center (FFRC, Wuxi, China). Healthy fish were selected and kept in a recirculating water system for 7 days to adapt to the experimental conditions. After domestication, the carp were randomly divided into 4 groups and fed diets containing different amounts of Pro (0, 0.2, 0.4, and 0.8 g/kg) for 10 weeks (Figure S1). Each group consisted of 30 fish, distributed evenly across 3 tanks (10 fish/tank). The 0 g/kg Pro group served as the normal control (NC). The fish were fed experimental diets (Table S1) twice a day (8:30 a.m. and 4:00 p.m.) to satiation. The Pro from grape seed (CAS 4852-22-6; purity, 95%) used in this study was purchased from Absin Biotechnology Co., Ltd (Shanghai, China).

After 10 weeks of farming, all fish were weighed under anesthesia using MS-222 (100 mg/L, Sigma, St. Louis, MO, USA). Subsequently, blood, liver, gills, muscle, and intestines were collected for the measurement of antioxidant parameters. From each group, 9 fish (3 fish/tank) were randomly selected for sampling. All collected samples were stored at −80 °C until further use. For analysis of metabolome, transcriptome, and 16S rRNA sequencing, as well as amino acid and fatty acid composition, muscle and intestinal tissues were collected from 12 randomly selected fish (4 fish/tank) in the NC and 0.8 g/kg Pro groups. These tissues obtained from three individuals were pooled together to form one sample [31,32]. The samples were flash-frozen in liquid nitrogen and sent with dry ice to Gene Denovo (Guangzhou, China) for sequencing. The experiment was conducted with consideration for animal welfare, and the animals used in this study were approved by the FFRC. During the experiment, one fish died in each of the NC, 0.2 g/kg Pro, and 0.8 g/kg Pro groups, while no fish died in the 0.4 g/kg Pro group.

2.2. Measurement of Antioxidant Parameters

The blood was centrifuged at 5000 r/min for 10 min at 4 °C to obtain the serum. The liver, gill, muscle, and intestine tissues were homogenized with 9 times (*w*/*v*) normal saline at 4 °C. In serum, liver, gill, muscle, and intestine tissues, 9 samples from each group were used to detect the antioxidant parameters, including malondialdehyde (MDA), superoxide dismutase (SOD), glutathione peroxidase (Gpx), total antioxidant capacity (T-AOC), and glutathione (GSH). Kits for measuring T-AOC, SOD, and GSH were supplied by Nanjing Jiancheng Bioengineering Institute (Nanjing, China). The Gpx kit was provided from Suzhou Grace Biotechnology Co., Ltd. (Suzhou, China), while the MDA kit was ordered from the Beyotime Institute of Biotechnology (Nantong, China). SOD activity was measured at an absorbance of 450 nm utilizing the WST-1 method, with calculation based on the production of WST-1 formazan. The T-AOC level was assessed using the FRAP assay at 593 nm, which quantifies the reduction of the Fe^{3+} to the Fe^{2+} form. The GSH content was assessed based on the intensity of the yellow color produced by its reaction with 5,5′-dithiobis(2-nitrobenzoic acid) (DTNB). The activity of GPx was assayed utilizing cumene hydroperoxide (Cum-OOH) as the substrate, with 5,5′-dithiobis(2-nitrobenzoic acid) (DTNB) serving as the chromogenic agent. The formation of MDA was evaluated using thiobarbituric acid (TBA) as a reactive substrate at 532 nm. The protein content in liver, gill, muscle, and intestine tissues was quantified using the bicinchoninic acid (BCA) assay and the OD value was measured at 562 nm. During the data analysis, the NC (normal control) group served as the control.

2.3. Determination of Amino Acids and Fatty Acids in Muscle

The wet muscle tissue (100 mg) was added to 10 mL of 6 mol/L hydrochloric acid solution. The mixture was then subjected to hydrolysis at 110 °C for 22 h. Following hydrolysis, the resulting hydrolysate was filtered into a 50 mL volumetric flask. Next, 1.0 mL of the filtrate was transferred into a 15 mL test tube and concentrated under reduced pressure at 40 °C. After drying, the concentrated mixture was dissolved using 1.0 mL of sodium citrate buffer (pH 2.2). Finally, the solution was filtered through a 0.22 µm filter membrane to analyze amino acids using automatic amino acid analyzer (HITACHI, Japan).

The muscle tissue (200 mg) was hydrolyzed by adding 10 mL of HCl (8.3 M) at 70–80 °C for 40 min. The resulting hydrolysate was used for total lipid extraction by adding 30 mL of a mixture of diethyl ether and petroleum ether (1:1, v/v) according to the method in the national standard of China (GB5009.168–2016) [33]. To convert the extracted lipid into methyl esters, it was subjected to methyl esterization using a 14% boron trifluoride–methanol solution at 45 °C for 20 min. The fatty acid methyl esters (FAMEs) were then analyzed using a gas chromatography instrument (Agilent 7890A, Agilent Technologies, Santa Clara, CA, USA) equipped with an HP-88 Agilent column (100 m × 0.25 mm × 0.20 µm). The injector and detector temperatures were set at 250 °C and 260 °C, respectively. The fatty acid composition was identified by comparison with 37 kinds of FAME standards (Sigma, St. Louis, MO, USA).

2.4. Non-Targeted Metabolome Sequencing in Muscle

The muscle metabolites were extracted using a solution (acetonitrile: methanol = 1:1) via centrifugation (12,000 rpm, 15 min, and 4 °C) from the NC group (named MNC, 3 pooled samples) and the 0.8 g/kg Pro group (named MPro, 3 pooled samples). The resulting supernatant was analyzed using an UHPLC system (Vanquish, San Jose, CA, USA) and a QE-MS (Orbitrap MS, San Jose, CA, USA). The raw data were converted into the mzXML format and metabolite annotation was performed using an in-house MS2 database (BiotreeDB, V2.1). To improve metabolite coverage, the metabolites were detected in both positive and negative ion modes. Principal component analysis (PCA) was used to evaluate the preliminary differences between groups of samples. Orthogonal projection to latent structures-discriminant analysis (OPLS-DA) was applied to distinguish the metabolomics profile of the two groups. The OPLS-DA model was further evaluated through cross-validation and permutation tests. Differential metabolites between the NC and 0.8 g/kg Pro groups were identified by comparing the VIP score of the OPLS-DA model using the following threshold values: VIP score ≥ 1 and p-value < 0.05 (t-test). The differential metabolites were mapped to the KEGG database to identify significantly enriched metabolic pathways.

2.5. Transcriptome Sequencing in Intestines

Total RNA was extracted from the intestinal tissue of the NC (4 pooled samples) and 0.8 g/kg Pro (4 pooled samples) groups using the TRIzol reagent kit (Invitrogen, San Diego, CA, USA) according to the manufacturer's instructions. The mRNA was reverse-transcribed into cDNA after enrichment and fragmentation. The purified double-stranded cDNA was used to construct a library via PCR amplification, which was sequenced using Illumina Nova 6000 system (Gene Denovo, Guangzhou, China).

To obtain clean data, the raw data were filtered using fastp (version 0.18.0). After the removal of residual ribosomal RNA, the clean data were mapped to the reference genome of *Cyprinus carpio* (NCBI: GCF_000951615.1) using HISAT2. 2.4. PCA was performed to evaluate the distance relationship between samples. DESeq2 software (version 3.0) was used to identify the differentially expressed genes (DEGs) between the NC and Pro groups using the following threshold values: FDR < 0.05 and $|\log FC| \geq 1$. To identify biological functions and key signaling pathways, the DEGs were mapped onto the GO (gene ontology) and KEGG databases. Furthermore, we performed a gene set enrichment analysis (GSEA) to discover distinctive pathways and GO terms between the NC and Pro groups, and threshold values for significance were set as a |normalized enrichment score (ES)| > 1, a nominal p-value < 0.05, and an FDR < 0.25.

The transcriptome sequencing was further validated via quantitative real-time PCR (qPCR) analysis, with the specific primers utilized in this study listed in Table S2. Total RNA was isolated from intestinal tissue using RNAiso Plus reagent (Takara, Beijing, China). The RNA was then used to synthesize cDNA via reverse transcription using the PrimeScript™ RT reagent kit (Takara). The cDNA served as a template to perform qPCR using a TB Green Premix Ex Taq II kit (Takara, RR820A). The resulting Cq value was used to calculate the relative expression of each gene using the $2^{-\Delta\Delta Cq}$ method, with β-actin used as the housekeeping gene.

2.6. 16S rRNA Sequencing in Intestinal Bacteria

Microbial DNA from the intestinal content of fish in the NC (4 pooled samples) and 0.8 g/kg Pro (4 pooled samples) groups was isolated using HiPure Stool DNA Kits (Meiji Biotechnology, Guangzhou, China) in accordance with the manufacturer's protocols. The target region of 16S rDNA was amplified by PCR using V3–V4 region primers (341F: CCTACGGGNGGCWGCAG, 806R: GGACTACHVGGGTATCTAAT). The amplicons were purified using AMPure XP Beads (Axygen, Union City, CA, USA), quantified using Real-Time PCR System (ABI, Foster City, CA, USA), and sequenced on an Illumina platform.

The raw data were subjected to a series of preprocessing steps, including merging, filtering, dereplication, denoising, and chimera removal, using the DADA2 R package (version 1.14). Following these procedures, the resulting clean tags were utilized to output the ASVs. The representative ASV sequences were classified into bacterial taxonomy using the RDP classifier (version 2.2) with reference to the SILVA database. After ASV annotation, the abundance statistics of each taxonomy were visualized using Krona (version 2.6). Alpha indices, including Chao1, Shannon, and Simpson, were calculated using the QIIME software (version 1.9.1), and the difference in these indices between the NC and 0.8 g/kg Pro groups was assessed by the Wilcoxon rank test. Principal coordinates analysis (PCoA) based on weighted Unifrac distances was plotted in R project, and the Anosim test was conducted using the Vegan package (version 2.5.3).

2.7. Statistical Analysis

SPSS was used to analyze the data in this study and the results are presented as the mean ± standard error of the mean (SEM). Differences in antioxidant parameters and growth parameters among groups were analyzed using ANOVA, followed by the LSD test. Differences in amino acid and fatty acid composition between the NC and Pro groups were analyzed using a t-test. The correlation between qPCR data and RNA-seq data was determined using the Pearson test. The level of significance was set at $p < 0.05$.

3. Results

3.1. Common Carp Growth Performance

During the experiment, one fish died in each of the NC, 0.2 g/kg Pro, and 0.8 g/kg Pro groups, while no fish died in the 0.4 g/kg Pro group (Table 1). After 10 weeks of feeding, there was a significant increase in the final weight and specific growth rate, but a significant decrease in the feed conversion ratio, in the groups fed 0.4 and 0.8 g/kg Pro compared to the NC group ($p < 0.05$; Table 1).

Table 1. Effects of dietary Pro on growth performance of *Cyprinus carpio* after 10 weeks of farming.

Growth Index	NC	0.2 g/kg Pro	0.4 g/kg Pro	0.8 g/kg Pro
Initial body weight (g)	42.5 ± 1.71	42.6 ± 1.63	42.6 ± 1.89	40.5 ± 2.01
Final body weight (g)	89.9 ± 3.48 ^{a}	86.5 ± 3.38 ^{a}	101.5 ± 2.80 ^{b}	103.5 ± 3.92 ^{b}
Specific growth rate (%/d)	1.07 ± 0.05 ^{a}	1.01 ± 0.06 ^{a}	1.24 ± 0.04 ^{b}	1.35 ± 0.01 ^{b}
Feed conversion ratio	2.13 ± 0.02 ^{a}	2.07 ± 0.02 ^{a}	1.83 ± 0.04 ^{b}	1.85 ± 0.03 ^{b}
Survival ratio (%)	96.7	96.7	100	96.7

The results are expressed as the mean ± SEM. Different letters in the same line indicate significant differences among groups ($p < 0.05$). Specific growth rate (SGR) = 100 × [Ln (average final weight) − Ln (average initial weight)]/number of days, feed conversion ratio (FCR) = food consumption/biomass increment, and survival ratio = final number of fish/initial number of fish.

3.2. Antioxidant and Lipid Peroxidation Parameters in Different Tissues

Antioxidant capacity was assessed by measuring the levels of MDA, SOD, T-AOC, GSH, and Gpx in different tissues (Figure 1). In the serum (Figure 1A), following Pro treatment, MDA content exhibited a declining trend, with significant reductions observed in the groups supplemented with 0.4 and 0.8 g/kg of Pro compared to the NC group ($p < 0.05$). Conversely, Gpx activity displayed an increasing trend, and it was significantly enhanced in the 0.4 and 0.8 g/kg Pro-supplemented groups relative to the NC group ($p < 0.05$). The GSH content also showed an increasing trend and a significant increase was observed in the 0.8 g/kg Pro-supplemented group relative to the NC group ($p < 0.05$). However, the SOD and T-AOC levels were not impacted by dietary Pro supplementation ($p > 0.05$).

In the liver (Figure 1B), following Pros administration, levels of SOD, Gpx, and T-AOC uniformly demonstrated an upward trend. For SOD and Gpx activities, there were significant differences in the groups supplemented with 0.4 and 0.8 g/kg Pro compared to the NC group ($p < 0.05$). Additionally, the T-AOC level was significantly higher in the 0.8 g/kg Pro-supplemented group than in the NC group ($p < 0.05$). However, there was no change in MDA and GSH levels among the different groups ($p > 0.05$).

In the muscle (Figure 1C), Pro treatment improved the levels of SOD, T-AOC, and GSH and lowered the MDA content. Notably, there was a marked decrease in the MDA content and a marked increase in the levels of SOD, T-AOC, and GSH in the groups fed 0.4 and 0.8 g/kg Pro compared with the NC group ($p < 0.05$). A similar increase in SOD level was also observed in the 0.2 g/kg Pro-treated group ($p < 0.05$). However, Gpx activities were not significantly changed by Pro treatment.

In the gills (Figure 1D), Pro treatment inhibited MDA formation but enhanced GSH production. Significant changes were observed in the 0.4 and 0.8 g/kg Pro treatments compared to the NC treatment ($p < 0.05$). Moreover, the levels of SOD, T-AOC, and Gpx were not altered by dietary Pro feeding ($p > 0.05$).

In the intestines (Figure 1E), the content of GSH was significantly increased in the three Pro-treated groups compared to the NC group ($p < 0.05$). However, no significant differences were observed in other parameters among the different treatments ($p > 0.05$).

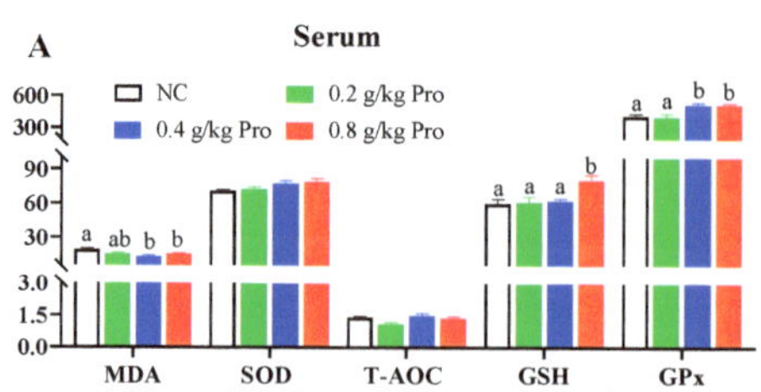

Figure 1. *Cont.*

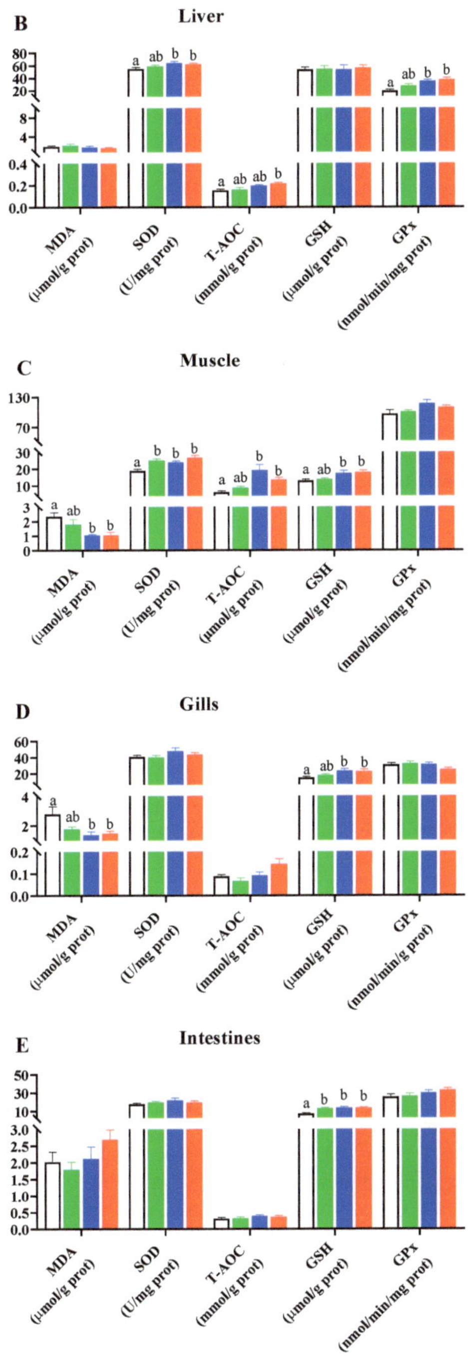

Figure 1. Effects of dietary Pro on antioxidant capacity of *Cyprinus carpio* after 10 weeks of farming. (**A–E**) Antioxidant parameters in the serum, liver, muscle, gill, and intestine, respectively. The results are expressed as the mean ± SEM (n = 9). Different letters above the bars indicate significant differences for each parameter between groups ($p < 0.05$).

3.3. Amino Acid and Fatty Acid Composition in Muscle

There were 17 amino acids identified in the muscle tissue of common carp, including 9 essential amino acids (EAAs) and 8 non-essential amino acids (NEAAs) (Table S3). EAAs consisted of Thr, Val, Met, Ile, Leu, Phe, Lys, His, and Arg, whereas NEAAs included Asp, Ser, Glu, Gly, Ala, Gys, Tyr, and Pro. However, there were no differences in the levels of these amino acids between the NC and 0.8 g/kg Pro groups.

Eleven fatty acids were detected in the muscle tissue of common carp, including two saturated fatty acids (SFA) and nine unsaturated fatty acids (NFA) (Table 2). The levels of C16:0, C18:0, C18:3n3, C20:1, C20:2, C22:1n9, and C20:4n6 were significantly higher in the 0.8 g/kg Pro group than in the NC group ($p < 0.05$). Similarly, Pro treatment exhibited higher levels of total polyunsaturated fatty acids (PUFA), including n-3 and n-6 PUFA ($p < 0.05$). Additionally, the n-6/n-3 ratio was slightly reduced, while the PUFA/SFA ratio was slightly increased in the Pro group compared to the NC group.

Table 2. Hydrolyzed fatty acid composition in muscle of *C. carpio* fed on normal diet (NC) and Pro-supplemented diet (Pro).

Fatty Acids (% of Total Fatty Acid)	Groups		
	NC	0.8 g/kg Pro	Sig.
C16:0	19.92 ± 0.41	22.41 ± 0.18	**
C18:0	7.72 ± 0.3	9.49 ± 0.15	**
C18:1n9c	22.67 ± 2.63	31.03 ± 2.79	ns
C18:2n6c	21.92 ± 2.05	29.5 ± 2.06	ns
C18:3n3	0.99 ± 0.04	1.72 ± 0.1	**
C20:1	1.04 ± 0.04	1.32 ± 0.07	*
C20:2	0.94 ± 0.08	1.20 ± 0.02	*
C20:3n6	2.53 ± 0.11	2.97 ± 0.13	ns
C22:1n9	0.84 ± 0.02	1.10 ± 0.03	**
C20:4n6	5.73 ± 0.35	7.81 ± 0.35	*
C22:6n3	3.24 ± 0.39	3.95 ± 0.11	ns
Total SFA	27.64 ± 0.7	31.91 ± 0.32	**
Total MUFA	24.55 ± 2.65	33.44 ± 2.84	ns
Total PUFA	35.35 ± 1.4	47.15 ± 1.81	**
n-3 PUFA	4.23 ± 0.41	5.67 ± 0.07	*
n-6 PUFA	30.18 ± 1.74	40.28 ± 1.73	*
n-6/n-3	7.61 ± 1.3	7.09 ± 0.24	ns
PUFA/SFA	1.28 ± 0.06	1.48 ± 0.08	ns

SFA, saturated fatty acid; MUFA, monounsaturated fatty acid; PUFA, polyunsaturated fatty acid. All values are expressed as the mean ± SEM (n = 4). * and ** indicate significant differences ($p < 0.05$ and $p < 0.01$) between the NC and 0.8 g/kg Pro groups; ns indicates no significant difference.

3.4. Metabolomics Analysis in Muscle

One sample from both the NC group and the 0.8 g/kg Pro group was excluded during metabolome sequencing due to sample degradation; thus, the metabolomic analysis was conducted with three pooled samples from each group. Quality control (QC) analysis showed that the sequencing data were acceptable based on the well-clustered QC samples (Figure S2). In the muscle, we identified 3086 metabolites in positive ion mode and 3025 metabolites in negative ion mode (Figure S3A,B). The unsupervised PCA displayed a clear separation between the NC and 0.8 g/kg Pro groups in both positive and negative ion modes (Figure S3C,D). The OPLS-DA results also showed that the metabolites between the NC and 0.8 g/kg Pro groups exhibited different classifications in both positive and negative (Figure S3E,F) ion modes. In addition, cross-validation and permutation test results revealed that the OPLS-DA model for the metabolites was reliable (Figure S3G,H).

Dietary 0.8 g/kg Pro supplementation resulted in an increase in 53 metabolites and a decrease in 21 metabolites in positive ion mode, while it led to an increase in 78 metabolites and a decrease in 17 metabolites in negative ion mode, compared to the NC group (Figure 2A,B). In positive ion mode (Figure 2C), the differential metabolites primarily belonged to lipids and lipid-like molecules (16), phenylpropanoids and polyketides (8), and organic acids and derivatives (7). In negative ion mode (Figure 2D), the differential metabolites were mainly organoheterocyclic compounds (34), lipids and lipid-like molecules (13), and phenylpropanoids and polyketides (4). KEGG enrichment analysis revealed that the differential metabolites were primarily associated with α-linolenic acid (α-LA) metabolism (q value = 0.024), glycerophospholipid (GP) metabolism (q value = 0.025), arachidonic acid (ARA) metabolism (q value = 0.025), and biosynthesis of UFA (q value = 0.038) (Figure 2E).

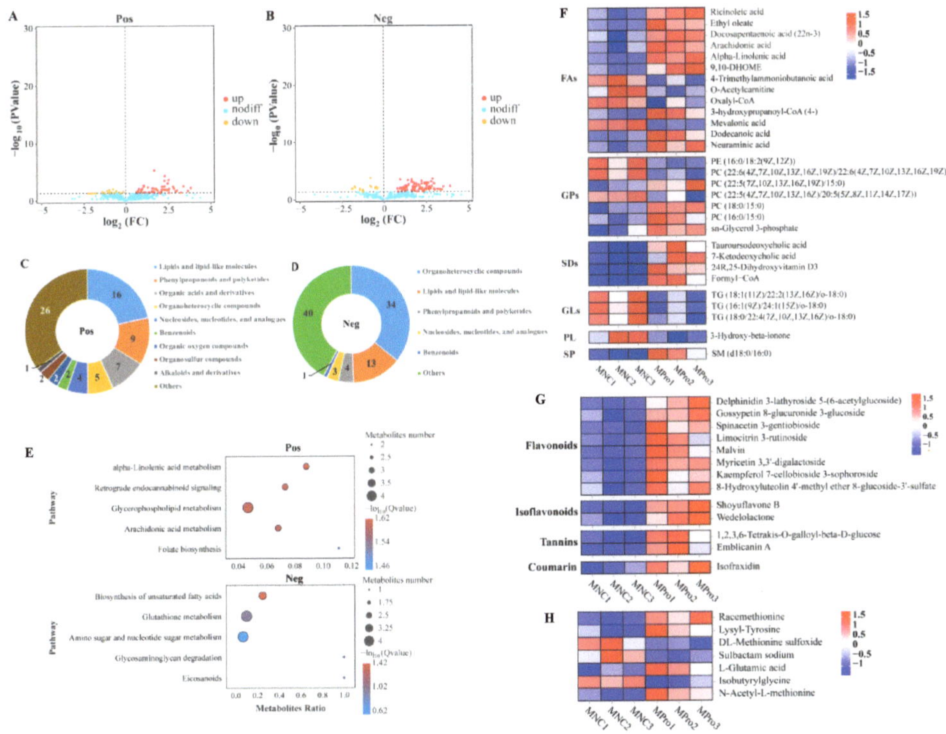

Figure 2. Differential metabolites in the muscle of *C. carpio* between the NC (MNC) and 0.8 g/kg Pro-fed groups (MPro). (**A**,**B**) Volcano plot of the differential metabolites in positive (Pos) and negative (Neg) ion modes. (**C**,**D**) Numbers and classification of the differential metabolites in Pos and Neg ion modes. (**E**) Main KEGG pathways enriched by the differential metabolites in Pos and Neg ion modes. (**F**) Differential metabolites related to lipids and lipid-like molecules (FA, fatty acyls; GP, glycerophospholipids; SD, steroids and steroid derivatives; GL, glycerolipids; SP, sphingolipid; and PL, prenol lipid). (**G**) Differential metabolites related to phenylpropanoids and polyketides. (**H**) Differential metabolites related to amino acids.

The analysis of lipid-related metabolites revealed significant alterations in 13 fatty acyls (FAs), 7 GPs, 4 steroids and steroid derivatives (SDs), 3 glycerolipids (GLs), 1 sphingolipid (SP), and 1 prenol lipid (PL) between the NC and 0.8 g/kg Pro groups (Figure 2F). Among FAs, nine metabolites (e.g., ARA, α-LA, ricinoleic acid, and docosapentaenoic acid) were significantly increased by 0.8 g/kg Pro treatment, while four metabolites were significantly decreased ($p < 0.05$). In GPs, 0.8 g/kg Pro treatment resulted in a marked increase in the levels of PC (16:0/15:0), PC (22:5/15:0), and PC (18:0/15:0), but a decrease in the levels of PC (22:6/22:6) and PC (22:5/20:5) ($p < 0.05$). In GLs, the three types of triglyceride (TG) levels were suppressed by 0.8 g/kg Pro treatment ($p < 0.05$).

Furthermore, 0.8 g/kg Pro treatment significantly increased the levels of various bioactive compounds, including eight flavonoids, two isoflavonoids, two tannins, and one coumarin (Figure 2G; $p < 0.05$). As shown in Figure 2H, Pro treatment also resulted in alterations in several amino acid derivatives. In detail, racemethionine (DL-methionine), lysyl-tyrosine, L-glutamic acid, and N-acetyl-L-methionine increased, while DL-methionine sulfoxide, sulbactam sodium, and isobutyrylglycine decreased, compared to the NC group ($p < 0.05$).

3.5. Transcriptomic Analysis in Intestines

The transcriptomic analysis was performed using four pooled samples from each group. After filtering, transcriptome sequencing obtained a total of 5,392,441,669–7,005,632,038 bp of clean reads (Table S4). The quality control results indicated that the sequencing data obtained from the intestines of the NC and 0.8 g/kg Pro groups were highly reliable (Table S4). The PCA demonstrated a distinct separation between the NC and 0.8 g/kg Pro groups, indicating that Pro treatment had a significant impact on the gene expression patterns in the intestinal tissue (Figure S4A). A total of 1909 DEGs were identified between the NC and 0.8 g/kg Pro groups, with 1035 upregulated genes and 874 downregulated genes in the Pro group compared to the NC group (Figure S4B,C).

To gain a deeper understanding of the effects of Pro treatment on biological function in the intestines, we conducted a GO enrichment analysis, focusing on three main categories: molecular function, cellular component, and biological process (Figure S5A). In the biological process, the DEGs were highly associated with lipid metabolic process (p.adjust < 0.001) and anion transport (p.adjust < 0.001) (Figure S5B). In the molecular function, the DEGs were primarily involved in exopeptidase activity (p.adjust < 0.001) and organic acid transmembrane transporter activity (p.adjust < 0.001) (Figure S5C). In the cellular component, the DEGs were primarily enriched in lysosome (p.adjust < 0.001) and vacuole (p.adjust < 0.001) (Figure S5D).

In the lipid metabolic process, sphingolipid catabolic process, glycolipid catabolic process, and lipid catabolic process were significantly affected by dietary 0.8 g/kg Pro supplementation (p.adjust < 0.001; Figure 3A). Furthermore, GSEA showed that these processes exhibited a high enrichment in 0.8 g/kg Pro group. Specifically, the sphingolipid catabolic process and glycolipid catabolic process showed statistically significant differences (Figure 3B). However, intestinal cholesterol absorption exhibited significantly lower enrichment in 0.8 g/kg Pro treatment (Figure 3A,B). In GO terms related to ion transport, specifically, organic anion transport, carboxylic acid transport, anion transmembrane transport, and carboxylic acid transmembrane transport were noticeably altered by 0.8 g/kg Pro treatment (Figure 3C). Further analysis using GSEA confirmed that these processes exhibited a higher enrichment in the 0.8 g/kg Pro group, with a significant difference in organic anion transport, anion transmembrane transport, and carboxylic acid transmembrane transport (Figure 3D).

To further investigate the key signaling pathways, we performed KEGG enrichment analysis using the DEGs (Figure S6). The DEGs were found to be enriched in five KEGG A classes: metabolism, organismal systems, cellular processes, genetic information processing, and environmental information processing (Figure S6A). The top 10 pathways were primarily associated with lysosome and metabolism function, specifically, sphingolipid metabolism and glycosphingolipid biosynthesis (Figure S6B,C). In the lysosome pathway, 34 genes were upregulated and 2 genes were downregulated following 0.8 g/kg Pro treatment ($q < 0.0001$; Figure 4A). GSEA confirmed the strong enrichment of the lysosome pathway in the 0.8 g/kg Pro group (Figure 4B). Further, we validated the expression of key genes involved in the lysosome pathway, including *ctsα*, *galcα*, *gba*, *asah1b*, *acp5b*, and *pla2g15*, using qPCR (Figure 4C), and the data revealed a clear positive correlation with the RNA-seq results ($r = 0.864$, $p = 0.027$; Figure 4D).

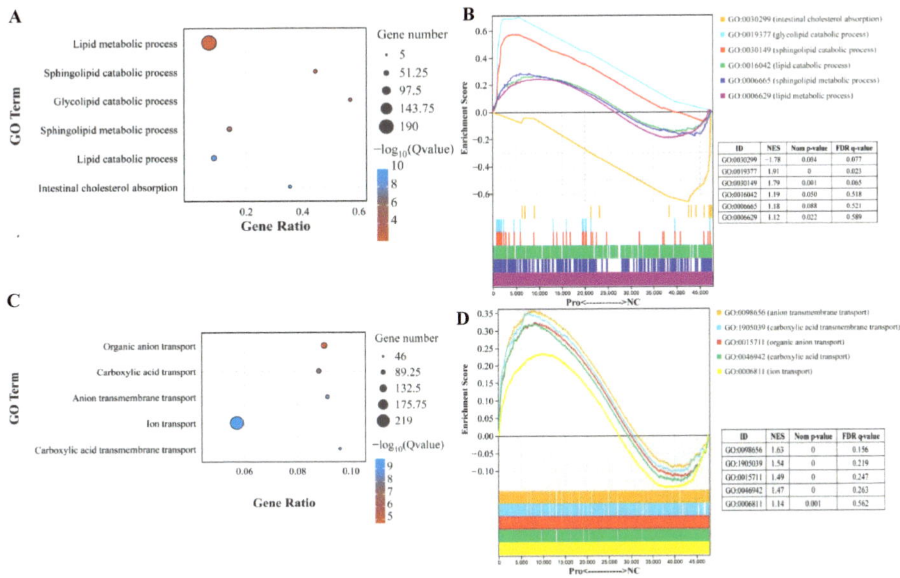

Figure 3. Effects of dietary Pro supplementation on lipid metabolism and ion transport in the intestines of *C. carpio*. (**A**) Differential GO terms related to lipid metabolism. (**B**) GSEA for the GO terms related to lipid metabolism. (**C**) Differential GO terms related to ion transport. (**D**) GSEA for the GO terms related to ion transport.

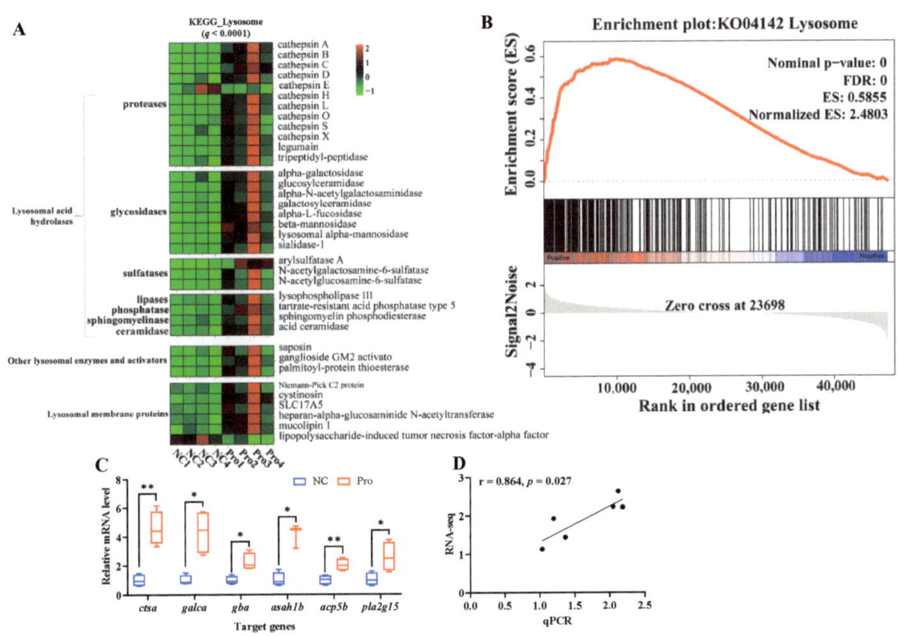

Figure 4. Changes in the lysosome pathway in the intestines of *C. carpio* between the NC and Pro groups. (**A**) DEGs in the KEGG lysosome pathway. (**B**) GSEA for the KEGG lysosome pathway. (**C**) Key genes expression in the lysosome pathway measure by qPCR, with values expressed as the mean ± SEM (n = 4), * $p < 0.05$ and ** $p < 0.01$. (**D**) Correlation between the qPCR and RNA-seq data.

It is important to highlight that 0.8 g/kg Pro treatment had a significant impact on the PPAR signaling pathway (q = 0.0029, Figure 5A). Specifically, 11 genes including *pparα* (a master in the PPARα signaling pathway) were downregulated, while 1 gene was upregulated, following 0.8 g/kg Pro treatment. Meanwhile, the GSEA results also indicated that the PPAR signaling pathway tended to be downregulated in the 0.8 g/kg Pro group (Figure 5B). Furthermore, the qPCR results showed a significantly positive correlation with RNA-seq, indicating the credibility of the transcriptome results (r = 0.915, p = 0.004; Figure 5C,D).

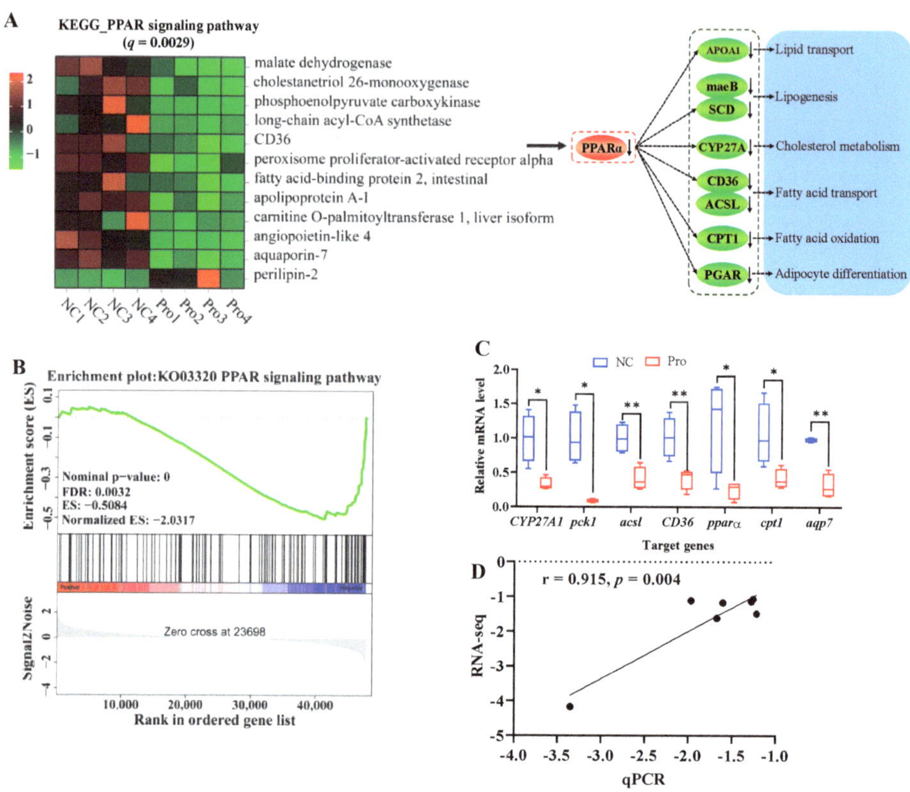

Figure 5. Changes in the PPAR signaling pathway in the intestines of *C. carpio* between the NC and Pro groups. (**A**) DEGs in the PPAR signaling pathway and possible mechanism regulating lipid metabolism. (**B**) GSEA for the PPAR signaling pathway. (**C**) Key genes expression in the PPARα signaling pathway measure by qPCR, with values expressed as the mean ± SEM (n = 4), * $p < 0.05$ and ** $p < 0.01$. (**D**) Correlation between the qPCR and RNA-seq data.

3.6. Intestinal Microbiota Characteristics

Microbiota characteristics were analyzed using four pooled samples from each group. Venn diagram analysis indicated that the total number of ASVs in the two groups was 6049 (Figure S7A). The Pro group had a higher total number of ASVs (3618) compared to the NC group (2899) ($p > 0.05$). Furthermore, there were 468 ASVs that were shared between the NC and 0.8 g/kg Pro groups. At the phylum level, 23 out of the total 29 phyla were found to be common to both the NC and 0.8 g/kg Pro groups (Figure S7B). At the genus level, 167 out of 336 genera were shared between the NC and 0.8 g/kg Pro groups (Figure S7C).

Microbial composition analysis revealed that the five most predominant phyla in the NC group were Proteobacteria (52.34%), Firmicutes (30.26%), Bacteroidota (6.97%), Fusobacteriota (2.73%), and Desulfobacterota (1.77%) (Figure 6A). In contrast, the 0.8 g/kg Pro group displayed a different pattern, with the top five phyla being Fusobacteriota (31.08%), Firmicutes (21.84%), Proteobacteria (18.93%), Bacteroidota (16.39%), and Planctomycetota (3.81%) (Figure 6A). At the genus level, *Aeromonas* (39.94%), *ZOR0006* (25.89%), *Bacteroides* (4.14%), *Vibrio* (2.97%), and *Cetobacterium* (2.58%) were the top five microbes in the NC group, while *Cetobacterium* (30.85%), *ZOR0006* (15.98%), *Bacteroides* (12.66%), *Aeromonas* (10.01%), and *Enterococcus* (1.83%) dominated in the 0.8 g/kg Pro group (Figure 6B).

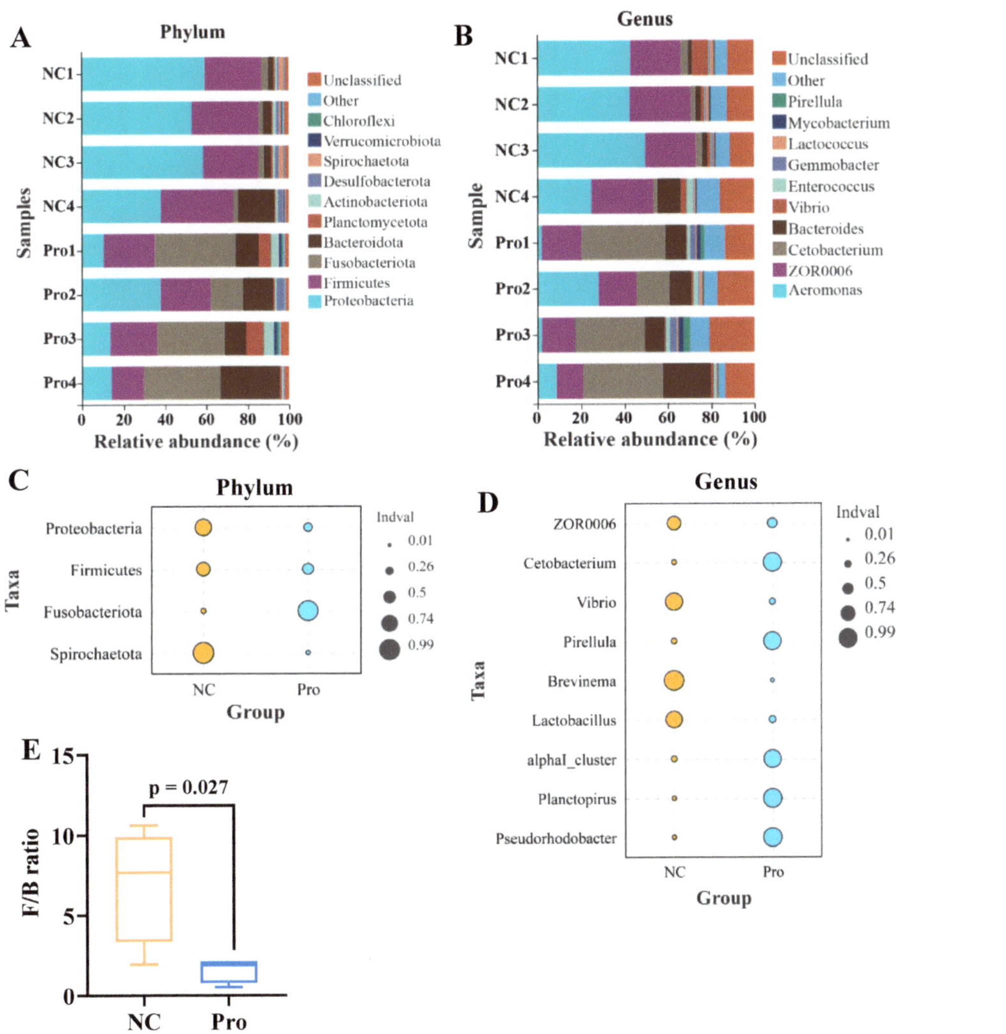

Figure 6. The proportions of intestinal microbiota in the NC and Pro groups. (**A**,**B**) The top 10 phyla and genera of intestinal microbiota in the two groups. (**C**,**D**) Differential phyla and genera in intestinal microbiota between the NC and Pro groups (Wilcoxon test, $p < 0.05$). (**E**) Firmicutes/Bacteroidetes (F/B) ratio in the intestinal microbiota collected from the NC and Pro groups.

Indicator species analysis showed that Fusobacteriota has a higher relative abundance, while Proteobacteria, Firmicutes, and Spirochaetota exhibited lower relative abundance in the 0.8 g/kg Pro group, compared to the NC group (Figure 6C). Similarly, at the genus level, the relative abundance of *Cetobacterium, Pirellula, alphaI_cluster, Planctopirus*, and *Pseudorhodobacter* were higher, whereas the relative abundance of *ZOR0006, Vibrio, Brevinema*, and *Lactobacillus* were lower in the 0.8 g/kg Pro group, compared to the NC group (Figure 6D). Notably, the Firmicutes/Bacteroidetes (F/B) ratio showed a decrease in the 0.8 g/kg Pro group compared with the NC group (Figure 6E).

According to α-diversity analysis, we observed a non-significant elevation in the Chao1 and Shannon indices ($p > 0.05$; Figure 7A,B) and a significant increase in the Simpson's index of diversity ($p < 0.05$; Figure 7C) in the 0.8 g/kg Pro group compared to the NC group. As for β-diversity, both PCA and PCoA analyses indicated that the samples of the NC and 0.8 g/kg Pro groups showed a distinct cluster (Figure 7D,E). Furthermore, Anosim analysis revealed a significant difference in the microbial community composition between the NC and 0.8 g/kg Pro groups (r = 0.75, $p = 0.026$; Figure 7F).

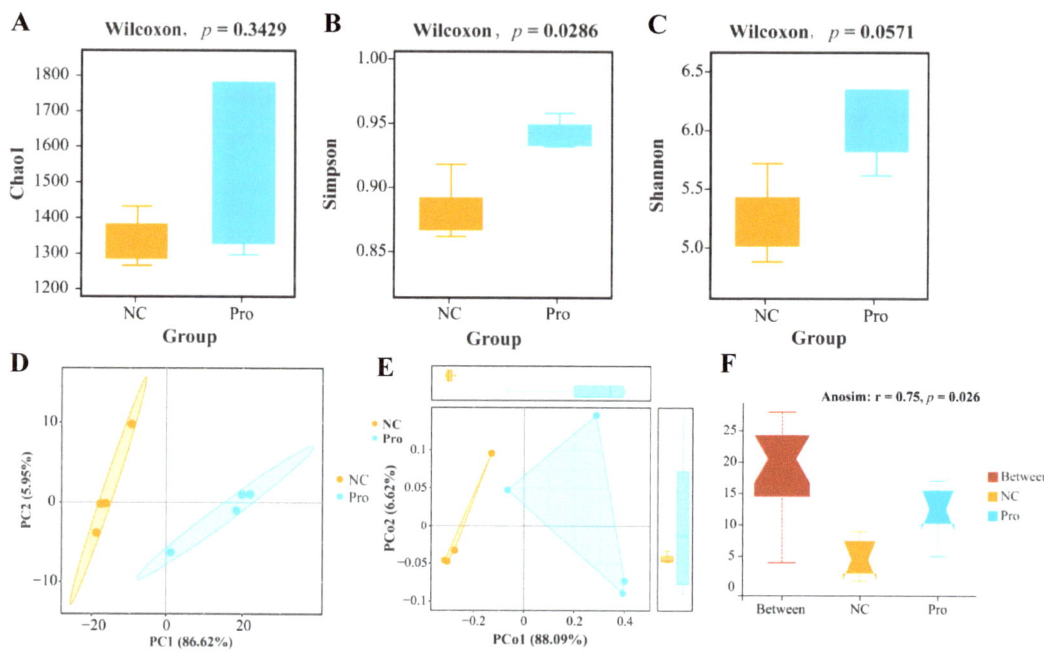

Figure 7. The α diversity (**A–C**) and β diversity (**D–F**) analysis of intestinal microbiota between the NC and Pro groups. (**A**) Chao1 index. (**B**) Simpson's index of diversity. (**C**) Shannon index. (**D**) PCA score plot. (**E**) PCoA score plot. (**F**) Anosim test.

We further predicted and analyzed nine potential bacterial phenotypes in the NC and 0.8 g/kg Pro groups (Figure 8A). Compared with the NC group, the relative abundance of mobile element containing, facultatively anaerobic, potentially pathogenic, and stress-tolerant bacteria were significantly lower in the intestine of the 0.8 g/kg Pro group ($p < 0.05$; Figure 8B). Tax4Fun analysis revealed that in the level 2 KEGG pathway, signal transduction, cell motility, and metabolism of terpenoids and polyketides were decreased in the 0.8 g/kg Pro group compared to the NC group ($p < 0.05$; Figure 8C). At the level 3 KEGG pathway, several pathways were found to be enhanced in the 0.8 g/kg Pro group, including purine metabolism, aminoacyl-tRNA biosynthesis, terpenoid backbone biosynthesis, tetracycline biosynthesis, and primary bile acid biosynthesis ($p < 0.05$; Figure 8D), suggesting that Pro mediated microbe–microbe and microbe–host interactions.

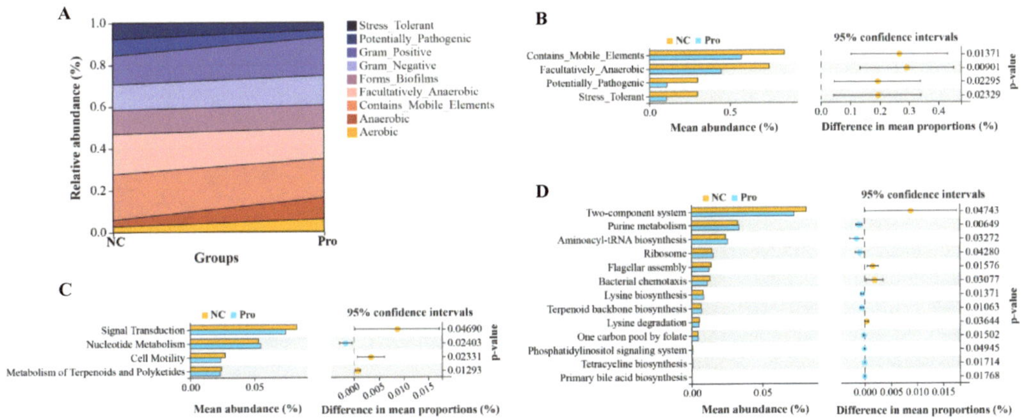

Figure 8. The analyses for potential phenotypes and functions of intestinal microbiota in the NC and Pro groups. (**A**) Relative abundance of nine potential phenotypes of intestinal microbiota. (**B**) Differential phenotypes of the intestinal microbiota between the NC and Pro groups. (**C**) Differential pathways at level 2 KEGG between the NC and Pro groups. (**D**) Differential pathways at level 3 KEGG between the NC and Pro groups. Statistical significance was identified by Welch's *t*-test with $p < 0.05$.

3.7. Interactions between Intestinal Microbes and Lipid Metabolism

To investigate the relationship between microbes and host genes in the intestines of common carp after 0.8 g/kg Pro feeding, we conducted a correlation analysis between 75 DEGs enriched in lipid metabolism and 15 microbial taxa at the genus level (Figure 9). As shown in Figure 9A, we observed significant positive correlations ($p < 0.05$) between the majority of DEGs involved in the lysosome pathway and sphingolipid metabolism and the relative abundance of *Gemmobacter*, *Mycobacterium*, *Pirellula*, and *Parabacteroides*. Additionally, several notable negative correlations ($p < 0.05$) were observed between dominant taxa, such as *Aeromonas* and *ZOR0006*, and specific genes associated with the lysosome pathway, including *cystinosin*, *lysosomal alpha-mannosidase*, and *acid ceramidase*.

We also specifically investigated the correlation between microbiota and the PPAR pathway, and visually depicted the significant gene–microbe correlations ($p < 0.05$; Figure 9B). The results demonstrated a statistically significant positive correlation between predominant taxa, specifically, *Aeromonas* and *ZOR0006*, and the majority of genes in the PPAR pathway. Conversely, a significant negative correlation was observed between *Cetobacterium*, which was significantly increased in the 0.8 g/kg Pro group, and the majority of genes.

In addition to the overall network, Figure 9C illustrates the correlation between representative gene expression and microbial taxa, both of which have previously been linked to host health and thus may be of interest. The *pparα*, a crucial regulator of lipid metabolism, was significantly negatively correlated with *Cetobacterium* (r = −0.834, $p = 0.01$). Similarly, *cpt1*, a vital enzyme involved in fatty acid oxidation, exhibited a negative correlation with *Cetobacterium* (r = −0.752, $p = 0.031$). Furthermore, both *gsta* (r = 0.874, $p = 0.008$) and *gba* (r = 0.949, $p < 0.001$) showed a positive correlation with *Pirllula*. CTSA (Cathepsin A) is a key proteolytic enzyme in the lysosome, while GBA (glucosylceramidase) is a crucial enzyme in the catabolic pathway for glucosylceramide, a membrane sphingolipid and a precursor for various glycolipids.

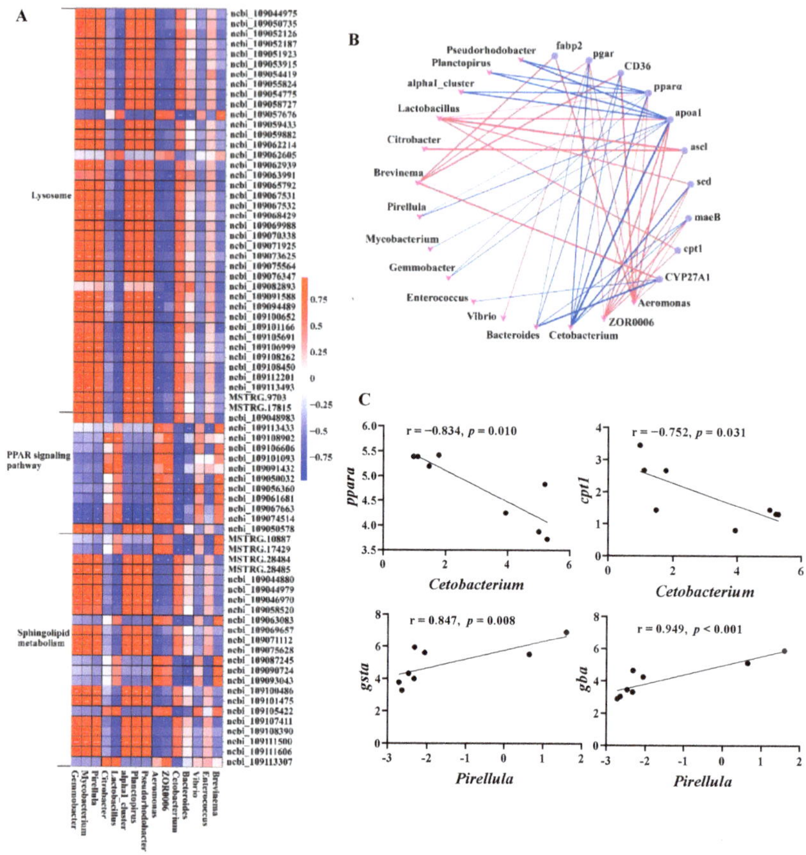

Figure 9. Interactions between the intestinal microbes and DEGs related to lipid metabolism. (**A**) Correlation heatmap of microbe–DEGs. (**B**) Network plot of significant microbe–gene (PPAR signaling pathway) correlations; red lines indicate positive correlation and blue lines indicate negative correlation. (**C**) Correlation plots of representative gene–microbe combinations.

4. Discussion

Pros are a type of polyphenolic compounds known for their potent antioxidant activity. They are abundantly present in various plant sources, such as grape seed, pinto bean, and blueberry. Emerging research has highlighted the important role of Pros in promoting animal growth, maintaining health, and preventing disease. Nevertheless, the effects of Pros on growth performance, muscle quality, and nutrients, as well as intestinal microbiota and function, may vary across different animal species.

4.1. The Effect of Pros on Growth Performance

Pros have reportedly been used as appetite stimulators and growth promoters in cultured fish. Adding Pros to the diet improved weight gain and feed utilization in juvenile American eel, European eel, and tilapia [20,21,34]. In addition, dietary Pro has been shown to ameliorate the growth retardation caused by histamine or cadmium stress in American eel, pearl gentian grouper, and tilapia [22,23,35]. In line with previous studies, the present study found that the common carp fed with 0.4 and 0.8 g/kg Pro exhibited higher growth performance, suggesting that Pro, as a feed additive, can effectively promote the growth of common carp. At present, the mechanism underlying the beneficial effect of Pros on fish growth remain unknown, but potential explanations have been mentioned in previous studies. Some studies have suggested that Pros can enhance the activities of intestinal digestive enzymes, thereby improving feed utilization rate [36,37]. Furthermore, other studies indicated that Pros could regulate intestinal microbial community composition, maintain intestinal health, and promote nutrient absorption, ultimately enhancing growth [38].

4.2. The Effect of Pros on Antioxidant Capacity

Animal experiments have shown that Pro treatment decreased the levels of reactive oxygen species (ROS) in different tissues or cells [10,39]. Pros also have the potential to improve cellular antioxidant systems, such as SOD, catalase, or Gpx systems [40]. The antioxidant properties of Pros could potentially lead to several beneficial effects, including anti-inflammatory, antimicrobial, anticarcinogenic, hypolipemic, and antihyperalgesic activities [10]. It has been reported that Pro treatment can effectively prevent the formation of H_2O_2, protein oxidation, and DNA damage in cells by enhancing antioxidant defense compounds, such as Gpx, SOD, catalase, and GSH [15]. In fish, Pro treatment has been found to increase antioxidant enzymes (e.g., SOD and Gpx) and non-enzymatic antioxidants (e.g., GSH) in the serum of European eels [34] and hybrid sturgeon [36]. Consistent with previous studies, our data also showed the increased levels of SOD (in liver and muscle), T-AOC (in liver and muscle), GSH (in serum, muscle, gills, and intestines), and Gpx (in liver) after 0.4 and 0.8 g/kg Pro feeding. These results demonstrate that Pros can significantly improve antioxidant ability in aquatic animals. Interestingly, in the muscle, almost all antioxidant parameters were enhanced after the administration of Pros.

The intrinsic antioxidant defense system of fish can be compromised under adverse conditions, resulting in the excessive production of ROS that cause lipid peroxidation via reacting with unsaturated fatty acids. Pros have been shown to scavenge free radicals and inhibit lipid peroxidation in the muscle of finishing pigs [41]. Similarly, in hybrid sturgeons, the administration of 50 and 100 mg/kg of Pro was found to effectively inhibit MDA formation [36]. Moreover, Pro treatment also mitigated the lipid peroxidation induced by Cd stress in the intestine of pearl gentian grouper [22]. In agreement with previous studies, our data showed that lipid peroxidation was inhibited by 0.4 and 0.8 g/kg Pro treatments. These results highlight the antioxidation ability of Pro to avoid lipid peroxidation. It is worth noting that Pro treatment was observed to effectively suppress lipid peroxidation in muscle, potentially leading to an enhancement in meat nutritional quality. Low lipid peroxidation may suppress the degradation of PUFA and vitamins, and formation of harmful substances [42].

4.3. The Effect of Pros on Muscle Nutrient Quality

The muscle of fish is the primary edible portion and a valuable source of nutriment for humans. Amino acid composition is a vital determinant of fish muscle nutritional quality. The addition of Pros to the diet was found to increase the crude protein content of the body in American eel [21] and tilapia [20], but the impact of Pros on amino acid composition in fish muscle remains unclear. This study showed that the amino acid composition including EAA and NEAA was not significantly changed in the muscle of common carp after Pro treatment. Similar effects of Pro were also observed in finishing pigs [41]. However, we found that dietary Pro supplementation significantly changed the levels of some amino acids analogs, such as increased DL-methionine and L-glutamic acid, in the muscle of common carp. DL-methionine is not only a vital source of methionine, but it also serves as a precursor to essential intermediates such as glutathione [43]. Glutamic acid was considered as a flavor-related amino acid, which would enhance the flavor of fish flesh [44]. Therefore, the increased levels of DL-methionine and L-glutamic acid may have a positive effect on flesh quality in the muscle of common carp.

The composition of fatty acids is also a crucial factor in evaluating the nutritional and health benefits of fish muscle. A high percentage of PUFAs in food has been linked to improved fetal development, brain health, and a reduced risk of coronary heart disease in humans [44]. In this study, the content of PUFAs, including both n-3 and n-6, in muscle was significantly increased by dietary supplementation of 0.8 g/kg Pro. More specifically, Pro treatment increased the levels of LA, docosapentaenoic acid (DPA), and ARA in the muscle. Similar findings were also observed in the muscle tissue of pigs treated with Pros [40]. n-3 PUFAs have been confirmed to possess properties that improve anti-inflammation, antioxidant capacity, and meat nutritional value [45]. LA, an essential fatty acid (EFA), exhibits cardiovascular-protective, neuro-protective, anti-osteoporotic, anti-inflammatory, and antioxidative effects [46]. ARA is also an EFA maintaining normal health, which plays a vital role in the functioning of all cells, especially in the nervous system, skeletal muscles, and immune system [47]. DPA is an important long-chain n-3 PUFA that can serve as a dietary source of eicosapentaenoic acid. It is also known to play a role in improving risk markers associated with cardiovascular and metabolic diseases [48]. Here, the increase in the levels of LA, DPA, ARA, and n-3 PUFAs suggests that dietary Pro supplementation may improve the nutritional and health benefits of common carp. Moreover, we found that α-LA metabolism, ARA metabolism, and biosynthesis of UFA pathways were significantly altered by Pro treatment, which may provide an explanation for the increase in the levels of these PUFAs.

PC and PE are the most abundant phospholipids in all types of mammalian cells and subcellular organelles, playing a crucial role in regulating lipid, lipoprotein, and energy metabolism [49]. They also acts as reservoirs for essential PUFAs like DHA and ARA [50]. Furthermore, changes in PC and PE are associated with the formation of volatile flavor compounds in muscle tissue [51]. In this study, metabolome analysis revealed an increase in the levels of PC (16:0/15:0), PC (22:5/15:0), and PC (18:0/15:0), but a decrease in the levels of PE (16:0/18:2) and PC (22:6/22:6) in muscle tissue after Pro treatment. We hypothesize that these changes in PC and PE levels may impact the nutritional value and volatile flavor of muscle, but the detailed mechanism regarding this phenomenon still requires further investigation. In addition, the decreased PE may prevent coalescence of the lipid droplets in muscle tissue [52].

Lipids are vital factors that affect meat quality, and excessive lipid content can reduce both meat quality and feed efficiency. Among the various types of lipids, triglycerides (TGs) are the most abundant in the muscle of many fishes [53]. Modifying TG content can reduce the risk of cardiovascular disease in consumers [54]. Previous studies have suggested that flavonoids can decrease the TG content of meat and increase the proportions of total PUFA in the breast muscle [55]. Consistent with these findings, the content of TGs, including TG (18:1/22:2/o-18:0), TG (16:1/24:1/o-18:0), and TG (18:0/22:4/o-18:0), in the muscle was reduced by Pro treatment. This reduction could potentially have positive impacts on the health of consumers who consume carp as a food. Animal models studies have demonstrated that Pros have a positive effect on the TG metabolism, such as reducing plasma TGs and controlling the endogenous liver lipid production (reviewed in [56]). Based on these findings, we hypothesize that this is a possible explanation for the decrease of TG levels in muscle tissue.

It is intriguing to note that the levels of 13 bioactive compounds, including 8 flavonoids, 2 isoflavonoids, 2 tannins, and 1 coumarin, significantly increase in the muscle of common carp after being fed a diet containing Pro. These bioactive compounds have been proven to possess health benefits, such as antioxidant, anti-inflammatory, neuroprotective, and hepatoprotective effects, in animals [57]. They can also improves meat quality by regulating lipid metabolism and antioxidant capacity [58]. The increase of bioactive compounds in common carp muscle may be attributed to the absorption, metabolism, and accumulation of Pro. It has been reported that Pros may undergo direct absorption in the proximal intestinal tract or be absorbed in the gastrointestinal tract after being metabolized by gut microbiota in mammals [59,60]. Absorbed Pros and their metabolites can be transported to other organs through the circulation, exerting health-promoting effects [38]. However, the mechanism of absorption and metabolism of Pros in fish remains unclear due to significant differences in intestinal structure between fish and mammals.

4.4. The Effect of Pros on Lipid Metabolism in Intestines

Intestinal barrier integrity is crucial for maintaining intestinal health, as it not only aids in nutrient absorption but also shields against harmful substances. Numerous studies have shown that Pros administration can ameliorate intestinal dysbiosis caused by dietary factors [61]. They have protective effects against inflammatory response in the intestinal barrier [62], and prevent metabolic syndrome by regulating intestinal function and energy metabolism [63]. Moreover, previous studies have reported that Pros administration regulates intestinal lipid homeostasis to improve cardiometabolic disorders [64]. Our study found that Pro treatment resulted in a high enrichment of lipid catabolic process, including glycolipid and sphingolipid catabolic processes, suggesting that Pro treatment may enhance lipid catabolic metabolism. Increasing lipid catabolic metabolism may inhibit the accumulation of lipids in the intestines and increase the availability of fatty acids in the body [65,66]. This process may also provide a source of energy and help maintain normal metabolic activity in intestines. Dietary polyphenols have been reported to improve glycolipid metabolism disorders in animals [67]. For instance, Xu et al., (2019) found that Pro treatment ameliorated intestinal barrier dysfunction induced by a high-fat diet in rats by modulating glycolipid digestion [68]. In this study, the upregulation of glycolipid catabolic process may indicate a beneficial effect of Pros in inhibiting cellular glycolipid accumulation. Sphingolipids are one of the most important membrane lipids, participating in the formation of membrane microdomains. However, the abnormal accumulation of sphingolipids in cells has been associated with metabolic disorders [69]. In this study, the upregulation in sphingolipid catabolic process may have an influence on cell signaling, cellular homeostasis, and immune regulation in intestines. Notably, the changes in the sphingolipid catabolic process were consistent with alterations in the lysosomal pathway, indicating that sphingolipids were possibly degraded via the lysosomal catabolic pathways [70]. Lysosomes, small organelles that contain various hydrolytic enzymes such as proteases, lipases, and nucleases, can promote lipid catabolism and transport [71]. Disruption of lysosome function is considered a key factor leading to metabolic derangement and neurodegeneration [72]. Therefore, the upregulation of the lysosomal pathway in this study indicates that Pro treatment may contribute to maintaining cellular lipid homeostasis and preventing metabolic diseases.

It has been confirmed that Pro has a hypolipidemic effect, and one of its possible mechanisms may be related to the delay of cholesterol and lipid absorption in intestines [56]. Pros supplementation reduced cholesterol absorption by increasing the excretion of neutral steroids and bile acids [73,74]. Treatment with red wine polyphenolics (containing Pros) has been found to reduce free cholesterol and total cholesterol in Caco-2 cells [75]. In this study, Pro treatment caused lower enrichment of gene expression in intestinal cholesterol absorption, suggesting that Pros interfered in this process. Meanwhile, we also observed a significant downregulation in the mRNA level of CYP27A (an important enzyme regulating cholesterol metabolism), suggesting that Pro treatment may exert inhibitory effects on cholesterol metabolism in the intestines.

It is worth noting that our results also indicated a downregulation of the PPARα signaling pathway after Pros administration, which may further affect lipid absorption. PPARα has been identified as a key regulator in lipid metabolism, involved in various processes including fatty acid transport, synthesis, and oxidation, as well as lipogenesis [76]. In Caco2 cells, Pro treatment repressed *pparα* and *acsl3* gene expression to decrease TG secretion [77]. However, contradictory results have also been reported, in which Pro treatment upregulated PPARa and CPT1 and downregulated ACC and SREBP1 to modify intestinal lipid homeostasis [64]. In this study, the decreased expression of *pparα* further inhibited its target genes, such as *apoa1*, *maeB*, *scd*, *CD36*, *acsl*, and *cpt1*, suggesting that lipogenesis, fatty acid transport, fatty acid oxidation, and lipid transport may be suppressed by Pro treatment in the intestines of common carp.

4.5. The Effect of Pros on Intestinal Microbiota

Intestinal microbiota is increasingly being linked to fish health, playing a crucial role in regulating intestinal immunity, nutrient absorption, and overall host health. Dietary Pro has been studied in relation to intestinal microbiota in animals, but the findings are inconsistent and contentious, possibly due to variations in Pros sources or types, as well as differences in animal models [37]. In a piglet model, Pro treatment increased the abundance and diversity of intestinal bacteria [78]. Similarly, Pro treatment improved the decrease in α-diversity induced by a high fat diet in C57BL/6J mice [79]. In this study, the Pro group had a higher number of ASVs, and a significantly higher Simpson's index of diversity in intestinal bacteria, compared to the NC group. Meanwhile, in conjunction with the β-diversity analysis, we speculate that Pro treatment may enhance the diversity of the intestinal microbiota in common carp. Interestingly, we also found that Pro treatment altered the relative abundance in the top four phyla and genus of the intestinal microbiota and resulted in a more balanced distribution among dominant taxa.

In animal intestines, the Firmicutes and Bacteroidetes phyla are dominant and play essential roles in promoting host health, boosting immunity, and maintaining homeostasis [80]. An elevated ratio of F/B is linked with some metabolic syndrome, such as obesity and chronic inflammation [81]. A marked reduction in the ratio of F/B was observed in ovariectomized mice after treatment with Pros (grape seed extract) [82]. Moreover, Pro treatment has been shown to alter the gut microbiota by increasing Bacteroidetes and decreasing Firmicutes, effectively alleviating metabolic syndrome induced by a high-fat diet (reviewed by Redondo-Castillejo, et al. [83]). Consistent with previous findings, our data also revealed a significant decrease in Firmicutes abundance and a 2.35-fold increase (but $p > 0.05$) in Bacteroidota abundance in the Pro group. Meanwhile, the ratio of F/B was significantly decreased by Pro treatment. These data suggest that dietary Pro may prevent metabolic syndrome in the intestines of common carp. In addition, some studies have found that dysbiotic Proteobacteria expansion is associated with epithelial dysfunction and inflammation [83,84]. Our results showed a lower Proteobacteria abundance in the Pro group, suggesting that Pro may have potential effects against epithelial dysfunction and inflammation in intestines.

Further, we observed the impact of Pro on intestinal microbiota at the genus. Here, compared with the NC group, there was a higher relative abundance in *Cetobacterium* and *Pirellula*, and a lower relative abundance in *Vibrio*, at the genus level. *Cetobacterium* is a dominant taxa of the intestinal microbiota, being involved in maintaining fish health, enhancing nutrition, and providing protection against pathogenic bacteria [85,86]. *Pirellula*, an important bacterial genus in fish intestines, has the potential to act as a probiotic, positively influencing fish growth [87]. *Vibrio* is also an important and diverse genus of bacteria, being widely distributed in fish intestines and the aquatic environment. However, a specific group within the *Vibrio* genus consists of several highly pathogenic species, such as *V. anguillarum*, *V. harveyi*, and *V. alginolyticus*, that can cause diseases in aquatic animals [88]. Based on previously published results and our data, we hypothesize that the increase in *Cetobacterium* and *Pirellula*, coupled with the decrease in *Vibrio*, in the Pro group may indicate positive effects on the resistance of common carp to pathogenic bacteria infection. Of interest, our phenotype analysis also showed a significant decrease in potential pathogenicity in the Pro group. It is worth noting that the changes of the intestinal microbiota showed a significant correlation with intestinal lipid metabolism. Taken together, it can be deduced that dietary Pro may enhance disease resistance, as well as regulate intestinal lipid metabolism in common carp, by altering the diversity and abundance of intestinal microbiota. However, further research is needed to elucidate the detailed underlying mechanism. In addition, Pro treatment also resulted in changes in some microbiota at the genus level, such as *ZOR0006*, *Brevinema*, *Planctopirus*, and *Pseudorhodobacter*. However, the potential effects of these changes on the host remain uncertain, as there is limited knowledge regarding these bacteria.

5. Conclusions

In summary, dietary Pro supplementation improved growth performance, antioxidant ability, and muscle nutrients in common carp. Dietary Pro may regulate lipid absorption and accumulation by upregulating lipid catabolic metabolism, inhibiting cholesterol absorption, and downregulating the PPARα pathway in the intestines. Meanwhile, Pro supplementation improved intestinal microbiota diversity and induced alterations in intestinal microbial composition and functions. Specifically, the alterations in microbiota may participate in the regulation of lipid metabolism, enhance disease resistance, and improve muscle nutrients through the interaction between intestinal microbiota and host metabolism. Our study not only provides novel insights into the action of Pro in enhancing growth, improving muscle nutrients, and regulating lipid metabolism, but also strongly supports the application of Pro as a feed additive in aquaculture.

Supplementary Materials: The following supporting information can be downloaded at: https://www.mdpi.com/article/10.3390/antiox12122095/s1, Figure S1: Schematic diagrams for experimental design and sampling; Figure S2: PCA score plot assessing quality of metabolomics data; Figure S3: Metabolome analysis of muscle between NC and Pro groups; Figure S4: DEGs in intestines of *C. carpio* between the NC and Pro groups; Figure S5: GO enrichment analysis for the DEGs in the intestines of *C. carpio* between the NC and Pro groups; Figure S6: KEGG enrichment analysis for the DEGs in the intestines of *C. carpio* between the NC and Pro groups; Figure S7: Venn diagram in ASVs (A), phylum (B), and genus (C) of intestinal microbiota in NC and Pro groups; Table S1: Formation and nutrient of the experimental diets [89]; Table S2: Specific primer sequences for qPCR in the study; Table S3: The hydrolyzed amino acid composition in muscle of C. carpio fed on normal diet (NC) and Pro-supplemented diet (Pro); Table S4: Valid data used in transcriptome analysis.

Author Contributions: Conceptualization, R.J., B.L. and J.Z.; methodology, R.J. and Y.H.; software, W.F.; validation, W.F. and M.N.; formal analysis, Y.H.; investigation, R.J. and M.N.; resources, B.L.; data curation, Y.H.; writing—original draft preparation, R.J.; writing—review and editing, B.L. and J.Z.; visualization, R.J.; supervision, B.L.; project administration, J.Z.; funding acquisition, J.Z. All authors have read and agreed to the published version of the manuscript.

Funding: This research was funded by the earmarked fund for CARS (CARS-45), Central Public-interest Scientific Institution Basal Research Fund, CAFS (2023TD64), Young Science-Technology Talents Support Project of Jiangsu Association Science and Technology (TJ-2021-076), the National Key R&D Program of China (2019YFD0900305), and Wuxi Modern Industry Development Fund Project (K20221053).

Institutional Review Board Statement: All animals in this study were approved by the Animal Care and Use Ethics Committee of the Freshwater Fisheries (2020TD60, 11 May 2022) and all procedures were performed according to Jiangsu Laboratory's Animal Management Guidelines (014000319/2008-00079).

Informed Consent Statement: Not applicable.

Data Availability Statement: The raw data of 16S rRNA and transcriptome sequencing used in this study have been submitted to open database NCBI Sequence Read Archive (SRA) and the accession numbers were PRJNA1012120 and PRJNA1012106, respectively. All other data are contained within the main manuscript and supplementary material.

Conflicts of Interest: The authors declare no conflict of interest.

References

1. Bondad-Reantaso, M.G.; Subasinghe, R.P.; Arthur, J.R.; Ogawa, K.; Chinabut, S.; Adlard, R.; Tan, Z.; Shariff, M. Disease and health management in Asian aquaculture. *Vet. Parasitol.* **2005**, *132*, 249–272. [CrossRef]
2. Rico, A.; Phu, T.M.; Satapornvanit, K.; Min, J.; Shahabuddin, A.M.; Henriksson, P.J.G.; Murray, F.J.; Little, D.C.; Dalsgaard, A.; Van den Brink, P.J. Use of veterinary medicines, feed additives and probiotics in four major internationally traded aquaculture species farmed in Asia. *Aquaculture* **2013**, *412–413*, 231–243. [CrossRef]
3. Chen, J.; Sun, R.; Pan, C.; Sun, Y.; Mai, B.; Li, Q.X. Antibiotics and Food Safety in Aquaculture. *J. Agric. Food Chem.* **2020**, *68*, 11908–11919. [CrossRef] [PubMed]

4. Bondad-Reantaso, M.G.; MacKinnon, B.; Karunasagar, I.; Fridman, S.; Alday-Sanz, V.; Brun, E.; Le Groumellec, M.; Li, A.; Surachetpong, W.; Karunasagar, I.; et al. Review of alternatives to antibiotic use in aquaculture. *Rev. Aquac.* **2023**, *15*, 1421–1451. [CrossRef]
5. Pereira, W.A.; Mendonca, C.M.N.; Villasante Urquiza, A.; Marteinsson, V.P.; Guy LeBlanc, J.; Cotter, P.D.; Figueroa Villalobos, E.; Romero, J.; Oliveira, R.P.S. Use of Probiotic Bacteria and Bacteriocins as an Alternative to Antibiotics in Aquaculture. *Microorganisms* **2022**, *10*, 1705. [CrossRef] [PubMed]
6. Tadese, D.A.; Song, C.Y.; Sun, C.X.; Liu, B.; Liu, B.; Zhou, Q.L.; Xu, P.; Ge, X.P.; Liu, M.Y.; Xu, X.D.; et al. The role of currently used medicinal plants in aquaculture and their action mechanisms: A review. *Rev. Aquac.* **2022**, *14*, 816–847. [CrossRef]
7. Reverter, M.; Bontemps, N.; Lecchini, D.; Banaigs, B.; Sasal, P. Use of plant extracts in fish aquaculture as an alternative to chemotherapy: Current status and future perspectives. *Aquaculture* **2014**, *433*, 50–61. [CrossRef]
8. Kuebutornye, F.K.A.; Abarike, E.D. The contribution of medicinal plants to tilapia aquaculture: A review. *Aquac. Int.* **2020**, *28*, 965–983. [CrossRef]
9. Pu, H.; Li, X.; Du, Q.; Cui, H.; Xu, Y. Research Progress in the Application of Chinese Herbal Medicines in Aquaculture: A Review. *Engineering* **2017**, *3*, 731–737. [CrossRef]
10. Rippin; Sharma, A.K.; Beniwal, V. Biosynthesis and medicinal applications of proanthocyanidins: A recent update. *Biocatal. Agric. Biotechnol.* **2022**, *45*, 102500. [CrossRef]
11. Wang, P.; Ma, G.; Zhang, L.; Li, Y.; Fu, Z.; Kan, X.; Han, Y.; Wang, H.; Jiang, X.; Liu, Y. A sucrose-induced MYB (SIMYB) transcription factor promoting proanthocyanidin accumulation in the tea plant (Camellia sinensis). *J. Agric. Food Chem.* **2019**, *67*, 1418–1428. [CrossRef]
12. Rauf, A.; Imran, M.; Abu-Izneid, T.; Iahtisham Ul, H.; Patel, S.; Pan, X.; Naz, S.; Sanches Silva, A.; Saeed, F.; Rasul Suleria, H.A. Proanthocyanidins: A comprehensive review. *Biomed. Pharmacother.* **2019**, *116*, 108999. [CrossRef] [PubMed]
13. Cai, Y.-Z.; Mei, S.; Jie, X.; Luo, Q.; Corke, H. Structure–radical scavenging activity relationships of phenolic compounds from traditional Chinese medicinal plants. *Life Sci.* **2006**, *78*, 2872–2888. [CrossRef] [PubMed]
14. Gil-Cardoso, K.; Ginés, I.; Pinent, M.; Ardévol, A.; Arola, L.; Blay, M.; Terra, X. Chronic supplementation with dietary proanthocyanidins protects from diet-induced intestinal alterations in obese rats. *Mol. Nutr. Food Res.* **2017**, *61*, 1601039. [CrossRef] [PubMed]
15. Ferreira, Y.A.M.; Jamar, G.; Estadella, D.; Pisani, L.P. Proanthocyanidins in grape seeds and their role in gut microbiota-white adipose tissue axis. *Food Chem.* **2023**, *404*, 134405. [CrossRef]
16. Kruger, M.J.; Davies, N.; Myburgh, K.H.; Lecour, S. Proanthocyanidins, anthocyanins and cardiovascular diseases. *Food Res. Int.* **2014**, *59*, 41–52. [CrossRef]
17. Strathearn, K.E.; Yousef, G.G.; Grace, M.H.; Roy, S.L.; Tambe, M.A.; Ferruzzi, M.G.; Wu, Q.-L.; Simon, J.E.; Lila, M.A.; Rochet, J.-C. Neuroprotective effects of anthocyanin- and proanthocyanidin-rich extracts in cellular models of Parkinson's disease. *Brain Res.* **2014**, *1555*, 60–77. [CrossRef] [PubMed]
18. Jiang, X.; Liu, J.; Lin, Q.; Mao, K.; Tian, F.; Jing, C.; Wang, C.; Ding, L.; Pang, C. Proanthocyanidin prevents lipopolysaccharide-induced depressive-like behavior in mice via neuroinflammatory pathway. *Brain Res. Bull.* **2017**, *135*, 40–46. [CrossRef]
19. Zhu, F. Proanthocyanidins in cereals and pseudocereals. *Crit. Rev. Food Sci. Nutr.* **2019**, *59*, 1521–1533. [CrossRef]
20. Zhai, S.-W.; Lu, J.-J.; Chen, X.-H. Effects of dietary grape seed proanthocyanidins on growth performance, some serum biochemical parameters and body composition of tilapia (*Oreochromis niloticus*) fingerlings. *Ital. J. Anim. Sci.* **2014**, *13*, 3357. [CrossRef]
21. Wang, Y.; Chen, X.-H.; Wu, X.-Y.; Cai, G.-H.; Zhai, S.-W. Effects of Dietary Supplementation of Peanut Skin Proanthocyanidins on Growth Performance and Lipid Metabolism of the Juvenile American Eel (*Anguilla rostrata*). *Animals* **2022**, *12*, 2375. [CrossRef] [PubMed]
22. Jia, Z.Y.; Tan, Y.T.; Liu, Y.M.; Cai, G.H.; Chen, X.H.; Zhai, S.W. Grape seed proanthocyanidins alleviate the negative effects of dietary cadmium on pearl gentian grouper (Epinephelus fuscoguttatus female x Epinephelus lanceolatus male). *Isr. J. Aquac. -Bamidgeh* **2021**, *73*, 1427267. [CrossRef]
23. Zhai, S.; Wang, Y.; He, Y.; Chen, X. Oligomeric proanthocyanidins counteracts the negative effects of high level of dietary histamine on american eel (*Anguilla rostrata*). *Front. Mar. Sci.* **2020**, *7*, 549145. [CrossRef]
24. Lu, R.H.; Qin, C.B.; Yang, F.; Zhang, W.Y.; Zhang, Y.R.; Yang, G.K.; Yang, L.P.; Meng, X.L.; Yan, X.; Nie, G.X. Grape seed proanthocyanidin extract ameliorates hepatic lipid accumulation and inflammation in grass carp (*Ctenopharyngodon idella*). *Fish Physiol. Biochem.* **2020**, *46*, 1665–1677. [CrossRef]
25. Ma, J.-L.; Qiang, J.; Tao, Y.-F.; Bao, J.-W.; Zhu, H.-J.; Li, L.-G.; Xu, P. Multi-omics analysis reveals the glycolipid metabolism response mechanism in the liver of genetically improved farmed Tilapia (GIFT, *Oreochromis niloticus*) under hypoxia stress. *BMC Genom.* **2021**, *22*, 105. [CrossRef]
26. Roh, H.; Kim, A.; Kim, N.; Lee, Y.; Kim, D.-H. Multi-omics analysis provides novel insight into immuno-physiological pathways and development of thermal resistance in rainbow trout exposed to acute thermal stress. *Int. J. Mol. Sci.* **2020**, *21*, 9198. [CrossRef] [PubMed]
27. Natnan, M.E.; Mayalvanan, Y.; Jazamuddin, F.M.; Aizat, W.M.; Low, C.-F.; Goh, H.-H.; Azizan, K.A.; Bunawan, H.; Baharum, S.N. Omics strategies in current advancements of infectious fish disease management. *Biology* **2021**, *10*, 1086. [CrossRef]
28. Li, Q.J.; Zheng, Y.; Sun, Y.; Xu, G.C. Resveratrol attenuated fatty acid synthesis through MAPK-PPAR pathway in red tilapia. *Comp. Biochem. Physiol. C-Toxicol. Pharmacol.* **2023**, *268*, 109598. [CrossRef] [PubMed]

29. Ahmadifar, E.; Mohammadzadeh, S.; Kalhor, N.; Yousefi, M.; Moghadam, M.S.; Naraballobh, W.; Ahmadifar, M.; Hoseinifar, S.H.; Van Doan, H. Cornelian cherry (*Cornus mas* L.) fruit extract improves growth performance, disease resistance, and serum immune-and antioxidant-related gene expression of common carp (*Cyprinus carpio*). *Aquaculture* **2022**, *558*, 738372. [CrossRef]
30. Paray, B.A.; Hoseini, S.M.; Hoseinifar, S.H.; Hien Van, D. Effects of dietary oak (*Quercus castaneifolia*) leaf extract on growth, antioxidant, and immune characteristics and responses to crowding stress in common carp (*Cyprinus carpio*). *Aquaculture* **2020**, *524*, 735276. [CrossRef]
31. Blüthgen, N.; Meili, N.; Chew, G.; Odermatt, A.; Fent, K. Accumulation and effects of the UV-filter octocrylene in adult and embryonic zebrafish (*Danio rerio*). *Sci. Total Environ.* **2014**, *476–477*, 207–217. [CrossRef]
32. Yang, E.-j.; Amenyogbe, E.; Zhang, J.-d.; Wang, W.-z.; Huang, J.-s.; Chen, G. Integrated transcriptomics and metabolomics analysis of the intestine of cobia (*Rachycentron canadum*) under hypoxia stress. *Aquac. Rep.* **2022**, *25*, 101261. [CrossRef]
33. *GB5009.168–2016*; National Food Safety Standard—Determination of Fatty Acid in Foods. National Health and Family Planning Commission of the China: Beijing, China, 2016.
34. He, Y.X.; Huang, L.X.; Xia, K.D.; Chen, X.H.; Zhai, S.W. Effect of grape seed proanthocyanidins on growth performance and antioxidant ability of European eels (*Anguilla anguilla*). *Feed Res.* **2019**, *7*, 20–23.
35. Zhai, S.-W.; Lu, J.-J.; Zhao, P.-Y.; Chen, X.-H. Effect of grape seed proanthocyanidins on alleviating dietary cadmium (Cd) induced growth retardation and oxidative stress in hepatopancreas of juvenile tilapia (Oreochromis niloticus). *Isr. J. Aquac.-Bamidgeh* **2018**, *70*, 1513. [CrossRef]
36. Xu, G.; Xing, W.; Li, T.; Yu, H.; Wei, S.; Jiang, N.; Ma, Z.; Luo, L. Dietary grape seed proanthocyanidins improved growth, immunity, antioxidant, digestive enzymes activities, and intestinal microbiota of juvenile hybrid sturgeon (Acipenser baeri Brandt ♀× A. schrenckii Brandt ♂). *Aquac. Nutr.* **2021**, *27*, 1983–1995. [CrossRef]
37. Wang, Y.; Liang, P.; Chen, X.-H.; Zhai, S.-W. Effects of dietary proanthocyanidins supplementation on growth performance, digestive enzymes activities and microbiota in the intestine of juvenile American eels (Anguilla rostrata) cultured in cement tanks. *Isr. J. Aquac.* **2020**, *70*, 1226208. [CrossRef]
38. Tao, W.; Zhang, Y.; Shen, X.; Cao, Y.; Shi, J.; Ye, X.; Chen, S. Rethinking the mechanism of the health benefits of proanthocyanidins: Absorption, metabolism, and interaction with gut microbiota. *Compr. Rev. Food Sci. Food Saf.* **2019**, *18*, 971–985. [CrossRef]
39. Bagchi, D.; Swaroop, A.; Preuss, H.G.; Bagchi, M. Free radical scavenging, antioxidant and cancer chemoprevention by grape seed proanthocyanidin: An overview. *Mutat. Res./Fundam. Mol. Mech. Mutagen.* **2014**, *768*, 69–73. [CrossRef] [PubMed]
40. Yang, L.Y.; Xian, D.H.; Xiong, X.; Lai, R.; Song, J.; Zhong, J.Q. Proanthocyanidins against Oxidative Stress: From Molecular Mechanisms to Clinical Applications. *Biomed Res. Int.* **2018**, *2018*, 8584136. [CrossRef]
41. Xu, M.; Chen, X.; Huang, Z.; Chen, D.; Li, M.; He, J.; Chen, H.; Zheng, P.; Yu, J.; Luo, Y. Effects of dietary grape seed proanthocyanidin extract supplementation on meat quality, muscle fiber characteristics and antioxidant capacity of finishing pigs. *Food Chem.* **2022**, *367*, 130781. [CrossRef]
42. Iglesias, J.; Pazos, M.; Torres, J.L.; Medina, I. Antioxidant mechanism of grape procyanidins in muscle tissues: Redox interactions with endogenous ascorbic acid and α-tocopherol. *Food Chem.* **2012**, *134*, 1767–1774. [CrossRef]
43. Jiang, S.-G.; Pan, N.-X.; Chen, M.-J.; Wang, X.-Q.; Yan, H.-C.; Gao, C.-Q. Effects of dietary supplementation with dl-methionine and dl-methionyl-dl-methionine in breeding pigeons on the carcass characteristics, meat quality and antioxidant activity of squabs. *Antioxidants* **2019**, *8*, 435. [CrossRef]
44. Wang, J.; Jiang, H.; Alhamoud, Y.; Chen, Y.; Zhuang, J.; Liu, T.; Cai, L.; Shen, W.; Wu, X.; Zheng, W. Integrated metabolomic and gene expression analyses to study the effects of glycerol monolaurate on flesh quality in large yellow croaker (Larimichthys crocea). *Food Chem.* **2022**, *367*, 130749. [CrossRef]
45. Canalis, M.B.; Baroni, M.; León, A.; Ribotta, P. Effect of peach puree incorportion on cookie quality and on simulated digestion of polyphenols and antioxidant properties. *Food Chem.* **2020**, *333*, 127464. [CrossRef] [PubMed]
46. Yuan, Q.; Xie, F.; Huang, W.; Hu, M.; Yan, Q.; Chen, Z.; Zheng, Y.; Liu, L. The review of alpha-linolenic acid: Sources, metabolism, and pharmacology. *Phytother. Res.* **2022**, *36*, 164–188. [CrossRef] [PubMed]
47. Tallima, H.; El Ridi, R. Arachidonic acid: Physiological roles and potential health benefits—A review. *J. Adv. Res.* **2018**, *11*, 33–41. [CrossRef] [PubMed]
48. Drouin, G.; Rioux, V.; Legrand, P. The n-3 docosapentaenoic acid (DPA): A new player in the n-3 long chain polyunsaturated fatty acid family. *Biochimie* **2019**, *159*, 36–48. [CrossRef] [PubMed]
49. van der Veen, J.N.; Kennelly, J.P.; Wan, S.; Vance, J.E.; Vance, D.E.; Jacobs, R.L. The critical role of phosphatidylcholine and phosphatidylethanolamine metabolism in health and disease. *Biochim. Et Biophys. Acta (BBA)—Biomembr.* **2017**, *1859*, 1558–1572. [CrossRef]
50. Cui, Z.; Houweling, M. Phosphatidylcholine and cell death. *Biochim. Et Biophys. Acta (BBA)—Mol. Cell Biol. Lipids* **2002**, *1585*, 87–96. [CrossRef]
51. Cui, Z.Y.; Liu, C.; Rao, W.X.; Chen, P.; Lei, K.K.; Mai, K.S.; Zhang, W.B. Dietary phospholipids improve growth performance and change the lipid composition and volatile flavor compound profiles in the muscle of abalone Haliotis discus hannai by affecting the glycerophospholipid metabolism. *Aquac. Rep.* **2023**, *30*, 101567. [CrossRef]
52. Hafez, I.M.; Cullis, P.R. Roles of lipid polymorphism in intracellular delivery. *Adv. Drug Deliv. Rev.* **2001**, *47*, 139–148. [CrossRef]

53. Wang, X.; Zhang, H.; Song, Y.; Cong, P.; Li, Z.; Xu, J.; Xue, C. Comparative lipid profile analysis of four fish species by ultraperformance liquid chromatography coupled with quadrupole time-of-flight mass spectrometry. *J. Agric. Food Chem.* **2019**, *67*, 9423–9431. [CrossRef]
54. Cavallini, D.C.; Bedani, R.; Bomdespacho, L.Q.; Vendramini, R.C.; Rossi, E.A. Effects of probiotic bacteria, isoflavones and simvastatin on lipid profile and atherosclerosis in cholesterol-fed rabbits: A randomized double-blind study. *Lipids Health Dis.* **2009**, *8*, 1–8. [CrossRef] [PubMed]
55. Kamboh, A.; Leghari, R.; Khan, M.; Kaka, U.; Naseer, M.; Sazili, A.; Malhi, K. Flavonoids supplementation-An ideal approach to improve quality of poultry products. *World's Poult. Sci. J.* **2019**, *75*, 115–126. [CrossRef]
56. Bladé, C.; Arola, L.; Salvadó, M.J. Hypolipidemic effects of proanthocyanidins and their underlying biochemical and molecular mechanisms. *Mol. Nutr. Food Res.* **2010**, *54*, 37–59. [CrossRef]
57. Álvarez-Rodríguez, J.; Urrutia, O.; Lobón, S.; Ripoll, G.; Bertolín, J.R.; Joy, M. Insights into the role of major bioactive dietary nutrients in lamb meat quality: A review. *J. Anim. Sci. Biotechnol.* **2022**, *13*, 20. [CrossRef]
58. Guo, M.; Wang, Z.; Gao, Z.; Ma, J.; Huangfu, W.; Niu, J.; Liu, B.; Li, D.; Zhu, X.; Sun, H.; et al. Alfalfa leaf meal as a new protein feedstuff improves meat quality by modulating lipid metabolism and antioxidant capacity of finishing pigs. *Food Chem. X* **2023**, *19*, 100815. [CrossRef]
59. Wu, T.; Grootaert, C.; Voorspoels, S.; Jacobs, G.; Pitart, J.; Kamiloglu, S.; Possemiers, S.; Heinonen, M.; Kardum, N.; Glibetic, M. Aronia (Aronia melanocarpa) phenolics bioavailability in a combined in vitro digestion/Caco-2 cell model is structure and colon region dependent. *J. Funct. Foods* **2017**, *38*, 128–139. [CrossRef]
60. Raab, T.; Barron, D.; Vera, F.A.; Crespy, V.; Oliveira, M.; Williamson, G. Catechin glucosides: Occurrence, synthesis, and stability. *J. Agric. Food Chem.* **2010**, *58*, 2138–2149. [CrossRef]
61. Zhang, X.; Song, X.; Hu, X.; Chen, F.; Ma, C. Health benefits of proanthocyanidins linking with gastrointestinal modulation: An updated review. *Food Chem.* **2022**, *404*, 134596. [CrossRef]
62. Sheng, K.; Zhang, G.; Sun, M.; He, S.; Kong, X.; Wang, J.; Zhu, F.; Zha, X.; Wang, Y. Grape seed proanthocyanidin extract ameliorates dextran sulfate sodium-induced colitis through intestinal barrier improvement, oxidative stress reduction, and inflammatory cytokines and gut microbiota modulation. *Food Funct.* **2020**, *11*, 7817–7829. [CrossRef]
63. Casanova-Martí, À.; González-Abuín, N.; Serrano, J.; Blay, M.T.; Terra, X.; Frost, G.; Pinent, M.; Ardévol, A. Long Term Exposure to a Grape Seed Proanthocyanidin Extract Enhances L-Cell Differentiation in Intestinal Organoids. *Mol. Nutr. Food Res.* **2020**, *64*, 2000303. [CrossRef]
64. Koudoufio, M.; Feldman, F.; Ahmarani, L.; Delvin, E.; Spahis, S.; Desjardins, Y.; Levy, E. Intestinal protection by proanthocyanidins involves anti-oxidative and anti-inflammatory actions in association with an improvement of insulin sensitivity, lipid and glucose homeostasis. *Sci. Rep.* **2021**, *11*, 3878. [CrossRef]
65. Xiao, W.; Wang, X.; Wang, T.; Chen, B.; Xing, J. HAO2 inhibits malignancy of clear cell renal cell carcinoma by promoting lipid catabolic process. *J. Cell Physiol.* **2019**, *234*, 23005–23016. [CrossRef]
66. Thomes, P.G.; Rasineni, K.; Yang, L.; Donohue, T.M.; Kubik, J.L.; McNiven, M.A.; Casey, C.A. Ethanol withdrawal mitigates fatty liver by normalizing lipid catabolism. *Am. J. Physiol.-Gastrointest. Liver Physiol.* **2019**, *316*, G509–G518. [CrossRef]
67. Wang, Z.; Zeng, M.; Wang, Z.; Qin, F.; Chen, J.; He, Z. Dietary luteolin: A narrative review focusing on its pharmacokinetic properties and effects on glycolipid metabolism. *J. Agric. Food Chem.* **2021**, *69*, 1441–1454. [CrossRef]
68. Xu, H.; Zhao, C.; Li, Y.; Liu, R.; Ao, M.; Li, F.; Yao, Y.; Tao, Z.; Yu, L. The ameliorative effect of the Pyracantha fortuneana (Maxim.) HL Li extract on intestinal barrier dysfunction through modulating glycolipid digestion and gut microbiota in high fat diet-fed rats. *Food Funct.* **2019**, *10*, 6517–6532. [CrossRef]
69. Sandhoff, K. Neuronal sphingolipidoses: Membrane lipids and sphingolipid activator proteins regulate lysosomal sphingolipid catabolism. *Biochimie* **2016**, *130*, 146–151. [CrossRef]
70. Schulze, H.; Sandhoff, K. Sphingolipids and lysosomal pathologies. *Biochim. Et Biophys. Acta (BBA)—Mol. Cell Biol. Lipids* **2014**, *1841*, 799–810. [CrossRef]
71. Thelen, A.M.; Zoncu, R. Emerging Roles for the Lysosome in Lipid Metabolism. *Trends Cell Biol.* **2017**, *27*, 833–850. [CrossRef]
72. Ballabio, A.; Gieselmann, V. Lysosomal disorders: From storage to cellular damage. *Biochim. Et Biophys. Acta (BBA)—Mol. Cell Res.* **2009**, *1793*, 684–696. [CrossRef]
73. Tebib, K.; Besançon, P.; Rouanet, J.-M. Dietary grape seed tannins affect lipoproteins, lipoprotein lipases and tissue lipids in rats fed hypercholesterolemic diets. *J. Nutr.* **1994**, *124*, 2451–2457. [CrossRef]
74. Osada, K.; Suzuki, T.; Kawakami, Y.; Senda, M.; Kasai, A.; Sami, M.; Ohta, Y.; Kanda, T.; Ikeda, M. Dose-dependent hypocholesterolemic actions of dietary apple polyphenol in rats fed cholesterol. *Lipids* **2006**, *41*, 133–139. [CrossRef]
75. Pal, S.; Ho, S.S.; Takechi, R. Red wine polyphenolics suppress the secretion of ApoB48 from human intestinal CaCo-2 cells. *J. Agric. Food Chem.* **2005**, *53*, 2767–2772. [CrossRef]
76. Wang, Y.P.; Nakajima, T.; Gonzalez, F.J.; Tanaka, N. PPARs as Metabolic Regulators in the Liver: Lessons from Liver-Specific PPAR-Null Mice. *Int. J. Mol. Sci.* **2020**, *21*, 2061. [CrossRef]
77. Quesada, H.; Pajuelo, D.; Fernández-Iglesias, A.; Díaz, S.; Ardevol, A.; Blay, M.; Salvadó, M.; Arola, L.; Blade, C. Proanthocyanidins modulate triglyceride secretion by repressing the expression of long chain acyl-CoA synthetases in Caco2 intestinal cells. *Food Chem.* **2011**, *129*, 1490–1494. [CrossRef]

78. Han, M.; Song, P.X.; Huang, C.; Rezaei, A.; Farrar, S.; Brown, M.A.; Ma, X. Dietary grape seed proanthocyanidins (GSPs) improve weaned intestinal microbiota and mucosal barrier using a piglet model. *Oncotarget* **2016**, *7*, 80313–80326. [CrossRef]
79. Yu, Y.; Chen, P.; Li, X.; Shen, S.; Li, K. Persimmon Proanthocyanidins with Different Degrees of Polymerization Possess Distinct Activities in Models of High Fat Diet Induced Obesity. *Nutrients* **2022**, *14*, 3718. [CrossRef]
80. Wang, E.; Zhou, Y.; Liang, Y.; Ling, F.; Xue, X.; He, X.; Zhai, X.; Xue, Y.; Zhou, C.; Tang, G. Rice flowering improves the muscle nutrient, intestinal microbiota diversity, and liver metabolism profiles of tilapia (Oreochromis niloticus) in rice-fish symbiosis. *Microbiome* **2022**, *10*, 231. [CrossRef]
81. Canfora, E.E.; Jocken, J.W.; Blaak, E.E. Short-chain fatty acids in control of body weight and insulin sensitivity. *Nat. Rev. Endocrinol.* **2015**, *11*, 577–591. [CrossRef]
82. Jin, G.; Asou, Y.; Ishiyama, K.; Okawa, A.; Kanno, T.; Niwano, Y. Proanthocyanidin-rich grape seed extract modulates intestinal microbiota in ovariectomized mice. *J. Food Sci.* **2018**, *83*, 1149–1152. [CrossRef] [PubMed]
83. Redondo-Castillejo, R.; Garcimartín, A.; Hernández-Martín, M.; López-Oliva, M.E.; Bocanegra, A.; Macho-González, A.; Bastida, S.; Benedí, J.; Sánchez-Muniz, F.J. Proanthocyanidins: Impact on Gut Microbiota and Intestinal Action Mechanisms in the Prevention and Treatment of Metabolic Syndrome. *Int. J. Mol. Sci.* **2023**, *24*, 5369. [CrossRef] [PubMed]
84. Litvak, Y.; Byndloss, M.X.; Tsolis, R.M.; Bäumler, A.J. Dysbiotic Proteobacteria expansion: A microbial signature of epithelial dysfunction. *Curr. Opin. Microbiol.* **2017**, *39*, 1–6. [CrossRef]
85. Qi, X.; Zhang, Y.; Zhang, Y.; Luo, F.; Song, K.; Wang, G.; Ling, F. Vitamin B12 produced by Cetobacterium somerae improves host resistance against pathogen infection through strengthening the interactions within gut microbiota. *Microbiome* **2023**, *11*, 135. [CrossRef] [PubMed]
86. Zhang, Z.; Fan, Z.; Yi, M.; Liu, Z.; Ke, X.; Gao, F.; Cao, J.; Wang, M.; Chen, G.; Lu, M. Characterization of the core gut microbiota of Nile tilapia (Oreochromis niloticus): Indication of a putative novel Cetobacterium species and analysis of its potential function on nutrition. *Arch. Microbiol.* **2022**, *204*, 690. [CrossRef] [PubMed]
87. Zhang, Y.; Ji, T.; Jiang, Y.; Zheng, C.; Yang, H.; Liu, Q. Long-term effects of three compound probiotics on water quality, growth performances, microbiota distributions and resistance to Aeromonas veronii in crucian carp Carassius auratus gibelio. *Fish Shellfish Immunol.* **2022**, *120*, 233–241. [CrossRef]
88. Nurliyana, M.; Amal, M.; Zamri-Saad, M.; Ina-Salwany, M. Possible transmission routes of Vibrio spp. in tropical cage-cultured marine fishes. *Lett. Appl. Microbiol.* **2019**, *68*, 485–496. [CrossRef]
89. He, Q.; Jia, R.; Cao, L.; Du, J.; Gu, Z.; Galina, J.; Xu, P.; Yin, G. Effects of Ginkgo biloba leaf extract on growth performance, antioxidant function and immune-related gene expression of Cyprinus Carpio. *Chin. Fish. Sci.* **2021**, *028*, 326–336.

antioxidants

MDPI

Article

The Enhanced Growth Performance and Antioxidant Capacity of Juvenile *Procambarus clarkii* Fed with Microbial Antioxidants

Zeyi Cheng [1,2,3], Jie Shi [1,2,3], Chen Qian [1,2,3], Jinghao Li [1,2,3], Xugan Wu [2,3,4], Ieong Kong [5] and Jiayao Li [1,2,3,*]

[1] Key Laboratory of Integrated Rice-Fish Farming Ecosystem, Ministry of Agriculture and Rural Affairs, Shanghai Ocean University, Shanghai 201306, China; m230150468@st.shou.edu.cn (Z.C.); shijie199767@163.com (J.S.); qianchenjoy@163.com (C.Q.); yxhdyhtc@163.com (J.L.)
[2] National Demonstration Center for Experimental Fisheries Science Education, Shanghai Ocean University, Shanghai 201306, China; xgwu@shou.edu.cn
[3] Shanghai Engineering Research Center of Aquaculture, Shanghai Ocean University, Shanghai 201306, China
[4] Key Laboratory of Freshwater Aquatic Genetic Resources, Ministry of Agriculture and Rural Affairs, Shanghai Ocean University, Shanghai 201306, China
[5] Shanghai Chuangbo Ecological Engineering, Shanghai 201108, China; kongieong1988@gmail.com
* Correspondence: jy-li@shou.edu.cn

Abstract: Given the economic significance of *Procambarus clarkii* in freshwater aquaculture and the lack of microbial antioxidants in *Procambarus clarkii* diet research, this study aimed to investigate the optimal supplementation level and feeding duration of microbial antioxidants in *Procambarus clarkii* diets. A series of three experiments were conducted to assess the long-term effects of different MA levels on crayfish and evaluate the palatability of the diets by observing feeding behavior and examining the short-term effects of high levels of MA. Our results indicate that long-term feeding using 1.5% MAs markedly increased the activities of antioxidant enzymes (T-AOC, T-SOD, and GSH-PX) and decreased the malondialdehyde (MDA) content in the hepatopancreas and hemolymph, with the crayfish showing significantly higher survival rates due to better antioxidant capacity after 24 h of air exposure stress. Under the condition of long-term feeding, the appropriate level of addition of MAs that can promote the growth of crayfish is 0.62–0.66%. The feeding behavior results indicate that the lower willingness and food intake of the crayfish in the high MA group may be the main reason affecting their growth. Conversely, short-term feeding using MAs alleviated the adverse effects on growth associated with the reduced palatability of the diet. The results indicate that the inclusion of 1.5% MAs in the diet for a period of 21 d optimized crayfish growth, accompanied by an improvement in antioxidant capacity and survival during transportation. This study demonstrates that diets supplemented with microbial antioxidants (MAs) can improve growth performance, antioxidant capacity, and resistance to air exposure stress in *Procambarus clarkii*. These results provide valuable insights into the potential benefits of MA supplementation in crayfish aquaculture.

Keywords: microbial antioxidant; growth performance; antioxidant capacity; resistance to air exposure stress; *Procambarus clarkii*

Academic Editor: Erchao Li

Received: 6 December 2024
Revised: 27 December 2024
Accepted: 2 January 2025
Published: 23 January 2025

Citation: Cheng, Z.; Shi, J.; Qian, C.; Li, J.; Wu, X.; Kong, I.; Li, J. The Enhanced Growth Performance and Antioxidant Capacity of Juvenile *Procambarus clarkii* Fed with Microbial Antioxidants. *Antioxidants* **2025**, *14*, 135. https://doi.org/10.3390/antiox14020135

1. Introduction

The red swamp crayfish *Procambarus clarkii* is one of the most significant freshwater crustaceans in China [1,2]. It is a popular consumer item, due to its rapid growth, adaptability, high nutritional value, and delicious taste [3,4]. At present, the annual yield of China's crayfish has reached 3,161,022 tons [5]. The primary method of crayfish transportation is via off-water means, which subjects crayfish to the dual stressors of dehydration and hypoxia. This disrupts their respiratory metabolism, prompting the body to produce a substantial quantity of reactive oxygen species (ROS) [6]. The ROS inflict oxidative damage to the tissues and diminish the survival rate if antioxidant enzymes cannot be promptly eliminated, resulting in greater economic losses [7]. Therefore, improving the health status of juvenile crayfish and reducing oxidative stress during transport is essential for the sustainable development of aquaculture [8,9].

In recent years, the use of probiotics and their metabolites in aquaculture to protect the health of farmed animals has become a trend [10,11]; various probiotic micro-organisms have been used as feed additives in aquaculture with some positive results [12,13], where *yeasts*, *Bacillus*, *Lactobacillus*, *Micrococcus*, *Enterococcus*, and *Lactobacillus* are widely used as probiotics in animal husbandry and aquaculture [8,12,14,15]. Microbial antioxidants (MAs) are novel compounds fermented by probiotics that are known to promote growth and improve nutrient absorption, as well as feed utilization and antioxidant capacity [16,17]. They can also replenish vitamins and amino acids. Furthermore, they facilitate the production of a range of active substances during the metabolic process, including proteases, amylases, and lipases. These substances enhance the digestion and absorption of nutrients, thereby improving the feed conversion ratio (FCR) and ensuring the healthy growth of aquatic animals [18,19]. A recent study reported that the dietary addition of MA increased lipase activity, promoted growth performance, and enhanced the immunity of *Eriocheir sinensis* by promoting antioxidant capacity against the negative effects of unfavorable environmental factors [20].

It can be reasonably proposed that the utilization of MAs as a dietary supplement for crayfish with the objective of enhancing its antioxidant capacity represents a viable solution to the issue of oxidative stress during the transportation process. This has the potential to significantly enhance the overall benefits of crayfish farming. MAs, as a feed additive, can maximize the improvement of the physiological functions of farmed animals and reduce the cost of farming, and are widely used in livestock diets. Numerous studies of mammals have found that the addition of MAs to diets can improve the growth performance in *Sus scrofa domestica* [17], *mus musculus* [16], and *gallus domestics* [14], but currently, there are few applications of MAs in aquatic animals [20]. Therefore, it is important to determine the performance of crayfish and the optimal MA supplementation level and feeding duration. The aim of this study was to investigate the effects of MAs as a feed additive on the growth performance and antioxidant capacity of crayfish, and to provide a theoretical basis for the use of MAs as a feed additive for crayfish, as well as a new way to alleviate oxidative stress during the transport of crayfish.

2. Materials and Methods

2.1. Experimental Diets

The experimental feeds used in this study were made by adding different doses of microbial antioxidants (MAs) to a commercial compound feed (Wuhan Dabeinong group, Wuhan, China) for crayfish. The MAs were made from the fruits of sea buckthorn and Rosa roxburghe, which were fermented by beneficial microbes, such as *Bacillus subtilis*, *Lactobacillus*, and *Saccharomyces*, through a compound fermentation process (The main components were as follows: *Bacillus subtilis* $\geq 1.0 \times 10^{10}$ cfu/g, *Lactobacillus* $\geq 1.0 \times 10^{7}$ cfu/g, and *Saccharomyces* $\geq 1.0 \times 10^{7}$ cfu/g). It is then extracted, concentrated, inactivated, lyophilized, and subjected to other processes, and it contains carotenoids, vitamins B1, B2, B12, and C, Quercetin-3-D-glucopyranose (quercetin), Quercetin (flavonoids), inositol, and metal derivatives of various trace elements (Shanghai Chuangbo Ecological Engineering, Shanghai, China). For the specific antioxidant content of the antioxidant components, see the study on microbial antioxidant on antioxidant performance and immune function in mice [21]. The diet preparation procedure was as follows: the pelleted diets were recrushed and sieved through a 60-mesh sieve with the addition of MAs at levels of 0% (A1), 0.25% (A2), 0.5% (A3), 1% (A4), and 1.5% (A5). To reduce the loss of vitamins due to comminution, the diets of each group were supplemented with 1% vitamin premix, weighed according to the ratio, and all ingredients were thoroughly mixed and then made into a 2 mm particle size diet by adding an appropriate amount of distilled water. The diets were dried and stored at -40 °C and protected from light. The experimental diets were analyzed for their nutrient composition according to standard procedures [22,23]. The crude protein (34%, dry matter) and crude fat (8.7%, dry matter) contents of the experimental diets were able to meet the growth requirements of juvenile crayfish [24]. The specific dietary ingredients and nutrient composition are shown in Table 1.

Table 1. Feed configuration and regular nutrient composition (dry matter).

Items	A1	A2	A3	A4	A5
Ingredients (%)					
Commercial feed	95	95	95	95	95
Vitamin premix	1	1	1	1	1
Carboxymethyl cellulose	2	2	2	2	2
Wheat meal	2	1.75	1.5	1	0.5
Microbial antioxidants [1]	0	0.25	0.5	1	1.5
Proximate composition (%)					
Moisture	7.17	6.68	8.03	7.96	8.08
Crude protein	34.22	33.93	33.75	34.48	34.30
Crude lipid	8.81	8.92	8.72	8.99	8.88
Ash	16.40	17.17	17.43	15.79	16.90

[1]: MAs were composed of 503 mg/kg Fe, 367 mg/kg Mn, 1.07 mg/kg Cu, 0.18 mg/kg Se, 194,000 U/100 g SOD, 322 mg/100 g vitamin C, 908 μg/100 g vitamin E. 4.43% total flavones, 1.37% *isoflavones*, 886 mg/100 g glutathione, 82.4 mg/100 g total saponins, 4.21% total amino acids, and 0.146% *taurine*.

2.2. *Experimental Design*

2.2.1. Trial 1: Different Levels of Microbial Antioxidant Feed Experiment

The experiment was conducted in September 2022, and the crayfish used in this study were vigorous juvenile crayfish with healthy appendages, selected from the Chongming base of Shanghai Ocean University. They were kept in an indoor recirculating water monoculture system containing dechlorinated tap water (temperature: 24.0 ± 1 °C; pH: 8.0 ± 0.5; dissolved oxygen: >7 mg/L; total ammonia nitrogen: below 0.2 mg/L; nitrite content: below 0.01 mg/L), and were fed commercial diets for 7 d of acclimatization. Subsequently, 240 crayfish of a similar size (initial average weight: 3.95 ± 0.6 g) were selected and randomly divided into five groups, with 48 crayfish in each group. The crayfish were fed the A1–A5 diets at 3–5% of their body weight daily, at 18:00, during the feeding trial. The following morning, every day of the experiment, all tanks were cleaned by siphoning out feed residues and feces to maintain water quality. Crayfish molting, water temperature, and dissolved oxygen were recorded daily; water quality indicators were measured twice a week, and the water was changed or added as appropriate, according to water quality. At the end of the 70 d feeding trial, the crayfish were fasted for 24 h. The surface water of the crayfish was then dried with absorbent paper, weighed with an electronic balance (accuracy of 0.01 g), and measured for body length and width, using a vernier caliper (accuracy of 0.01 mm). The hepatopancreas and hemolymph of 18 crayfish from each group were collected for the weighing and determination of oxidase activity.

After the feeding trial, 15 juveniles from each dietary treatment were randomly selected, removed from their tanks, and maintained in individual plastic boxes (8.5 cm × 6.5 cm × 12 cm; volume = 450 mL). The temperature and relative humidity of the room were maintained at 21 ± 0.7 °C and 39.25 ± 0.89%, respectively. After 24 h, all the surviving crayfish in each group were counted and sampled for subsequent determination of respiratory metabolic enzymes.

2.2.2. Trial 2: Behavioral Experiments with Microbial Antioxidant Diets

The feeding behavior experiments were conducted in round polyethylene barrels (diameter: 50 cm), and a fixed infrared camera recording system (Hikvision, DS-2DC24021W-DE3) was placed directly above each drum to record the feeding behavior of the crayfish during the experimental process. Considering the behavior of crayfish in darkness, the behavioral observation was carried out from 18:00 to 21:00, and each crayfish was used only once during the whole experiment. The water and crayfish were replaced after the completion of one shot to start a new round of feeding behavior observation.

The temporary culture conditions for 1 week were the same as above, and then 52 juvenile crayfish of a similar size (initial average weight: 7.04 ± 0.43 g) were divided into four groups that were fed with only A1 and A5 diets under satiation or starvation (for 48 h), with 10 replicates for each of the two feeding groups. Due to the small size and limited food intake of 3 g of juvenile crayfish, which also tend to become satiated more quickly, 7 g of juvenile crayfish with a longer feeding time were used in the feeding behavior trial. One crayfish was placed in the experimental barrel and allowed to acclimatize to the water environment for 2 h. The 20 experimental diets were weighed and placed in the PC tube in the center of the barrel, and the crayfish were fixed in the PC tube area directly in front of the diet to ensure that the crayfish in each group had the same distance of access to the diet. At the same time, the two PC tubes were removed, so that the feeding behavior of the crayfish could be automatically recorded for 3 h. After the experiment, the residual feed was immediately collected, dried at 65 °C for 24 h, and weighed.

2.2.3. Trial 3: Short-Term Optimization Experiments with Microbial Antioxidant Diets

The temporary culture conditions for 1 week were the same as above, and then 240 crayfish of a similar size (initial average weight: 5.62 ± 0.85 g) were selected and randomly divided into two groups, control group A1 (0% microbial source antioxidant) and antioxidant group A5 (1.5% microbial source antioxidant), with 150 crayfish in each group, and three replicates were set up. The experimental management was as described above. The experiment was carried out for 28 d. After feeding for 7 d, 14 d, 21 d, and 28 d, 18 crayfish were collected from each of the two feeding groups for growth statistics and antioxidant enzyme activity measurements (the methods used were the same as Trial 1), and 18 crayfish were randomly selected from each of the two feeding groups for 24 h air exposure stress experiments under the same conditions as above. After 24 h, all the surviving crayfish in each group were counted and sampled for the subsequent determination of respiratory metabolic enzymes.

2.3. Index Determination

The weight growth rate (WGR), specific growth rate (SGR), survival rate (SR), meat yield (MY), and hepatosomatic index (HSI) were calculated according to the following formulas:

$$\text{WGR (\%)} = (W_t - W_0)/W_0 \times 100 \quad (1)$$

$$\text{SGR (\%/d)} = (\ln W_t - \ln W_0)/D \times 100 \quad (2)$$

$$\text{SR (\%)} = N_f/N_i \times 100 \quad (3)$$

$$MY (\%) = W_m / W_t \times 100 \quad (4)$$

$$HSI (\%) = W_h / W_t \times 100 \quad (5)$$

where W_0 is the initial weight of the crayfish (g), W_t is the final weight of the crayfish (g), and D is the number of experimental days. N_f is the final crayfish numbers, and N_i is the initial crayfish numbers. W_m is the muscle weight (g), and W_h is the hepatopancreas weight (g).

For the determination of conventional nutrients, the moisture content was determined by drying the samples at 105 °C, using a constant weight method. The crude protein content was quantified by digesting the samples with concentrated sulfuric acid and then by employing the Kjeldahl method (TM 8200, Foss, Hoganas, Sweden). The total lipid content was determined using the Folch method [22], which employs a chloroform–methanol solution and a solution of 0.37 mol/L KCl for extraction. The samples were subjected to a muffle furnace (PCDE3000, Shanghai, China) analysis to determine the ash content in accordance with the AOAC methodology [23].

The hemolymph and hepatopancreas of 18 crayfish from each group after Trial 1 and Trial 3 were thawed, weighed, and homogenized, separately, in nine volumes (*v*/*v*) of a precooled physiological saline solution, using a T10B homogenizer (IKA Co., Staufen, Germany). The homogenates were centrifuged at 10,000× *g* for 10 min at 4 °C, and the supernatants were collected using a pipette for later analysis. Subsequently, the total superoxide dismutase (T-SOD), total antioxidant capacity (T-AOC), glutathione peroxidase (GSH-PX), and malondialdehyde (MDA) levels were detected in the supernatants of the hepatopancreas and hemolymph after Trial 1 and Trial 3. The hepatopancreas and muscle of six crayfish from each group after air exposure stress were thawed, and hepatopancreas treatment was the same as above. Muscle tissue was crushed and homogenized, separately, in nine volumes (*v*/*v*) of a precooled physiological saline solution, using a T10B homogenizer. Subsequently, MDA, lactic acid (LD), lactic dehydrogenase (LDH), and succinate dehydrogenase (SDH) levels were analyzed in the supernatants of the hepatopancreas and muscle after air exposure stress. All the above parameters were determined using the corresponding bio-kits (Nanjing Jiancheng Bioengineering Institute, Nanjing, China) with a T6 New Century spectrophotometer (Beijing Purkinje General Instrument Co., Ltd., Beijing, China), according to the manufacturer's instructions.

We watched and analyzed the behavior observation videos, recorded using an infrared monitoring system, and calculated the duration of various behaviors [25]. The statistical behavior parameters mainly included the following:

$$\text{Dissolution rate (\%)} = 100 \times (\text{Initial feed weight} - \text{dried feed weight})/\text{initial feed weight} \quad (6)$$

$$\text{Food intake (g)} = \text{Initial feed weight} - \text{dried feed weight}/(1 - \text{dissolution rate}) \quad (7)$$

$$\text{Searching time (min): Time of first exposure of crayfish to feed} \quad (8)$$

$$\text{Feeding time (min): Time for crayfish to sniff, test, and ingest feed} \quad (9)$$

$$\text{Willingness to feed (s): Time to start feed ingestion} - \text{time of first feed exposure} \quad (10)$$

where the initial feed weight is the weight of the feed that is not put into the water, and the dried feed weight is the weight of the feed taken out of the water after 3 h and dried.

2.4. Statistical Analysis

SPSS 22.0 software (SPSS, Michigan Avenue, Chicago, IL, USA) was used for statistical analysis. Levene's equal variance test and the Shapiro–Wilk test were used to test for the homogeneity of variance and normal distribution of all data, respectively. Under the same feeding period, one-way analysis of variance (ANOVA) was used to analyze whether there was a significant difference between crayfish that were fed different MA levels. If the one-way ANOVA was significant, a Duncan's range test was used to further analyze the significance of the different treatments. A Student's t-test was used to analyze the growth property differences between the control and MA treatments for short-term feeding, and it was also used to determine the significant differences between the behavioral indicators of crayfish in A1 and A5 in the same state of starvation or satiation. The significance level was set at $p < 0.05$. The data are presented as mean ± standard deviation (SD). Based on the actual amount of MAs in the diet, regression analysis was used to determine the optimal level of MA supplementation, using a quadratic curve model (Supplementary Table S1).

3. Results

3.1. Growth Performance and Whole-Body Proximate Composition

After 70 d of feeding using the A1–A5 group diets, the SR of juvenile crayfish in the diet group with added MAs was significantly higher than that of the A1 group, and with an increase in MAs in the diet, FBW, WGR, and SGR showed a trend of increasing and then decreasing, with the highest being in the A4 group, the second highest in the A3 group, and the lowest in the A5 group ($p < 0.05$; Table 2); the MY was highest in the A4–A5 groups, and the HSI was highest in the A5 group. There was no significant difference in the final length and width of juvenile crayfish between groups ($p > 0.05$; Table 2). In addition, the crude protein content in muscle was highest in group A4, and significantly lower in group A5 than in groups A1–A4 ($p < 0.05$); the ash content in muscle was lower in groups A4–A5 than in groups A1–A3 ($p < 0.05$; Table 3). The A5 group was significantly longer than the other groups in both molt cycles, and A3 was significantly shorter than the other groups in the second molt cycle ($p < 0.05$; Figure 1). A regression analysis of the FBW, WGR, and SGR on the MA level was used to determine the optimal level of MA supplementation, using a quadratic curve model. With increasing MA concentrations, the FBE, WGR, and SGR all showed a trend of increasing and then decreasing, to peak at 0.62%, 0.65%, and 0.66%, respectively (Figure 2).

Table 2. Effects of different levels of microbial antioxidants (in feed) on survival, growth performance, and the hepatopancreas index of juvenile crayfish ($n = 33$).

Items	A1	A2	A3	A4	A5
IBW (g)	3.81 ± 0.56	4.01 ± 0.49	3.97 ± 0.49	3.79 ± 0.51	3.92 ± 0.71
FBW (g)	8.88 ± 1.24 ^{ab}	9.35 ± 1.07 ^{b}	9.38 ± 0.94 ^{b}	9.16 ± 0.92 ^{b}	8.55 ± 1.30 ^{a}
WGR (%)	133.55 ± 15.46 ^{b}	133.87 ± 19.18 ^{b}	138.41 ± 24.22 ^{b}	143.81 ± 22.88 ^{b}	119.99 ± 17.02 ^{a}
Final length (mm)	51.55 ± 2.31 ^{a}	52.98 ± 1.99 ^{bc}	53.70 ± 2.33 ^{c}	53.19 ± 2.29 ^{bc}	52.14 ± 2.52 ^{ab}
Final width (mm)	15.05 ± 0.74	15.12 ± 0.66	15.00 ± 0.60	15.24 ± 0.57	15.05 ± 0.63
SR (%)	91.67 ± 5.89	97.92 ± 4.66	97.92 ± 4.66	97.92 ± 4.66	97.92 ± 4.66
SGR (%/d)	1.21 ± 0.09 ^{b}	1.21 ± 0.11 ^{b}	1.23 ± 0.15 ^{b}	1.27 ± 0.13 ^{b}	1.12 ± 0.11 ^{a}
MY (%)	14.58 ± 1.98	14.63 ± 1.23	14.43 ± 2.04	15.38 ± 1.85	15.44 ± 2.18
HSI (%)	7.58 ± 1.01	7.65 ± 0.93	7.92 ± 0.98	7.76 ± 1.01	8.13 ± 1.17

Notes: The numerical values represent the mean and standard deviation (SD). The letters indicate significant differences between groups ($p < 0.05$).

Table 3. Effects of different levels of microbial antioxidants on the routine biochemical composition of juvenile crayfish muscle (wet mass: $n = 6$).

Items	A1	A2	A3	A4	A5
Moisture	76.97 ± 1.31	78.68 ± 1.97	78.19 ± 1.45	78.20 ± 1.92	77.52 ± 1.71
Crude protein	16.63 ± 2.09 ^{ab}	17.05 ± 1.85 ^{b}	16.88 ± 1.20 ^{b}	17.39 ± 0.54 ^{b}	14.87 ± 0.65 ^{a}
Crude lipid	1.03 ± 0.15	0.89 ± 0.08	0.96 ± 0.09	0.98 ± 0.24	1.01 ± 0.11
Ash	3.07 ± 0.21	3.10 ± 0.31	3.06 ± 0.35	2.94 ± 0.32	2.98 ± 0.35

Notes: The numerical values represent the mean and standard deviation (SD). The letters indicate significant differences between groups ($p < 0.05$).

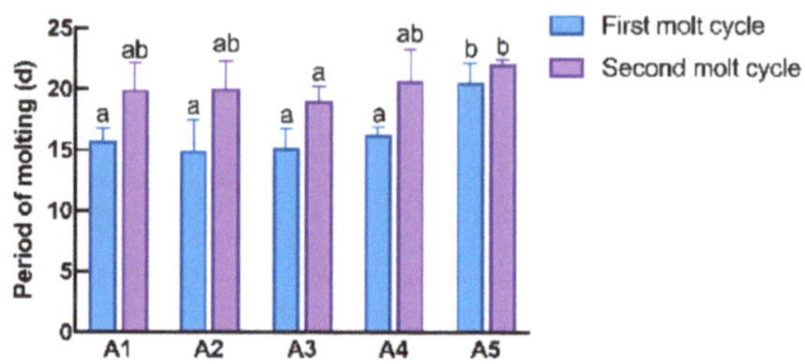

Figure 1. Effects of different levels of microbial antioxidants in the diet on the molting cycle of crayfish. Note: The data are presented as mean ± SD ($n = 48$). The superscripts in the figure indicate a significant difference ($p < 0.05$).

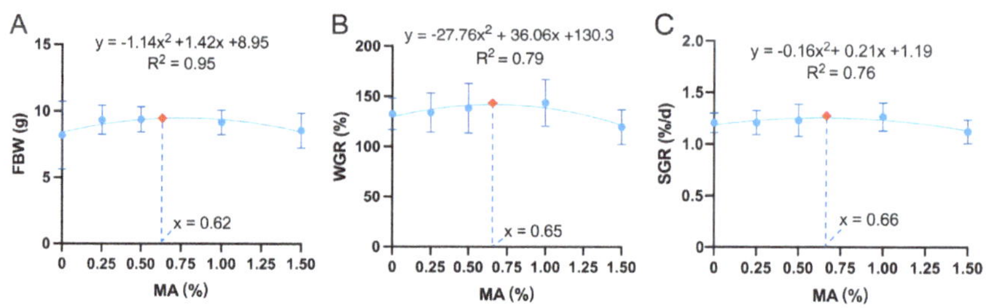

Figure 2. Quadratic relationship between dietary microbial antioxidant supplementation and the FBW (**A**), WGR (**B**), and SGR (**C**) of juvenile crayfish ($n = 33$).

3.2. Antioxidative Capacity

The parameters related to the antioxidant capacity of crayfish were significantly affected by the level of antioxidant supplementation from MAs, and the addition of antioxidants from MAs to the diet significantly increased the activities of T-AOC, T-SOD, and GSH-PX, and decreased the content of MDA in the hepatopancreas and hemolymph of juvenile crayfish ($p < 0.05$). In the hepatopancreas, the activities of T-AOC and GSH-PX showed a trend of increasing with increasing MA doses, T-AOC and GSH-PX activities were significantly higher in groups A4–A5 than in group A1 ($p < 0.05$; Figure 3A,E), and the MDA content was significantly lower than in group A1 ($p < 0.05$; Figure 3G). The activities of T-SOD were significantly higher in groups A3–A5 than in group A1, and peaked in group A4 ($p < 0.05$; Figure 3C). In the hemolymph, the contents of T-AOC, T-SOD, and GSH-PX showed an increasing trend with an increase in MA supplementation, and were all significantly higher in groups A4–A5 than in groups A1–A3 ($p < 0.05$; Figure 3B,D,F); the MDA contents of juvenile crayfish in the MA-supplemented groups were lower than in the control group, and showed a decreasing and then increasing trend, reaching the lowest level in group A4 ($p < 0.05$; Figure 3H).

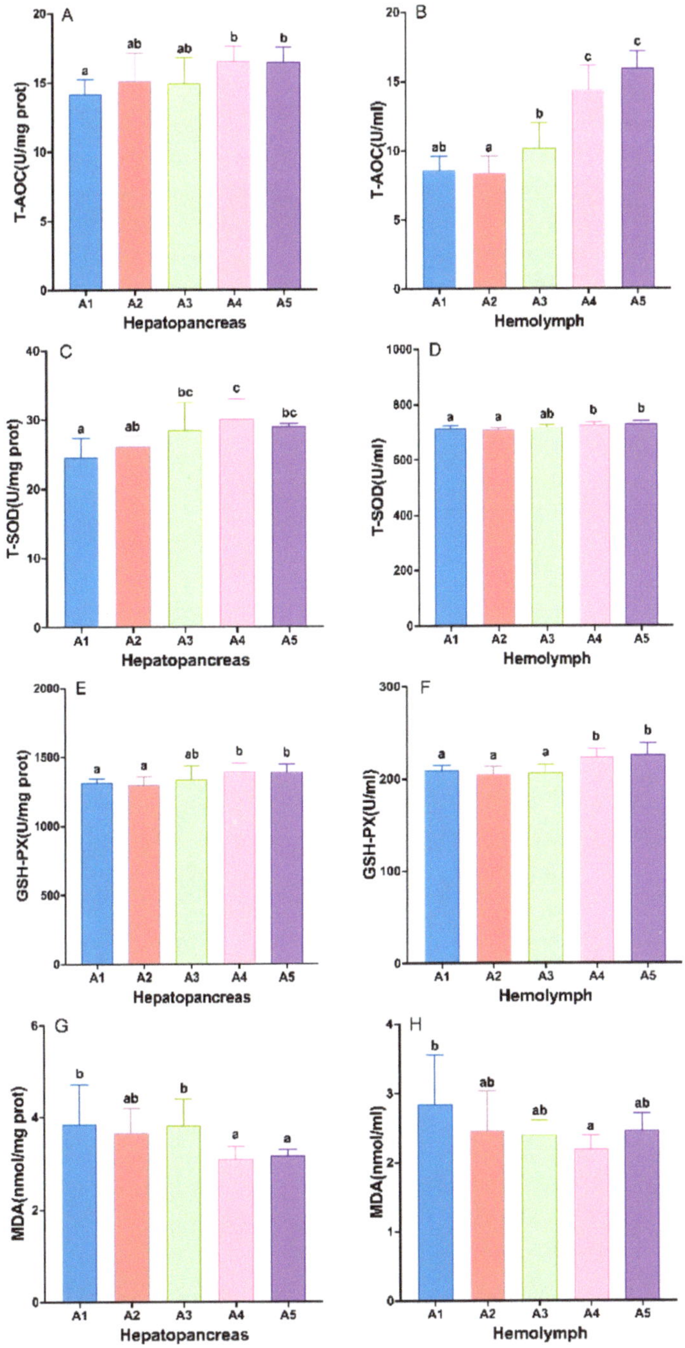

Figure 3. Effects of different levels of microbial antioxidants on hepatopancreas and hemolymph antioxidant enzyme activities in terms of T-AOC (**A**,**B**), T-SOD (**C**,**D**), GSH-PX (**E**,**F**), and MDA (**G**,**H**) content in juvenile crayfish. Note: The data are presented as mean ± SD (n = 6). The different letters on the top of the bars mean significant differences (p < 0.05).

3.3. Stress Response to Air Exposure

After 24 h of air exposure stress, the survival rate of the crayfish in each group gradually increased with increasing addition of MA to the diet, and the survival rate of juvenile crayfish in the MA-added group was significantly higher than that of the control group, with the highest survival rates in groups A4 and A5 (Figure 4).

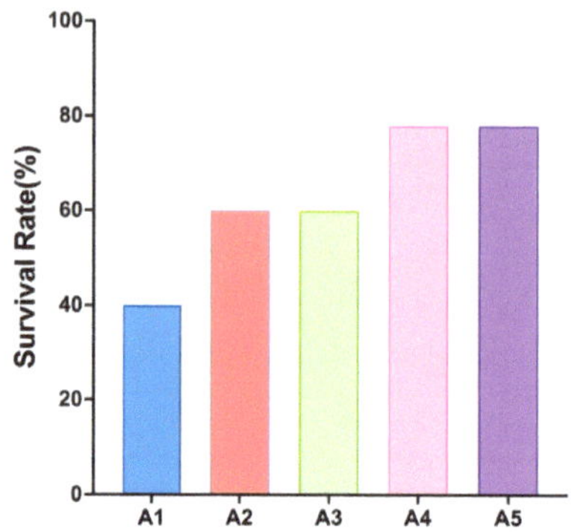

Figure 4. Effects of different levels of microbial antioxidants on the survival rate of juvenile crayfish after 24 h of air exposure stress.

The parameters related to the antioxidant capacity of juvenile crayfish exposed to air stress were significantly affected by the level of MA supplementation, and the addition of MAs to the diet significantly decreased LD content and LDH activity in the hepatopancreas and muscle of juvenile crayfish ($p < 0.05$). In the hepatopancreas, the MDA, LD, and LDH contents of juvenile crayfish in the MA-supplemented group were significantly lower than those in the control group, with the MDA content being lowest in the A4 and A5 groups, and significantly lower than that in the A1 group ($p < 0.05$; Figure 5A); LD content and LDH activity were highest in the A1 group and significantly lower in the A2–A5 groups, with no significant difference between groups ($p < 0.05$; Figure 5C,E); and SDH activity was significantly higher in the A2–A5 groups than in the A1 group ($p < 0.05$; Figure 5G). In the muscles, both LD content and LDH activity were significantly lower in groups A2–A5 than in group A1 ($p < 0.05$; Figure 5D,F), and SDH activity was significantly higher in groups A2–A5 than in group A1, with no significant difference between groups A2–A5 ($p > 0.05$; Figure 5H); MDA content among the groups was not significantly different between the A1 group and the A3–A5 groups ($p > 0.05$; Figure 5B).

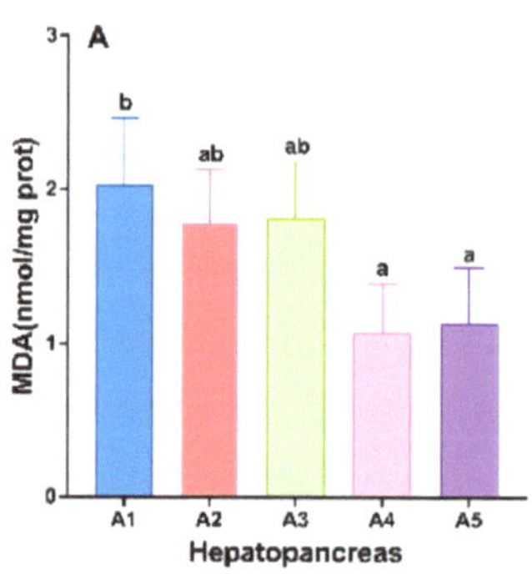

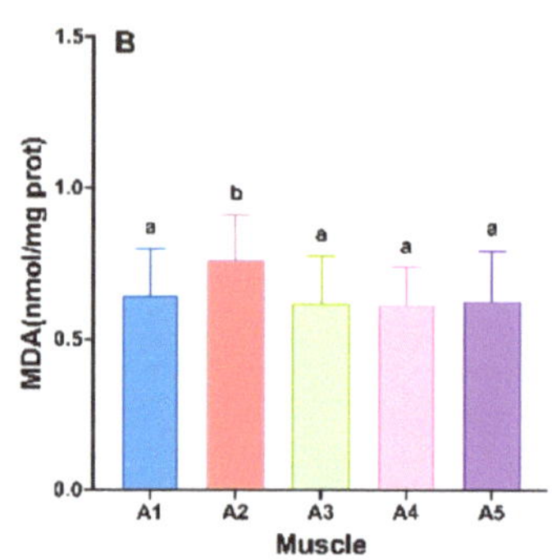

Figure 5. *Cont.*

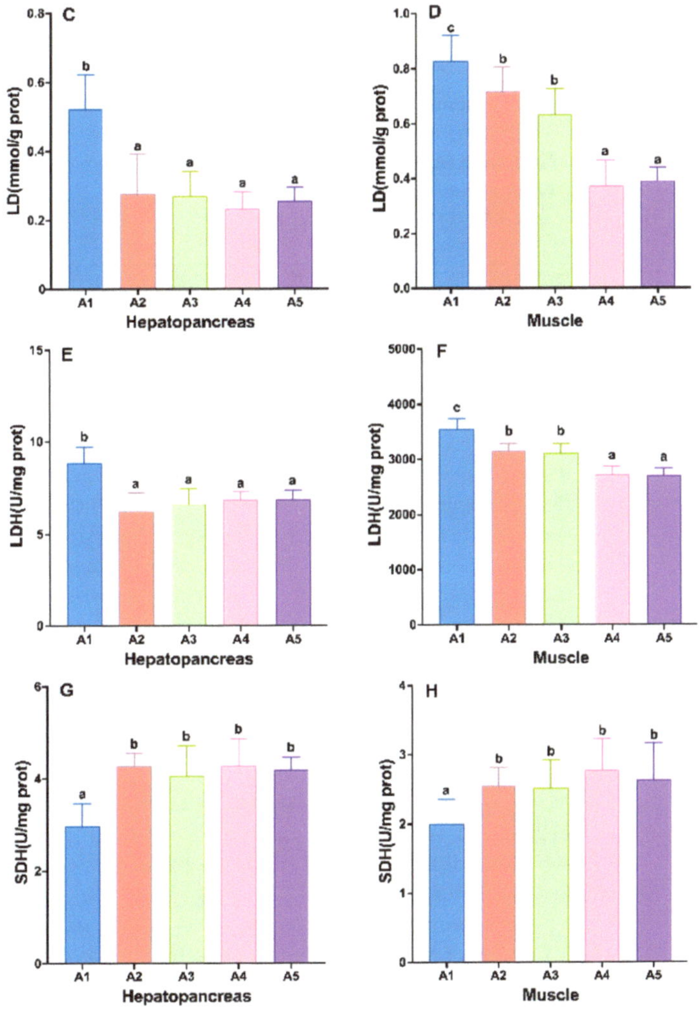

Figure 5. Effects of different levels of microbial antioxidants on the hepatopancreas and muscle in terms of the MDA (**A**,**B**) content and LD (**C**,**D**), LDH (**E**,**F**), and SDH (**G**,**H**) activities in crayfish after air exposure stress. Note: The data are presented as mean ± SD (n = 6). The different letters on the top of the bars mean significant differences ($p < 0.05$).

3.4. MA Feed Palatability

In the starvation condition, the time spent searching for food (5.6 min) and the willingness to feed (0.3 s) in group A1 was shorter than that in group A5 (8.4 min and 0.8 s), but there was no significant difference after the analysis of the results ($p > 0.05$, Table 4); In the satiation condition, the time spent searching for food (12. 9 min) in group A1 was significantly shorter than that in group A5 (23.2 min) ($p < 0.05$, Table 4). In the satiation condition, the time interval between exposure to A1 and the start of eating (0.7 s) was significantly shorter than in group A5 (4.8 s) ($p < 0.05$, Table 4), so the crayfish's willingness to feed on A5 was significantly weaker than that on A1.

Table 4. Effects of different levels of microbial antioxidants on the behavioral parameters of juvenile crayfish ($n = 10$).

Items	Starvation		Satiation	
	A1	A5	A1	A5
Searching time (min)	5.6 ^{a}	8.4 ^{a}	12.9 ^{a}	23.2 ^{b}
Willingness to feed (s)	0.3 ^{a}	0.8 ^{a}	0.7 ^{a}	4.8 ^{b}
Feeding time (min)	16.8 ^{a}	16.8 ^{a}	16.5 ^{a}	24.8 ^{b}
Food intake (g)	0.18 ^{b}	0.13 ^{a}	0.17 ^{b}	0.10 ^{a}

Note: The numerical values represent the mean and standard deviation (SD). The letters indicate significant differences between groups ($p < 0.05$).

When the crayfish were in the starved state, there was no significant difference between the feeding time of group A1 and group A5 ($p > 0.05$; Table 4); when in the satiated state, the feeding time of crayfish in group A5 (24.8 min) was significantly longer than that of the crayfish in group A1 (16.5 min) ($p < 0.05$; Table 4). During the 3 h of filming, whether the crayfish were starved or satiated, the intake of group A1 (0.18 g and 0.17 g) was higher than that of group A5 (0.13 g and 0.10 g) ($p < 0.05$; Table 4). In general, the crayfish ingested significantly less of the MA-supplemented diet than the unsupplemented group in both the starvation and satiation conditions, and the crayfish had a lower willingness to ingest the MA-supplemented diet for a longer period in the satiation condition.

3.5. Short-Term Feeding Optimization

The growth performance of the crayfish in groups A1 and A5 showed an increasing trend for FBW and WGR with increasing feeding time ($p < 0.05$; Table 5), which reached the highest at 14 d, but there was no significant difference in terms of WGR and SGR between the two groups, according to the growth performance results at 7, 14, and 21 d of feeding ($p > 0.05$). The SGR shows an increasing and then decreasing trend for the juvenile crayfish in group A5, which was significantly lower than that in group A1 ($p < 0.05$); there was no significant difference in HSI and MY between the two groups at the four feeding periods ($p > 0.05$; Table 5).

The parameters related to the antioxidant capacity of juvenile crayfish were significantly affected by the duration of MA feeding, as the activities of T-AOC, T-SOD, and GSH-PX in the hepatopancreas and hemolymph of the juvenile crayfish increased with increasing duration of MA feeding, while the content of MDA decreased ($p < 0.05$; Figure 6). In the hepatopancreas, the activities of T-SOD, T-AOC, and GSH-PX in group A5 were found to be significantly higher than those in group A1 after 21 d and 28 d of feeding, reaching their peak at 28 d of feeding ($p < 0.05$; Figure 6A,C,E); the MDA content in group A5 showed a tendency to increase and then decrease, with the lowest antioxidant content at 28 d of feeding, which was significantly lower than group A1 ($p < 0.05$; Figure 6G). In the hemolymph, the activities of T-AOC and GSH-PX in the A5 group were significantly higher than those in the control group after 21 d and 28 d of feeding, reaching a peak at 21 d ($p < 0.05$; Figure 6B,F). The MDA content of the juvenile crayfish at 28 d of feeding was significantly lower than that of the A1 group ($p < 0.05$; Figure 6H). In the hepatopancreas and hemolymph, the indicators of the four antioxidant enzymes of the crayfish in the two groups at the time periods of 7 d and 14 d of feeding did not show any statistical differences ($p > 0.05$).

Table 5. Effects of the different feeding times of microbial antioxidant feed on the growth performance and hepatopancreas index of juvenile crayfish (n = 18).

Items	7 d		14 d		21 d		28 d	
	A1	A5	A1	A5	A1	A5	A1	A5
IBW (g)	5.51 ± 0.91	5.34 ± 0.73	5.08 ± 0.87	5.30 ± 0.94	5.57 ± 0.71	5.85 ± 0.83	5.45 ± 1.00	5.73 ± 0.98
FBW (g)	5.79 ± 0.95	5.65 ± 0.77	6.96 ± 1.22	7.20 ± 1.18	7.83 ± 1.07	8.16 ± 1.21	8.46 ± 1.55	8.50 ± 1.46
Final length (mm)	48.06 ± 2.36	46.86 ± 1.61	50.58 ± 2.10	50.71 ± 2.45	51.50 ± 2.17	51.81 ± 2.43	53.49 ± 2.28	53.05 ± 2.04
Final width (mm)	13.70 ± 0.91	13.73 ± 0.73	14.34 ± 0.61	14.28 ± 0.66	14.79 ± 0.75	15.04 ± 0.80	14.86 ± 1.05	14.85 ± 0.83
WGR (%)	5.04 ± 1.69	5.65 ± 1.42	36.92 ± 3.45	36.38 ± 5.70	40.41 ± 4.04	39.37 ± 3.82	55.09 ± 2.81 [b]	48.34 ± 2.04 [a]
SGR (%/d)	0.70 ± 0.23	0.78 ± 0.19	2.24 ± 0.18	2.21 ± 0.30	1.61 ± 0.14	1.58 ± 0.13	1.57 ± 0.06 [b]	1.41 ± 0.05 [a]
HSI (%)	7.08 ± 0.86	7.28 ± 0.96	6.86 ± 0.88	6.52 ± 0.76	6.67 ± 0.81	6.29 ± 0.68	7.13 ± 0.68	7.21 ± 0.86
MY (%)	16.53 ± 1.30	16.92 ± 1.66	14.38 ± 1.33	14.74 ± 1.61	12.86 ± 1.37	12.55 ± 1.42	15.06 ± 1.62	15.81 ± 2.30

Notes: The numerical values represent the mean and standard deviation (SD). The letters indicate significant differences between groups ($p < 0.05$).

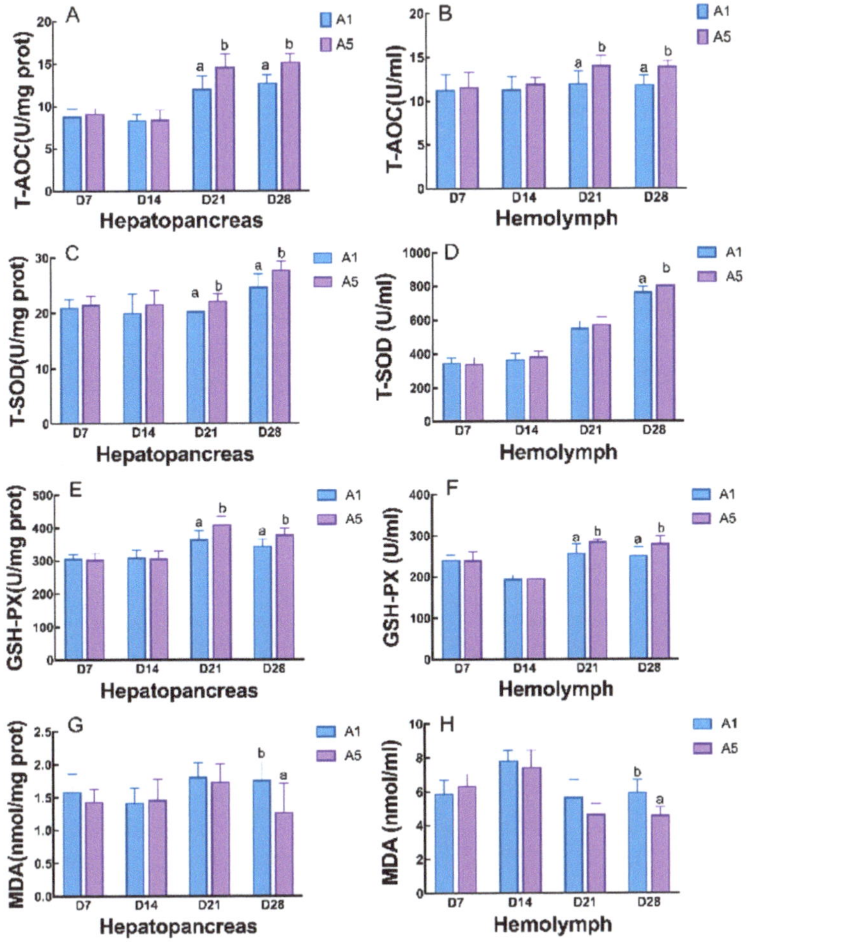

Figure 6. Effects of different microbial antioxidant feeding times on the antioxidant parameters in the hepatopancreas and hemolymph of juvenile crayfish, in terms of T-AOC (**A**,**B**), T-SOD (**C**,**D**), GSH-PX (**E**,**F**), and MDA (**G**,**H**) content. Note: The data are presented as mean ± SD (n = 6). The different letters on the top of the bars mean significant differences ($p < 0.05$).

After 24 h of air exposure stress, the crayfish in the A5 group showed the highest survival rates after 21 d and 28 d of feeding, which were significantly higher than that of the control group ($p < 0.05$; Figure 7). The parameters related to the antioxidant capacity of air exposure stress in the juvenile crayfish were significantly affected by the antioxidant feeding period, and the addition of MAs to the diet both significantly reduced LD content and LDH activity in the hepatopancreas and muscle of the crayfish ($p < 0.05$; Figure 8). In the hepatopancreas, the MDA, LD, and LHD contents of the crayfish in group A5 were significantly lower than those in group A1, with the MDA content of the crayfish at 28 d of feeding being significantly lower than that of group A1 ($p < 0.05$; Figure 8A). The LD content and LDH activity of group A5 were significantly lower than those of group A1 at 21 d and 28 d of feeding ($p < 0.05$; Figure 8C,E); the SDH content of group A5 was higher than that of group A1 after 21 d and 28 d of feeding ($p < 0.05$; Figure 8G). In the muscles, the LD content and LDH activity began to show a decreasing trend after 14 d of feeding, and were significantly lower in group A5 than in group A1, at 21 d and 28 d ($p < 0.05$; Figure 8D,F); the SDH activity increased significantly at 21 d and 28 d of feeding, and it was significantly higher in group A5 than in group A1, reaching a peak at 21 d ($p < 0.05$; Figure 8H); and there was no significant difference between the two groups for the MDA content ($p > 0.05$; Figure 8B).

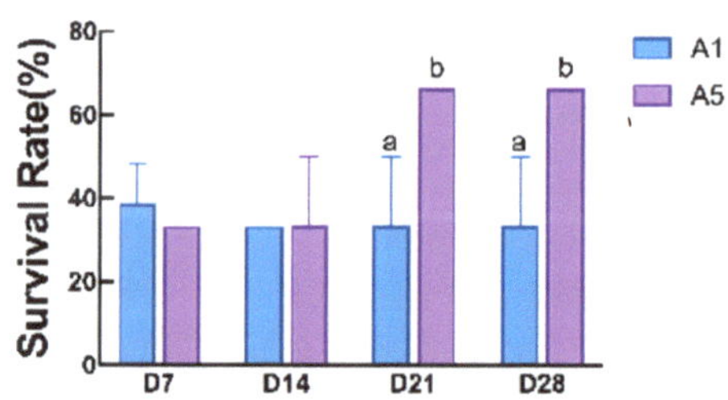

Figure 7. Effects of different microbial antioxidant feeding times on the survival rate of juvenile crayfish after 24 h of air exposure stress. The different letters on the top of the bars mean significant differences ($p < 0.05$).

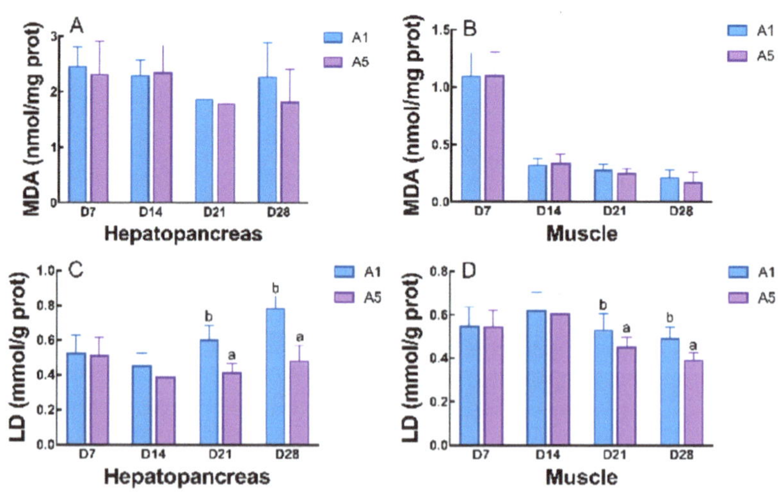

Figure 8. *Cont.*

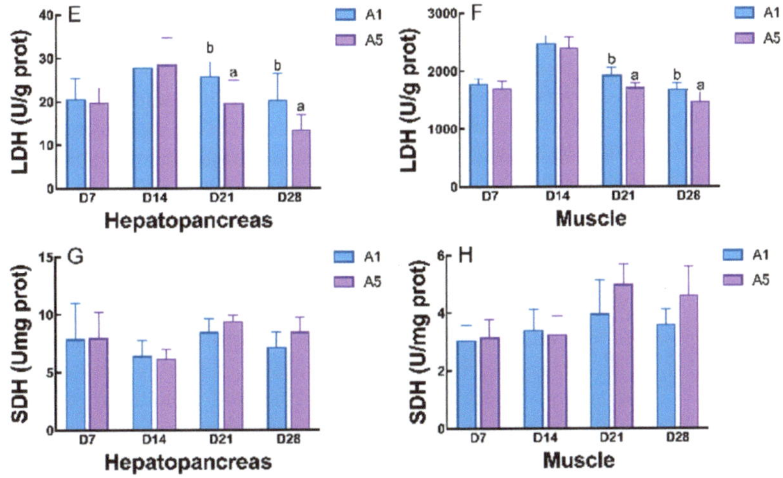

Figure 8. Effects of different microbial antioxidant feeding times on the hepatopancreas and muscle of crayfish, in terms of MDA (**A**,**B**) content and LD (**C**,**D**), LDH (**E**,**F**), and SDH (**G**,**H**) activities, after 24 h of air exposure stress. The data are presented as mean ± SD (n = 6). The different letters on the top of the bars mean significant differences ($p < 0.05$).

4. Discussion

The components of microbial antioxidant additives have been demonstrated to enhance the ecological balance of intestinal flora, facilitate growth and development, prevent and control diseases, and improve production performance [19]. Previous studies on mammals found that daily weight gain was significantly increased by the addition of 0.5% MAs fed to piglets for 2 weeks [26] and 0.8% MAs fed to mice for 3 weeks [27]. In a study on crustaceans, the WG and SGR of *E. sinensis* were significantly higher than those of the control group after 56 d of adding 0.2% microbe-derived antioxidant feed [20]. In the present study, it was found that the addition of 0.25–1% MAs to the diet under long-term feeding significantly increased the weight gain rate of crayfish; however, the addition of 1% MAs could make the WGR and SGR of juvenile crayfish significantly higher than those of the other groups, and the content of crude protein and crude fat in the body was also higher than that of the other groups. The MAs include components such as VC, and numerous studies have shown that feed supplements of VC can significantly improve specific growth rates and reduce mortality in crustaceans [28,29]. The results of studies utilizing microbial-fermented feeds in aquatic animals have also demonstrated that growth performance can be enhanced. Panigrahi et al. [30] found that probiotic bacteria, such as *B. subtilis*, in bio-flocculent systems significantly increased *Penaeus indicus* WGR, MY, and FCR, and improved survival. Wang et al. [31] found that *Macrobrachium nipponense* consuming probiotic-supplemented diets containing *Lactobacillus* improved survival and optimized growth. It can thus be concluded that the fermentation components and metabolic yields of MAs can facilitate crayfish growth.

Studies on functional feed additives have highlighted that overdosing can lead to adverse effects, with dosage being a crucial factor influencing probiotic efficacy [32]. Several studies have demonstrated that excessive probiotic supplementation in diets impairs the growth performance of aquatic animals [19,33]. Zhang et al. [34] reported limited, or even negative, effects on the growth and physiological functions of crustaceans when astaxanthin was overdosed, and similarly, excess VC in the diet reduced weight gain in crayfish, compared to the control group [29]. Appropriate taurine levels (10 and 25 mg kg^{-1}) accelerated molting and improved survival in juvenile *L. vannamei* [35], whereas excessive taurine (>0.8%) negatively impacted growth, nutrient accumulation, and survival in *E. sinensis* [36]. In terms of MA fermentation components, sea buckthorn fermentation produces *γ-aminobutyric acid* (GABA), which is also present in the fermentation products of *Clostridium butyricum*, and may affect the effectiveness of the diet due to its content. Meng et al. [37] found that high concentrations of *C. butyricum* supplements (2%) inhibited growth in *Micropterus salmoides*, while low concentrations (0.5%) enhanced body weight and total fatty acid content in the gut and liver, positively modulating intestinal flora. In this study, the 1.5% MA group exhibited the longest molting cycles and reduced the final body weight (FBW), weight gain rate (WGR), and specific growth rate (SGR). Conversely, the 0.5% MA group had the shortest second molting cycle, aligning with regression analyses indicating optimal additions for growth performance. Therefore, the incorporation of microbial antioxidants in diets should adhere to the principle of moderation. Based on the conditions of this study, 0.65% MA feed is suitable for long-term feeding in order to optimize the growth performance of crayfish.

Although the addition of 1.5% MAs affected growth performance, the antioxidant capacity indices in both the hepatopancreas and hemolymph of juvenile crayfish increased with an increase in MAs in the diet, and the best antioxidant capacity was found in the crayfish that were fed an addition of 1–1.5% MAs. Gu et al. [38] also found that mice being gavaged with 1% MAs could significantly increase the vitality of GSH-PX and SOD in the hemolymph and reduce the MDA content at the same time so that the antioxidant defense system in the body was enhanced and the antioxidant capacity of the mice was improved. Shi et al. [19] also found that, when *E. sinensis* ingested diets supplemented with 0.2% MAs, T-AOC and T-SOD were significantly higher than that of the control group, and MDA was significantly reduced under ammonia nitrogen stress. This is due to the presence of *B. subtilis*, *Lactobacillus*, and *flavonoids* in MAs, which can improve the antioxidant capacity of juvenile crayfish. *B. subtilis* can effectively inhibit the formation of lipid peroxidation products [39]. Prasuna et al. [40] found that adding *B. subtilis* to the diet could increase the SOD and CAT activities in the blood of *Macrobrachium rosenbergii*, reduce the MDA content, and improve antioxidant capacity. The same results were found in *Penaeus vannamei* regarding ingesting fermented feed supplemented with *B. subtilis* [39]. *Lactobacillus* can produce SOD in the organism, which is also an important barrier against free radical attack [41]. In crayfish, Li et al. [2] showed that T-AOC and SOD activities significantly increased and MDA content reduced after the intake of probiotic biocontinues containing *Lactobacillus*, *Bacillus*, and other probiotic organisms, which improved the immune and antioxidant capacity of the crayfish. The same results were found in *Fennerpenaeus chinensis* [42]. *Flavonoids* can reduce lipid peroxidation and ROS, and play a role in lymphocyte protection [43]. Li et al. [44] found that, when crayfish ingested diets supplemented with different concentrations of fermented *Moringa oleifera* leaves (*flavonoids*, etc.), this increased the hepatopancreatic SOD and GSH-PX activities and reduced the MDA content compared to the basal diet, with the addition of 1% fermented *M. oleifera* leaves being the most effective in improving antioxidant capacity. The present study showed that different levels of MAs enhanced the antioxidant defense system in crayfish, and the addition of 1–1.5% MAs was the best for the antioxidant capacity of juvenile crayfish.

The dual stress of dehydration and hypoxia during transport disrupts respiratory metabolism, induces oxidative damage to tissues, and reduces the survival of crustaceans [6,45]. Air exposure stress generates a large amount of ROS in the organism, which can cause oxidative damage if antioxidant enzymes are unable to remove them in a timely manner [46]. Respiratory metabolic enzyme results further confirmed that different levels of MAs could potentially mitigate the oxidative damage caused by air exposure stress on the organism. Antioxidant components, such as VC, *taurine*, and *isoflavones*, in MA products can eliminate ROS produced by the organism due to air exposure stress and reduce the degree of anaerobic respiration [47], which improves the resistance of crayfish to desiccation. *Flavonoids* can reduce lipid peroxidation and ROS, and play a role in lymphocyte protection [43]. In the hepatopancreas and muscle of crayfish, 1.5% MAs showed the lowest LD content and LDH capacity and the highest SDH capacity. These results further indicate that MAs can prolong the transition from aerobic to anaerobic respiration under air exposure stress [48]. Furthermore, the lowest levels of MDA, reflecting free radical metabolism and internal oxidative damage in the form of lipid oxidation end products [49], were found in the 1–1.5% MA groups. Dong et al. [36] found that, when taurine was added to the diet of *E. sinensis* at different concentrations, the MDA content in the plasma and hepatopancreas of *E. sinensis* was significantly lower in crabs fed the 0.4% and 0.8% *taurine* diets than in those fed the control diet. Thus, *taurine* may improve the response of crayfish to dehydration and hypoxia-induced stress during off-water transport. Thus, the antioxidant components in MA products reduced the response to dehydration and hypoxia-induced stress, and improved survival to air exposure stress. Although the 1.5% MA diet was effective in improving the antioxidant capacity and resistance to air exposure stress of crayfish in the long-term diet, the growth performance was also significantly lower than the other additive groups.

Feeding behavior is a basic living activity of animals, and the main way to obtain nutrients [25]. The weight gain of aquatic animals is closely related to dietary nutrients, and the palatability of the diet is also directly related to the growth and development of aquatic animals [50]. In terms of dietary nutrient composition, the crude protein, crude fat, and ash contents did not differ significantly among the additive groups, and there was no significant difference in body composition (Table 3) between the juvenile crayfish that ingested the 1.5% MA diet for 70 d and the other additive groups in terms of crude fat and ash contents. In the diet design, there was no difference in nutrient composition between the diet groups, but there was a dose difference in the addition of MAs. Therefore, the weight gain and palatability of crayfish were dependent on the level of MAs. From the perspective of feed palatability, a number of studies have shown that the addition of attractants influences the feeding behavior of aquatic animals [51,52]. Suresh et al. [53] found that the addition of small peptides and nucleotides significantly increased the intake of *Litopenaeus stylirostris*, which can be used as an attractant and palatability enhancer to influence the feeding behavior of animals, since MAs are produced by the fermentation of *Lactobacillus*, *B. subtilis*, and *yeast*. The fermentation of *Lactobacillus* produces organic acid (*lactic acid*) [54], which results in lower pH and higher acid concentration in fermented diets, exacerbating the sour taste of the diets [15]. The results of the feeding behavior showed that the 1.5% MA-supplemented group was significantly different from the control group in terms of seeking time, willingness to feed, feeding time, and feeding amount, and the amount of food intake by the crayfish was significantly lower than that of the control group in both the satiated and starved conditions; the willingness to feed was weaker, suggesting that the excessive addition of MAs had, indeed, affected the palatability of the diets. This finding is in accordance with the results of Zhang et al. [55]. When evaluating the sensory index of fermented feeds (fermented with different concentrations of *Lactobacillus*, *Brewer yeast*, and *B. subtilis*) for *P. vannamei*, it was found that the fermented feeds produced a sour taste compared to the unfermented diets. The crustaceans usually have a strong ability to recognize chemical information, including food odors [25]. Therefore, it can be reasonably inferred that the increase in sour taste brought about by adding more MAs to the feed may be the main reason affecting the palatability of crayfish feed. The low intake under long-term feeding reduced the nutrient accumulation of the juvenile crayfish in the 1.5%

group, resulting in a prolonged molting period and lower growth performance than in other groups. Based on the positive effects of MAs on the growth and antioxidant capacity of crustacean aquatic animals, how to reduce the effect of MAs on feed palatability, and which attractant can improve the effect of MAs on palatability are also directions worthy of further research in the future.

Long-term feeding using the 1.5% MA diet has an effect on the growth performance of crayfish, as well as resulting in a significant improvement in the antioxidant capacity of the organism and resistance to air exposure stress. By optimizing the levels of MAs via feeding strategies to facilitate the better application of 1.5% MA supplements in crayfish production practice, using short-term feeding to determine the optimal feeding time for 1.5% MA supplements and improving resistance to air exposure stress with transport requirements can also be combined for cost reduction and efficiency. The short-term feeding results showed that the 1.5% MA diet had no negative effect on growth performance for 21 d (3 weeks), but after 28 d (4 weeks), the WGR and SGR of juvenile crayfish were significantly lower than those of the control group, indicating a negative effect on growth. The results were similar to those of *E. sinensis* fed on a diet supplemented with MAs, with a significant decrease in the weight gain rate after 28 d [19]. Some studies have shown that *lactic acid bacteria* and *yeast*-fermented diets are not suitable for long-term feeding [56,57]. Long-term feeding on *lactic acid bacteria* and *yeast* additive diets results in the continuous proliferation of live bacteria colonizing the gut [58], which can prey on and absorb nutrients in the gut of crayfish, affecting their growth performance. In terms of antioxidant capacity, the experimental results of Zhu [59], who added 1% MAs to the drinking water of rats, showed that the SOD and GSH-PX activities in serum and liver were significantly higher than those of the no-addition group after 28 d, and the MDA content reached the lowest level. The present study demonstrates that the application of 1.5% MAs enhances antioxidant capacity, with the activities of T-AOC and GSH-PX reaching a peak at 21 d, while the MDA content was also the lowest at 21 d. Therefore, feeding juvenile crayfish with a 1.5% MA diet for 21 d can enhance the antioxidant capacity and achieve the optimal antioxidant level. According to the indexes after air exposure stress, the survival rate of the 21 d and 28 d antioxidant groups was significantly higher than that of other groups, and the level of respiratory metabolic enzymes was also better than that of the other groups, indicating that the addition of 1.5% MAs to the diet for 21 d could reduce the degree of anaerobic respiration in the process of air exposure stress. Optimized feeding strategies may be a viable means of using high levels of MA-supplemented diets in crayfish production to meet transport requirements, reduce costs, and increase efficiency.

5. Conclusions

In conclusion, the addition of microbial antioxidants to a diet can improve the growth performance, antioxidant capacity, and immunity of juvenile crayfish, and reduce oxidative damage during transport. However, the addition of excessive amounts of microbial antioxidants to long-term diets was found to have limited beneficial effects on the growth and physiological functions of crayfish and negatively affect the palatability of the diets. By combining the indicators of growth, antioxidants, resistance to air exposure stress, and economic benefits, the addition of 1.5% microbial antioxidants to a diet for 21 d was effective in improving antioxidant capacity and reducing oxidative damage during transport.

Supplementary Materials: The following supporting information can be downloaded at: https://www.mdpi.com/article/10.3390/antiox14020135/s1, Supplement Table S1. Detailed data of quadratic relationship between dietary microbial antioxidant supplementation and FBW(A), WGR(B) and SGR(C) of juvenile crayfish.

Author Contributions: Conceptualization and design, J.L. (Jiayao Li) and Z.C.; feeding experiments, J.S.; contribution to the collection of samples, J.S., J.L. (Jinghao Li) and C.Q.; provision of essential materials and technical supports, I.K.; financial supports, X.W. and J.L. (Jiayao Li); data analysis and interpretation, Z.C., J.S. and J.L. (Jinghao Li); writing—review and editing, Z.C., J.L. (Jiayao Li)and C.Q. All authors contributed experimental assistance and intellectual input to this study. All authors have read and agreed to the published version of the manuscript.

Funding: This study was funded by the Asian Cooperation Fund project "Technical Cooperation in Rice–Fish and Personnel Training in the Lancang–Mekong Area"(No. 18240066) and the Special Fund of the Chinese Agriculture Research System from the Ministry of Agriculture of China, with grant number [CARS-48].

Institutional Review Board Statement: All animal care and handling procedures in this study were conducted under the guidance of the Standards for the Care and Use of Laboratory Animals in China, and strictly followed the Standard Operation Procedures (SOPs) detailed in the Guide for Use of Experimental Animals at Shanghai Ocean University.

Informed Consent Statement: Not applicable.

Data Availability Statement: The data that support the findings of this study are available from the corresponding author upon reasonable request.

Acknowledgments: The authors would like to thank the staff at the Key Laboratory of Integrated Rice–Fish Farming, the Ministry of Agriculture and Rural Affairs, Shanghai Ocean University, and KONG IEONG (Shanghai Chuangbo Ecological Engineering, Shanghai, China) for their assistance during the experiments.

Conflicts of Interest: The authors declare that the research was conducted in the absence of any commercial or financial relationships that could be construed as a potential conflict of interest.

References

1. Jin, S.; Jacquin, L.; Ren, Y.; Yu, J.; Li, W.; Lek, S.; Liu, J.; Li, Z.; Zhang, T. Growth performance and muscle composition response to reduced feeding levels in juvenile red swamp crayfish *Procambarus clarkii* (Girard, 1852). *Aquac. Res.* **2019**, *50*, 934–943. [CrossRef]
2. Li, J.; Huang, J.; Li, C.; Zhang, Y.; Wang, Y.; Hou, S.; Cheng, Y.; Li, J. Evaluation of the nutritional quality of edible tissues (muscle and hepatopancreas) of cultivated *Procambarus clarkii* using biofloc technology. *Aquac. Rep.* **2021**, *19*, 100586. [CrossRef]
3. Guo, K.; Ruan, G.L.; Fan, W.H.; Wang, Q.; Fang, L.; Luo, J.B.; Liu, Y.L. Immune response to acute heat stress in the intestine of the red swamp crayfish, *Procambarus clarkii*. *Fish Shellfish Immunol.* **2020**, *100*, 146–151. [CrossRef] [PubMed]
4. Wang, J.; Ye, J.; Zhang, Z.; An, Z.; Wang, T.; Dong, X. Comparison of the nutrient value, nonspecific immunity, and intestinal microflora of red swamp crayfish (*Procambarus clarkii*) in different culture modes. *Aquac. Rep.* **2023**, *31*, 101683. [CrossRef]
5. Ministry of Agriculture and Rural Fisheries Administration; National Fisheries Technology Extension Station; Chinese Fisheries Society. *China Fisheries Statistical Yearbook of 2024*; China Agriculture Press: Beijing, China, 2024; p. 159.
6. Paital, B. Antioxidant and oxidative stress parameters in brain of Heteropneustes fossilis under air exposure condition; role of mitochondrial electron transport chain. *Ecotoxicol Env. Saf.* **2013**, *95*, 69–77. [CrossRef] [PubMed]
7. Luo, Z.; Gao, Q.; Zhang, H.; Zhang, Y.; Zhou, S.; Zhang, J.; Xu, W.; Xu, J. Microbe-derived antioxidants attenuate cobalt chloride-induced mitochondrial function, autophagy and BNIP3-dependent mitophagy pathways in *BRL3A* cells. *Ecotoxicol. Environ. Saf.* **2022**, *232*, 113219. [CrossRef]
8. Lu, Y.P.; Zheng, P.H.; Zhang, X.X.; Li, J.T.; Zhang, Z.L.; Xu, J.R.; Meng, Y.Q.; Li, J.J.; Xian, J.A.; Wang, A.L. New insights into the regulation mechanism of red claw crayfish (*Cherax quadricarinatus*) hepatopancreas under air exposure using transcriptome analysis. *Fish Shellfish Immunol.* **2023**, *132*, 108505. [CrossRef] [PubMed]
9. Lei, X.; Yang, L.; Tan, L.; Yang, Q.; Zhou, F.; Jiang, S.; Huang, J. Effect of Air Exposure and Re-Submersion on the Histological Structure, Antioxidant Response, and Gene Expression of *Procambarus clarkii*. *Animals* **2023**, *13*, 462. [CrossRef] [PubMed]
10. Dawood, M.A.O.; Koshio, S.; Abdel-Daim, M.M.; Van Doan, H. Probiotic application for sustainable aquaculture. *Rev. Aquac.* **2019**, *11*, 907–924. [CrossRef]
11. Kuebutornye, F.K.A.; Abarike, E.D.; Lu, Y. A review on the application of *Bacillus* as probiotics in aquaculture. *Fish Shellfish Immunol.* **2019**, *87*, 820–828. [CrossRef] [PubMed]

12. Li, H.D.; Tian, X.L.; Dong, S.L. Growth performance, non-specific immunity, intestinal histology and disease resistance of *Litopenaeus vannamei* fed on a diet supplemented with live cells of *Clostridium butyricum*. *Aquaculture* **2019**, *498*, 470–481. [CrossRef]
13. Ringø, E.; Van Doan, H.; Lee, S.H.; Soltani, M.; Hoseinifar, S.H.; Harikrishnan, R.; Song, S.K. Probiotics, lactic acid bacteria and bacilli: Interesting supplementation for aquaculture. *J. Appl. Microbiol.* **2020**, *129*, 116–136. [CrossRef]
14. Lee, K.W.; Kim, D.K.; Lillehoj, H.S.; Jang, S.I.; Lee, S.-H. Immune modulation by *Bacillus subtilis*-based direct-fed microbials in commercial broiler chickens. *Anim. Feed Sci. Technol.* **2015**, *200*, 76–85. [CrossRef]
15. Zhang, M.; Pan, L.; Fan, D.; He, J.; Su, C.; Gao, S.; Zhang, M. Study of fermented feed by mixed strains and their effects on the survival, growth, digestive enzyme activity and intestinal flora of *Penaeus vannamei*. *Aquaculture* **2021**, *530*, 735703. [CrossRef]
16. Luo, Z.; Xu, X.; Zhao, S.; Sho, T.; Luo, W.; Zhang, J.; Xu, W.; Hon, K.; Xu, J. Inclusion of microbe-derived antioxidant during pregnancy and lactation attenuates high-fat diet-induced hepatic oxidative stress, lipid disorders, and *NLRP3* inflammasome in mother rats and offspring. *Food Nutr. Res.* **2019**, *63*, 3504. [CrossRef] [PubMed]
17. Yu, C.; Luo, Y.; Shen, C.; Luo, Z.; Zhang, H.; Zhang, J.; Xu, W.; Xu, J. Effects of microbe-derived antioxidants on growth performance, hepatic oxidative stress, mitochondrial function and cell apoptosis in weaning piglets. *J. Anim. Sci. Biotechnol.* **2024**, *15*, 128. [CrossRef]
18. Alvanou, M.V.; Feidantsis, K.; Staikou, A.; Apostolidis, A.P.; Michaelidis, B.; Giantsis, I.A. Probiotics, Prebiotics, and Synbiotics Utilization in Crayfish Aquaculture and Factors Affecting Gut Microbiota. *Microorganisms* **2023**, *11*, 1232. [CrossRef]
19. Tao, L.T.; Lu, H.; Xiong, J.; Zhang, L.; Sun, W.W.; Shan, X.F. The application and potential of postbiotics as sustainable feed additives in aquaculture. *Aquaculture* **2024**, *592*, 741237. [CrossRef]
20. Shi, A.; Ma, H.; Shi, X.; Zhou, W.; Pan, W.; Song, Y.; Chen, Q.; Yu, X.; Niu, C.; Yang, Y.; et al. Effects of microbe-derived antioxidants on growth, digestive and aminotransferase activities, and antioxidant capacities in the hepatopancreas of *Eriocheir sinensis* under ammonia nitrogen stress. *Aquac. Fish.* **2024**, *9*, 957–966. [CrossRef]
21. Wu, D.; An, R.; Wang, D.; Jiang, L.; Huang, L.; Lu, T.; Xu, W.; Xu, J.; Zhang, J. Regulatory Effects of Maternal Intake of Microbial-Derived Antioxidants on Colonization of Microbiota in Breastmilk and That of Intestinal Microbiota in Offspring. *Animals* **2024**, *14*, 2582. [CrossRef]
22. Folch, J.; Lees, M.; Stanley, G.H.S. A simple method for the isolation and purification of total lipides from animal tissues. *J. Biol. Chem.* **1957**, *226*, 497–509. [CrossRef] [PubMed]
23. Arlington, V. *Official Methods of Analysis of the Association of Official Analytical Chemists*, 16th ed.; Association of Official Analytical Chemists, AOAC: Rockville, MD, USA, 1995; p. 993.
24. Mo, N.; Zhu, D.D.; Liu, J.X.; Feng, T.Y.; Cui, Z.X. Metabolic responses to air-exposure stress of the Chinese mitten crab (*Eriocheir sinensis*) revealed by a combined analysis of metabolome and transcriptome. *Aquaculture* **2022**, *548*, 737710. [CrossRef]
25. Chen, Q.; Pang, Y.Y.; Zhang, J.Y.; Cheng, Y.X.; Yang, X.Z. The role of different sensory organs in the feeding behavior of Chinese mitten crab (*Eriocheir sinensis*). *Aquaculture* **2023**, *566*, 739231. [CrossRef]
26. Luo, Z.; Luo, W.; Zhao, S.; Li, S.; Zhang, J.; Xu, W.; Xu, J. The effects of micro-derived antioxidants on the growth performance and antioxidant capacity of weaned piglets. *Feed Ind.* **2017**, *38*, 15–17. [CrossRef]
27. Chen, X.; Gong, L.Z.; Xu, J.X. Study of the role of microbial-derived antioxidants on antioxidant capacity and hepatocyte DNA damage in oxidatively stressed rats. *Feed Ind.* **2012**, *33*, 14–17.
28. Kong, F.; Zhu, Y.; Yu, H.; Wang, X.; Abouel Azm, F.R.; Yuan, J.; Tan, Q. Effect of dietary vitamin C on the growth performance, nonspecific immunity and antioxidant ability of red swamp crayfish (*Procambarus clarkii*). *Aquaculture* **2021**, *541*, 736785. [CrossRef]
29. Song, Y.; Cai, X.; Bu, X.; Liu, S.; Song, M.; Yang, Y.; Wang, X.; Shi, Q.; Qin, J.; Chen, L. Effects of Iron and Vitamin C on Growth Performance, Iron Utilization, Antioxidant Capacity, Nonspecific Immunity, and Disease Resistance to Aeromonas hydrophila in Chinese Mitten Crab (*Eriocheir sinensis*). *Aquac. Nutr.* **2023**, *2023*, 7228854. [CrossRef]
30. Panigrahi, A.; Das, R.R.; Sivakumar, M.R.; Saravanan, A.; Saranya, C.; Sudheer, N.S.; Kumaraguru Vasagam, K.P.; Mahalakshmi, P.; Kannappan, S.; Gopikrishna, G. Bio-augmentation of heterotrophic bacteria in biofloc system improves growth, survival, and immunity of Indian white shrimp *Penaeus indicus*. *Fish Shellfish Immunol.* **2020**, *98*, 477–487. [CrossRef]
31. Wang, J.; Li, S.; Jian, Y.; Song, J.; Zheng, J.; Zhou, D.; Kong, Y.; Limbu, S.M.; Ye, J.; Ding, Z. Dietary postbiotics supplementation improves growth, survival rate, antioxidant capacity, non-specific immunity and gut health of juvenile oriental river prawn (*Macrobrachium nipponense*). *Aquac. Rep.* **2023**, *33*, 101771. [CrossRef]
32. Nikoskelainen, S.; Ouwehand, A.C.; Bylund, G.; Salminen, S.; Lilius, E.-M. Immune enhancement in rainbow trout (Oncorhynchus mykiss) by potential probiotic bacteria (*Lactobacillus rhamnosus*). *Fish Shellfish Immunol.* **2003**, *15*, 443–452. [CrossRef]
33. Ghosh, T. Recent advances in the probiotic application of the *Bacillus* as a potential candidate in the sustainable development of aquaculture. *Aquaculture* **2025**, *594*, 741432. [CrossRef]
34. Zhang, Y.; Qian, C.; Huang, J.; Li, J.; Jiang, X.; Li, Z.; Cheng, Y.; Li, J. Suitable natural astaxanthin supplementation with *Haematococcus pluvialis* improves the physiological function and stress response to air exposure of juvenile red swamp crayfish (*Procambarus clarkii*). *Aquaculture* **2023**, *573*, 739577. [CrossRef]

35. El-Sayed, A.-F.M. Is dietary *taurine* supplementation beneficial for farmed fish and shrimp? a comprehensive review. *Rev. Aquac.* **2014**, *6*, 241–255. [CrossRef]
36. Dong, J.; Cheng, R.J.; Yang, Y.H.; Zhao, Y.Y.; Wu, G.F.; Zhang, R.Y.; Zhu, X.C.; Li, L.; Li, X.D. Effects of dietary *taurine* on growth, non-specific immunity, antioxidative properties and gut immunity in the Chinese mitten crab *Eriocheir sinensis*. *Fish Shellfish Immunol.* **2018**, *82*, 212–219. [CrossRef] [PubMed]
37. Meng, X.; Cai, H.; Li, H.; You, F.; Jiang, A.; Hu, W.; Li, K.; Zhang, X.; Zhang, Y.; Chang, X.; et al. *Clostridium butyricum*-fermented Chinese herbal medicine enhances the immunity by modulating the intestinal microflora of largemouth bass (*Micropterus salmoides*). *Aquaculture* **2023**, *562*, 738768. [CrossRef]
38. Gu, J.; Chen, X.L.; Li, X.; Wang, Y.F.; Cai, X.; Gu, Y.Y.; Xu, J.X. Microbial antioxidant on antioxidant performance and immune function in mice. *Biotechnol. Bull.* **2012**, *9*, 99–103. [CrossRef]
39. Shen, W.Y.; Fu, L.L.; Li, W.F.; Zhu, Y.R. Effect of dietary supplementation with *Bacillus subtilis* on the growth, performance, immune response and antioxidant activities of the shrimp (*Litopenaeus vannamei*). *Aquac. Res.* **2010**, *41*, 1691–1698. [CrossRef]
40. Devi, D.P.; Hareesh, K.; Reddy, M.S. Enhancement of growth potentials in freshwater prawn *Macrobrachium rosenbergii* through supplementation of probiotic diets of *Bacillus subtilis* and *Lactobacillus rhamnosus*. *Int. J. Fish. Aquat. Stud.* **2015**, *3*, 124–131.
41. Daenen, K.; Andries, A.; Mekahli, D.; Van Schepdael, A.; Jouret, F.; Bammens, B. Oxidative stress in chronic kidney disease. *Pediatr. Nephrol.* **2019**, *34*, 975–991. [CrossRef]
42. Kim, M.S.; Min, E.Y.; Kim, J.H.; Koo, J.K.; Kang, J.C. Growth performance and immunological and antioxidant status of Chinese shrimp, *Fennerpenaeus chinensis* reared in bio-floc culture system using probiotics. *Fish Shellfish Immunol.* **2015**, *47*, 141–146. [CrossRef] [PubMed]
43. Yan, X.; Pang, P.; Qin, C.; Mi, J.; Yang, L.; Yang, B.; Nie, G. Improvement of sea buckthorn (*Hippophae rhamnoides* L.) *flavonoids* on the antioxidant and immune performance of Yellow River carp (*Cyprinus carpio* L.) fed high-carbohydrate diet. *Fish Shellfish Immunol.* **2024**, *144*, 109289. [CrossRef] [PubMed]
44. Li, Z.; Luo, W.; Zhou, Q.; Sun, C.; Zheng, X.; Liu, B.; Mpange, K.; Zhu, A.; Wang, A. Investigation of the Fermentation Process of *Moringa oleifera* Leaves and Its Effects on the Growth Performance, Antioxidant Capacity, and Intestinal Microbiome of *Procambarus clarkii*. *Antioxidants* **2024**, *13*, 1355. [CrossRef] [PubMed]
45. Duan, Y.F.; Zhang, J.S.; Dong, H.B.; Wang, Y.; Liu, Q.S.; Li, H. Effect of desiccation and resubmersion on the oxidative stress response of the *kuruma shrimp Marsupenaeus japonicus*. *Fish Shellfish Immunol.* **2016**, *49*, 91–99. [CrossRef] [PubMed]
46. Duan, Y.F.; Liu, P.; Li, J.T.; Li, J.; Chen, P. Expression profiles of selenium dependent glutathione peroxidase and glutathione S-transferase from Exopalaemon carinicauda in response to Vibrio anguillarum and WSSV challenge. *Fish Shellfish Immunol.* **2013**, *35*, 661–670. [CrossRef]
47. Choi, W.; Moniruzzaman, M.; Hamidoghli, A.; Bae, J.; Lee, S.; Lee, S.; Min, T.; Bai, S.C. Effect of Four Functional Feed Additives on Growth, Serum Biochemistry, Antioxidant Capacity, Gene Expressions, Histomorphology, Digestive Enzyme Activities and Disease Resistance in Juvenile Olive Flounder, *Paralichthys olivaceus*. *Antioxidants* **2023**, *12*, 1494. [CrossRef]
48. Lorenzon, S.; Giulianini, P.G.; Libralato, S.; Martinis, M.; Ferrero, E.A. Stress effect of two different transport systems on the physiological profiles of the *crab Cancer pagurus*. *Aquaculture* **2008**, *278*, 156–163. [CrossRef]
49. Liu, F.; Geng, C.; Qu, Y.-K.; Cheng, B.-X.; Zhang, Y.; Wang, A.-M.; Zhang, J.-H.; Liu, B.; Tian, H.-Y.; Yang, W.-P.; et al. The feeding of dietary Codonopsis pilosula polysaccharide enhances the immune responses, the expression of immune-related genes and the growth performance of red swamp crayfish (*Procambarus clarkii*). *Fish Shellfish Immunol.* **2020**, *103*, 321–331. [CrossRef]
50. Dworjanyn, S.A.; Pirozzi, I.; Liu, W. The effect of the addition of algae feeding stimulants to artificial diets for the sea urchin *Tripneustes gratilla*. *Aquaculture* **2007**, *273*, 624–633. [CrossRef]
51. Nunes, A.J.P.; Sá, M.V.C.; Andriola Neto, F.F.; Lemos, D. Behavioral response to selected feed attractants and stimulants in Pacific white shrimp, *Litopenaeus vannamei*. *Aquaculture* **2006**, *260*, 244–254. [CrossRef]
52. Peixoto, S.; Strebel, L.; Soares, R.; Davis, D.A. Acoustic feeding responses using marine chemoattractants in plant-based diets for naive and non-naive *Litopenaeus vannamei*. *Appl. Anim. Behav. Sci.* **2022**, *257*, 105792. [CrossRef]
53. Suresh, A.V.; Paramasivam, K.V.K.; Nates, S. Attractability and palatability of protein ingredients of aquatic and terrestrial animal origin, and their practical value for blue shrimp, *Litopenaeus stylirostris* fed diets formulated with high levels of poultry byproduct meal. *Aquaculture* **2011**, *319*, 132–140. [CrossRef]
54. Lin, Y.H.; Chen, Y.T. *Lactobacillus* spp. fermented soybean meal partially substitution to fish meal enhances innate immune responses and nutrient digestibility of white shrimp (*Litopenaeus vannamei*) fed diet with low fish meal. *Aquaculture* **2022**, *548*, 737634. [CrossRef]
55. Zhang, X.; Tian, X.; Jiang, W.; Zhao, K.; Dong, S.; Cai, Y.; Liang, J. Growth performance, non-specific immunity and Vibrio parahaemolyticus resistance of Pacific white shrimp, *Litopenaeus vannamei*, in response to various microbial-derived additives. *Aquac. Nutr.* **2021**, *27*, 666–678. [CrossRef]
56. Figueras, A.; Chizhayeva, A.; Amangeldi, A.; Oleinikova, Y.; Alybaeva, A.; Sadanov, A. *Lactic acid bacteria* as probiotics in sustainable development of aquaculture. *Aquat. Living Resour.* **2022**, *35*, 10. [CrossRef]

57. Jahan, N.; Islam, S.M.M.; Rohani, M.F.; Hossain, M.T.; Shahjahan, M. Probiotic yeast enhances growth performance of rohu (*Labeo rohita*) through upgrading hematology, and intestinal microbiota and morphology. *Aquaculture* **2021**, *545*, 737243. [CrossRef]
58. Gupta, S.; Fečkaninová, A.; Lokesh, J.; Koščová, J.; Sørensen, M.; Fernandes, J.; Kiron, V. *Lactobacillus* Dominate in the Intestine of Atlantic Salmon Fed Dietary Probiotics. *Front. Microbiol.* **2018**, *9*, 3247. [CrossRef]
59. Zhu, Y. Effects of Antioxidant from Microbe on Free Radical Metabolic, Liver Damage Repairs and Produce Performance of Animals. Master's Thesis, Shanghai Jiao Tong University, Shanghai, China, 2007.

antioxidants

Article

Dietary Lysophosphatidylcholine Improves the Uptake of Astaxanthin and Modulates Cholesterol Transport in Pacific White Shrimp *Litopenaeus vannamei*

Ziling Song [1,2], Yang Liu [1,2], Huan Liu [2], Zhengwei Ye [2], Qiang Ma [2], Yuliang Wei [2], Lindong Xiao [3], Mengqing Liang [2] and Houguo Xu [2,*]

1 College of Fisheries and Life Sciences, Shanghai Ocean University, 999 Huchenghuan Road, Shanghai 201306, China

2 State Key Laboratory of Mariculture Biobreeding and Sustainable Goods, Yellow Sea Fisheries Research Institute, Chinese Academy of Fishery Sciences, 106 Nanjing Road, Qingdao 266071, China

3 Weifang Key Laboratory of Precise Animal Nutrition, Weifang Kenon Biotechnology Co., Ltd., Weifang 261108, China

* Correspondence: xuhg@ysfri.ac.cn

Abstract: Astaxanthin (AST), functioning as an efficient antioxidant and pigment, is one of the most expensive additives in shrimp feeds. How to improve the uptake efficiency of dietary astaxanthin into farmed shrimp is of significance. The present study investigated the effects of lysophosphatidylcholine (LPC), an emulsifier, on dietary astaxanthin efficiency, growth performance, body color, body composition, as well as lipid metabolism of juvenile Pacific white shrimp (average initial body weight: 2.4 g). Three diets were prepared: control group, the AST group (supplemented with 0.02% AST), and the AST + LPC group (supplemented with 0.02% AST and 0.1% LPC). Each diet was fed to triplicate tanks, and each tank was stocked with 30 shrimp. The shrimp were fed four times daily for eight weeks. The AST supplementation improved the growth of white shrimp, while LPC further promoted the final weight of shrimp, but the whole-shrimp proximate composition and fatty acid composition were only slightly affected by AST and LPC. The LPC supplementation significantly increased the astaxanthin deposition in the muscle. The LPC supplementation significantly increased the shell yellowness of both raw and cooked shrimp compared to the AST group. Moreover, the dietary LPC increased the high-density lipoprotein-cholesterol content but decreased the low-density lipoprotein-cholesterol content in the serum, indicating the possible regulation of lipid and cholesterol transport. The addition of astaxanthin significantly up-regulated the expression of *npc2* in the hepatopancreas compared to the control group, while the addition of LPC down-regulated the expression of *mttp* compared to the AST group. In conclusion, the LPC supplementation could facilitate the deposition of dietary astaxanthin into farmed shrimp and further enlarge the beneficial effects of dietary astaxanthin. LPC may also independently regulate shrimp body color and cholesterol transportation. This was the first investigation of the promoting effects of LPC on dietary astaxanthin efficiency.

Keywords: carotenoid; lysophospholipid; lipid metabolism; antioxidant capacity; shrimp quality

Citation: Song, Z.; Liu, Y.; Liu, H.; Ye, Z.; Ma, Q.; Wei, Y.; Xiao, L.; Liang, M.; Xu, H. Dietary Lysophosphatidylcholine Improves the Uptake of Astaxanthin and Modulates Cholesterol Transport in Pacific White Shrimp *Litopenaeus vannamei*. *Antioxidants* **2024**, *13*, 505. https://doi.org/10.3390/antiox13050505

Academic Editors: Alessandra Napolitano and Erchao Li

Received: 3 April 2024
Revised: 18 April 2024
Accepted: 19 April 2024
Published: 23 April 2024

1. Introduction

Pacific white shrimp, *Litopenaeus vannamei*, also known as white-leg shrimp, is the most important aquaculture shrimp species all over the world [1]. The body color of shrimp is a very important quality trait and, consequently, pigments such as natural or artificial carotenoids are commonly added into shrimp feeds to regulate the body color. Meanwhile, the prominent antioxidative effects of carotenoids are also considered when they are used [2,3].

Astaxanthin, as a non-vitamin A-derived carotenoid, is the most commonly used anti-oxidant and pigment in aqua-feeds. The use of astaxanthin in shrimp feeds has been found to not only promote the shrimp growth, survival, stress resistance, and pigmentation but also improve the antioxidant capacity and immune response [4–6]. However, to date, astaxanthin, whether natural or artificial, is still expensive and constitutes a large proportion of the feed cost [7]. How to increase the efficiency of dietary astaxanthin is of significance to the shrimp feed industry. Measures such as spraying post-pelleting has been tried to decrease the loss of astaxanthin during pelleting.

Lysophosphatidylcholine (LPC), also known as lysolecithin, is being more and more commonly used as an emulsifier in animal feeds. Research on aquatic animals havs shown the positive effects of LPC on growth performance, lipid accumulation, and antioxidation status [8–11]. Because astaxanthin is lipid-soluble, LPC is assumed to be able to facilitate the uptake of dietary astaxanthin into aquatic animals. However, this speculation has not been validated. Also, little information has been available regarding the use of LPC in shrimp feeds. Therefore, the present study was aimed at investigating whether LPC supplementation can enhance the uptake and the subsequent biological activities of astaxanthin in shrimp feeds.

2. Materials and Methods

2.1. Experimental Diets and Feeding Trial

Three groups of experimental diets were prepared (Table 1). The control diet had a lipid content of 6.5%, which is suitable for Pacific white shrimp. The other two diets were prepared by adding 0.02% astaxanthin (AST, DSM Co., Ltd., Shanghai, China) or 0.02% AST + 0.1% LPC (Weifang Kenon Biotechnology Co., Ltd., Weifang, China) into the control diet. The selection of the AST dose was according to previous AST studies on the same shrimp species [4,12,13]. However, the selection of the LPC dose was according to a previous study of ours on turbot considering the lack of accurate information in shrimp [8]. In the diet preparation process, all the ingredients were firstly grinded and sieved through an 80-mesh sieve. Then, the dietary ingredients were thoroughly mixed. Oils were then added and thoroughly mixed into the ingredient mixture. Thereafter, water was added to make the dough. Pellets with 1.0 mm diameter were made using a single-screw pelleting machine. After that, the pellets were loaded into a tray and dried in an oven at 55 °C. The dry pellets were then packed and stored at −20 °C until use.

Table 1. The formulation and proximate composition of the experimental diets used in this study (% dry matter except moisture).

Ingredient	Control	AST	AST + LPC
Fish meal	20	20	20
Soybean meal	30	30	30
Peanut meal	14	14	14
Poultry by-product meal	4	4	4
Wheat meal	21	21	21
Mineral premix [a]	0.5	0.5	0.5
Vitamin premix [a]	1	1	1
Monocalcium phosphate	1	1	1
Vitamin C	0.2	0.2	0.2
Choline chloride	0.2	0.2	0.2
Ethoxyquin	0.02	0.02	0.02
Mold inhibitor	0.1	0.1	0.1
Betaine	0.3	0.3	0.3
Soya lecithin	1.5	1.5	1.5
Soybean oil	3	3	3
CarophyllPink [b]	0	0.2	0.2

Table 1. *Cont.*

Ingredient	Control	AST	AST + LPC
Lysophosphatidylcholine [c]	0	0	0.1
Y_2O_3	0.1	0.1	0.1
Bentonite	1.08	0.88	0.78
Alginate	2	2	2
Proximate composition			
Crude protein	43.9	44.0	44.5
Crude lipid	6.5	6.4	6.7
Ash	10.3	10.5	10.2
Moisture	8.3	8.1	8.5
Gross energy (MJ/kg)	17.8	17.8	17.7

[a] Vitamin premix and mineral premix, designed for marine fish, were purchased from Qingdao Master Biotech Co., Ltd., Qingdao, China. Generally, the vitamin premix contained retinyl acetate, vitamin D_3, DL-α-tocopherol acetate, menadione nicotinamide bisulfite, thiamine, riboflavin, vitamin B6, cyanocobalamin, D-calcium pantothenate, niacinamide, folic acid, D-biotin, L-ascorbate-2-phosphate, inositol, betaine hydrochloride, yeast hydrolysate, and rice hull powder; the mineral premix contained ferrous sulfate, zinc sulphate, manganese sulphate, cupric sulfate, cobaltous chloride, sodium selenite, calcium iodate, and zeolite powder. [b] Astaxanthin was added as Carophyllpink of DSM Co., Ltd. (Shanghai, China), which contained 10% astaxanthin. [c] The phosphatidylcholine supplied by Weifang Kenon Biological Technology Co., Ltd. (Weifang, China) had a purity of 98%, and the available LPC concentration was 5%.

2.2. Experimental Shrimp and Feeding Management

The feeding experiment was conducted at the Langya Experimental Base of Yellow Sea Fisheries Research Institute (Qingdao, Shandong, China). In this experiment, Pacific white shrimp with an average initial body weight of 2.4 g were firstly acclimatized to the experimental environment with a commercial feed for two weeks. At the beginning of the feeding experiment, after fasting for 24 h, the experimental shrimp were randomly assigned to nine polyethylene tanks (300 L). In total, 3 replicate tanks were set up for each group of diet, and 30 shrimp were reared in each tank. During the experimental period, the shrimp were fed to apparent satiation four times (7:00, 12:00, 17:00, and 22:00) each day. During the shrimp rearing for eight weeks, the recirculating seawater system was turned on for two hours each day to remove the suspended solids. The system was kept hydrostatic at the rest time of the day, considering the big tank volume. The water was changed by 1/2 per day. The residual feeds and feces were siphoned out every day, and the tanks were brushed regularly from inside. During the feeding, the water temperature was 27~31 °C; salinity 28~30; dissolved oxygen > 8 mg/L; pH 7.6 to 7.9; total ammonia nitrogen < 0.3 mg/L; non-ionic ammonia nitrogen < 0.01 mg/L. All shrimp handling protocols, as well as the sampling protocols described below, in this study were reviewed and approved by the Animal Care and Use Committee (ACUC) of Yellow Sea Fisheries Research Institute, Chinese Academy of Fishery Sciences (protocol code ACUC202306125274; date of approval, 12 June 2023).

2.3. Sample Collection

At the end of the feeding experiment, after anaesthetized with eugenol, all shrimp in each tank were bulk weighed and counted. In addition, eight shrimp were randomly selected from each tank for tissue sampling. The hemolymph was collected from the pericardiac coelom of the shrimp using a 1 mL syringe. Anticoagulant was added at a 1:2 (*v*/*v*) ratio into a 1.5 mL centrifuge tube, which was centrifuged (4000× *g*, 10 min, 4 °C) after four hours to separate the plasma. After the blood was taken, three shrimp per tank were selected for the subsequent analysis of body color and proximate composition, and another three shrimp per tank were dissected for the collection of hepatopancreas and muscle samples. All tissue samples were snap-frozen in liquid nitrogen immediately after sampling and then brought back to the laboratory and stored in a −80 °C refrigerator.

2.4. Analysis of the Proximate Composition and Fatty Acid Composition

The proximate composition analysis of experimental diets and whole shrimp (three individual shrimp per tank) was conducted according to the standard methods of the Association of Official Analytical Chemists (AOAC). The moisture, crude protein, crude lipid, and ash were determined by the drying to constant weight in an oven at 105 °C, Kjeldahl method (FOSS KJELTEC 2300, Hillerod, Denmark), chloroform-methanol method [14], and incineration by muffle furnace at 550 °C, respectively.

The fatty acid composition of the muscles was analyzed by gas chromatography (GC-2010 Pro, Shimadzu, Kyoto, Japan). The samples were first freeze-dried in a freeze-dryer (FDU-1100, Tokyo Rikakikai, Co., Ltd., Tokyo, Japan) for 48 h. After the lipid was extracted by the chloroform-methanol method, 30 μL of sample was placed into a 10 mL glass tube and 2 mL of 0.5 mol/L potassium hydroxide-methanol solution was added. The samples were placed into a water bath at 75 °C for 30 min. After cooling, 1 mL of boron trifluoride-methanol solution was added, and the samples were placed into a water bath again at 75 °C for 30 min. After 1 mL of water was added, 1 mL of n-hexane was added and vortexed. The solution was left on ice for 1 h, and then the supernatant was taken for the determination with gas chromatography, which was equipped with a fused silica capillary column (SH-RT-2560, 100 m × 0.25 mm × 0.20 μm, Shimadzu, Japan; dicyano-propyl-polysiloxane as stationary phase) and a flame ionization detector. The column temperature increase was programmed: from 150 °C up to 200 °C at 15 °C/min and then from 200 °C to 250 °C at 2°C/min. Both the injector and detector temperatures were 250 °C. Results were expressed as percentage of each fatty acid with respect to total fatty acids (TFAs).

2.5. Analysis of Astaxanthin Concentration

The muscle astaxanthin concentration was determined by Qingdao Yuanxin Testing Technology Co., Ltd. (Qingdao, China). A total of 1 g of sample was weighed precisely and placed into a 50 mL round-bottomed centrifuge tube. Then, 4 mL of extraction solution was added and homogenized for 2 min. The homogenizer was rinsed with 4 mL of extraction solution. The two batches of solution were merged and vortexed. Then, ultrasonic extraction was carried out at <15 °C for 10 min, and centrifugation was carried out at 7104 r/min for 5 min to collect all the supernatant (into a 50 mL centrifuge tube). The residues were also collected. After centrifugation, 4 mL of extraction solution was added into the residue, and the above processes were repeated. The extracts were combined. The extracts were dehydrated by filtration with anhydrous sodium sulfate into a 100 mL brown rotary evaporation flask. The anhydrous sodium sulfate was washed with 15 mL of extract 2 times, and the extract was combined into the evaporation flask. The extract was concentrated under reduced pressure on a rotary evaporator at 40 ± 2 °C to be nearly dried. The extracts were further blow-dried with N_2 and re-dissolved with 1.0 mL 0.1% BHT ethanol solution. The extract was filtered through 0.22 μm organic membrane before subjection to high-performance liquid chromatography (LC-2030 C 3 D, Shimadzu, Kyoto, Japan).

2.6. Biochemical Parameters in Serum and Hepatopancreas

The levels of total cholesterol (TC), triglycerides (TG), high-density lipoprotein cholesterol (HDL-C), low-density lipoprotein cholesterol (LDL-C), malondialdehyde (MDA), hepatic lipase (HL), and lipoproteins esterase (LPL) in serum were measured with commercial kits. The TC, HL, and LPL kits were purchased from Beijing Solarbio Science & Technology Co., Ltd. (Beijing, China). The other kits were purchased from Nanjing Jiancheng Bioengineering Institute, China (Nanjing, China). Homogenates of hepatopancreas were prepared in 0.9% saline or 1:1 n-heptane:isopropanol according to the instructions and then centrifuged to separate the supernatant.

2.7. Analysis of Body Color

The raw shrimp were first wiped to clean the body surface, and then the body color was measured with a high-quality colorimeter (NR60CP+, Shenzhen ThreeNH Technology

Co., Ltd., China). Two spots (Figure 1), i.e., the first one near the head (A1) and the second one near the tail (A2), were selected for the color determination. The indices L* (brightness degree), a* (redness degree), and b* (yellowness degree) were recorded. The shrimp were then placed in a self-sealing bag and boiled in 95 °C water for 3 min for the color analysis of cooked shrimp. Three shrimp per tank were used for this analysis.

Figure 1. Schematic diagram of points (A1 and A2) for the shrimp body color measurement.

2.8. Quantitative Real-Time Polymerase Chain Reaction (qRT-PCR)

The qRT-PCR methods (see Table 2 for primers), regents (Accurate Biotechnology and TsingKe Biological Technology, Qingdao, China), and equipment (Roche LightCycler 96, Basel, Switzerland) were the same as our previous publications [15]. The relative mRNA expression was evaluated with the $2^{-\Delta\Delta CT}$ method [16]. Two reference genes, *ef-1α* and *β-actin*, which were screened according to our previous methods, were used in the calculation of mRNA expression of target genes. The geometrical mean of Ct values of these two references was used in the calculation [17].

Table 2. Sequences of the PCR primers used in this work.

Primer	Forward Primer (5′–3′)	GenBank Reference	Product Length (bp)
npc1-F	CGAAGGGGAAAAGCCAGAGT	XM_027363410.1	87
npc1-R	TTGAGGAGGAAGGGAGCGTA		
npc2-F	CGCAGTATCGGCAGTCAAGA	XM_027358057.1	149
npc2-R	GTGGTGTGAAAGGCAACGAC		
abca1-F	ACCTGAAGGCAGGACGAAAG	XM_027375794.1	73
abca1-R	GGCATACTCGCGCTATCTGT		
abcg1-F	TCCCGGAAGGCAGGAATAGA	XM_027356334.1	121
abcg1-R	TTCATCAGCGTCGACTTCCC		
abcg5-F	TCCGTTTGCCCCGATAACAA	XM_027382028.1	74
abcg5-R	TAGCGCTCGAGCAGGTAGTA		
abcg8-F	CCAACGATTCCGAAGGGTCT	XM_027368984.1	93
abcg8-R	CGTTGAGGATGAAGTCCCCC		
ldlr-F	CGTCACATGCCCAGCCATAA	XM_027363319.1	86
ldlr-R	GATCCGTCATGGCACTCGAA		
srb1-F	GTTCGACATCTACCCGGACC	XM_027352601.1	132
srb1-R	AACCAGAAGATGGGCAGGAC		
mttp-F	GCTGCTAAGGAAAGTGCGTG	XM_027380336.1	215
mttp-R	AAGGATGCGTCGCTAAGGAG		
apod-F	CAACGCGGTAACAGGGAAAG	XM_027368898.1	85
apod-R	GACAACCAGCTTGGCTTCAC		
ef-1α-F	GTATTGGAACAGTGCCCGTG	GU136229.1	143
ef-1α-R	ACCAGGGACAGCCTCAGTAAG		
β-actin-F	CGAGGTATCCTCACCCTGAA	AF300705.2	176
β-actin-R	GTCATCTTCTCGCGGTTAGC		

npc: NPC intracellular cholesterol transporter; *abca*: ATP-binding cassette sub-family A; *abcg*: ATP-binding cassette sub-family G; *ldlr*: low-density lipoprotein receptor; *srb1*: scavenger receptor class B member 1; *mttp*: microsomal triglyceride transfer protein large subunit; *apod*: apolipoprotein D.

2.9. Statistical Methods

Calculations are according to the following equations:

$$\text{Weight gain g} = \text{final weight} - \text{initial weight}$$

$$\text{Weight gain \%} = (\text{final weight} - \text{initial weight})/\text{IBW} \times 100;$$

$$\text{Feed conversion ratio} = \text{feed intake}/\text{weight gain};$$

$$\text{Feed intake \%} = \text{feed dry weight}/[\text{experimental days} \times (\text{initial weight} + \text{final weight})/2] \times 100;$$

$$\text{Survival \%} = \text{final shrimp number}/\text{initial shrimp number} \times 100.$$

All data were subjected to one-way analysis of variance (ANOVA) in SPSS 16.0 for Windows. Tukey's multiple range test was used to detect the significant differences between the means. The significance was accepted when $p < 0.05$. The results are presented as means of triplicate tanks ± standard error.

3. Results

3.1. Growth Performances and Somatic Indices

The final weight and weight gain of the AST + LPC group was significantly ($p < 0.05$) higher than that of the control group (Table 3). There was no significant difference ($p > 0.05$) in weight gain %, feed conversation ratio, feed intake, and survival among the groups.

Table 3. Growth performance of experimental shrimp (mean ± standard error).

Parameter	Control	AST	AST + LPC	*p* Value
Initial weight g	2.41 ± 0.09	2.34 ± 0.04	2.58 ± 0.05	0.162
Final weight g	10.35 ± 0.03 ^{a}	10.96 ± 0.25 ^{ab}	11.45 ± 0.34 ^{b}	0.046
Weight gain g	7.94 ± 0.07 ^{a}	8.62 ± 0.21 ^{ab}	8.87 ± 0.29 ^{b}	0.041
Weight gain %	330.33 ± 14.31	367.67 ± 3.28	344.00 ± 5.00	0.229
Feed conversion ratio	1.35 ± 0.06	1.24 ± 0.05	1.24 ± 0.04	0.573
Feed intake %	2.93 ± 0.14	2.83 ± 0.10	2.82 ± 0.01	0.792
Survival %	95.33 ± 2.33	98.00 ± 1.00	98.50 ± 1.50	0.460

Data in the same row not sharing a superscript letter were significantly ($p < 0.05$) different.

3.2. Body Proximate Composition

There was no significant ($p > 0.05$) difference in the proximate composition of whole shrimp and muscle among all groups (Table 4). The whole-shrimp crude lipid content in the AST and AST + LPC groups showed increasing trends compared to the control group.

Table 4. Body proximate composition of experimental shrimp (% wet weight, mean ± standard error).

Parameter	Control	AST	AST + LPC	*p* Value
Whole shrimp				
Moisture	77.20 ± 0.61	76.71 ± 0.68	76.76 ± 1.03	0.894
Crude protein	17.59 ± 0.47	17.72 ± 0.42	17.83 ± 0.58	0.943
Crude lipid	0.68 ± 0.07	1.32 ± 0.26	1.38 ± 0.38	0.583
Ash	3.10 ± 0.12	2.90 ± 0.10	2.90 ± 0.20	0.582
Muscle				
Moisture	74.74 ± 0.40	74.36 ± 1.11	74.33 ± 0.32	0.902
Crude protein	22.65 ± 0.29	22.80 ± 82	22.58 ± 0.33	0.959
Crude lipid	1.15 ± 0.08	1.13 ± 0.07	1.17 ± 0.02	0.873

3.3. Muscle Astaxanthin Concentration

The astaxanthin content of the AST + LPC group was significantly ($p < 0.05$) higher than those of the other two groups (Figure 2). The astaxanthin concentration in the AST group was higher numerically compared to the control group, but there was no significant statistical difference between the two groups.

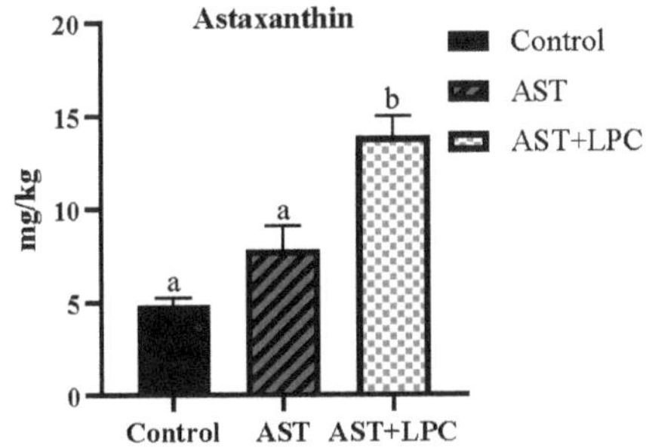

Figure 2. Muscle astaxanthin concentration of experimental shrimp (mean ± standard error). Data bars not sharing the same letter were significantly different ($p < 0.05$).

3.4. Body Color

The L* value of shrimp body significantly ($p < 0.05$) decreased after the addition of AST (Figures 3 and 4). The a* value was significantly ($p < 0.05$) increased by the addition of AST, but the addition of LPC slightly decreased the a* value compared to the AST group. The b* values significantly ($p < 0.05$) ranked as follows: AST + LPC > AST > control.

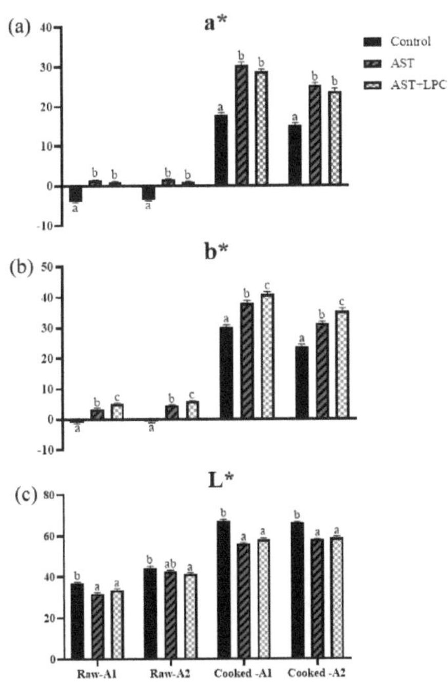

Figure 3. Body color parameters of experimental shrimp (mean ± standard error). For each sampling point, data bars not sharing the same letter were significantly different ($p < 0.05$). (**a**) a* value; (**b**) b* value; (**c**) L* value. The a*, b*, and L* indicate the redness (red–green), yellowness (yellow–blue), and brightness (white–black), respectively. A1 and A2 indicate the sampling points near the head and tail, respectively, as shown in Figure 1.

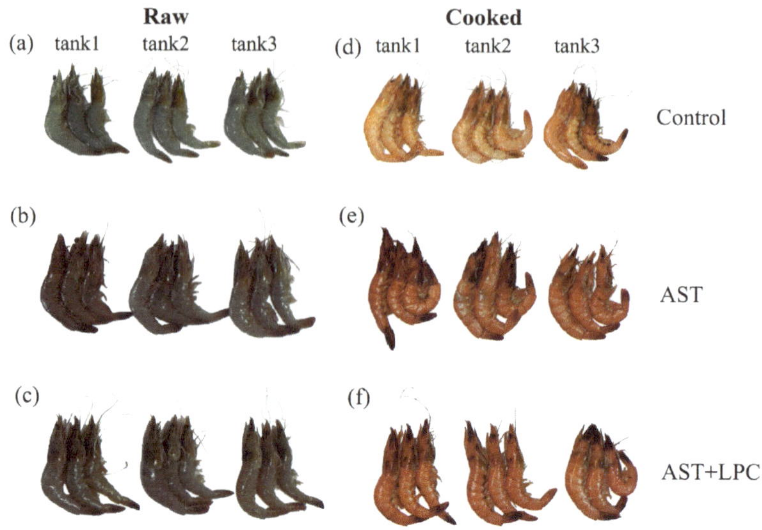

Figure 4. Photo pictures of raw and cooked shrimp fed experimental diets AST and LPC. (**a**) Raw—control group; (**b**) raw—AST group; (**c**) raw—AST + LPC group; (**d**) cooked—control group; (**e**) cooked—AST group; (**f**) cooked—AST + LPC group.

3.5. Fatty Acid Profiles of the Hepatopancreas

The contents of 14:1n-5, 16:1n-7, 18:1n-9, and mono-unsaturated fatty acids (MUFAs) were significantly ($p < 0.05$) lower in the AST group than in the other two groups, and there was no significant ($p > 0.05$) difference between the control and LPC + AST groups (Table 5). The contents of 18:2n-6 and n-6 polyunsaturated fatty acid (PUFA) were significantly ($p < 0.05$) higher in the LPC + AST group than those in the control group, and there was no significant ($p > 0.05$) difference between the control and AST groups. The n-3 PUFA content and n-3/n-6 ratio were lower in the LPC + AST group than those in the other two groups, and there was no significant ($p > 0.05$) difference between the control and AST groups.

Table 5. Muscle fatty acid composition of experimental shrimp (%TFA, mean ± standard error).

Fatty Acid	Control	AST	LPC + AST	*p* Value
14:0	0.33 ± 0.04	0.21 ± 0.01	0.33 ± 0.04	0.093
16:0	19.54 ± 0.29	19.49 ± 0.18	20.07 ± 1.00	0.764
18:0	9.21 ± 0.09	9.96 ± 0.24	9.32 ± 0.26	0.098
20:0	0.19 ± 0.00	0.19 ± 0.00	0.19 ± 0.01	0.527
SFA	29.26 ± 0.28	29.85 ± 0.06	29.91 ± 0.86	0.648
14:1n-5	0.20 ± 0.02 ^{b}	0.14 ± 0.01 ^{a}	0.19 ± 0.02 ^{ab}	0.035
16:1n-7	1.10 ± 0.08 ^{b}	0.76 ± 0.04 ^{a}	1.07 ± 0.06 ^{b}	0.013
18:1n-9	13.44 ± 0.13 ^{ab}	12.83 ± 0.15 ^{a}	13.90 ± 0.21 ^{b}	0.011
22:1n-9	0.16 ± 0.00	0.17 ± 0.01	0.18 ± 0.01	0.530
MUFA	14.90 ± 0.11 ^{b}	13.90 ± 0.19 ^{a}	15.35 ± 0.25 ^{b}	0.005
8:2n-6	18.58 ± 0.32 ^{a}	18.88 ± 0.42 ^{ab}	19.99 ± 0.04 ^{b}	0.038
20:2n-6	1.54 ± 0.03	1.72 ± 0.08	1.59 ± 0.08	0.242
n-6 PUFA	21.82 ± 0.31 ^{a}	22.45 ± 0.41 ^{ab}	23.20 ± 0.17 ^{b}	0.055
18:3n-3	1.05 ± 0.03	0.99 ± 0.06	1.10 ± 0.04	0.292
20:3n-3	1.69 ± 0.03	1.84 ± 0.08	1.62 ± 0.13	0.295
20:5n-3	9.39 ± 0.15	9.75 ± 0.22	8.61 ± 0.37	0.057

Table 5. *Cont.*

Fatty Acid	Control	AST	LPC + AST	*p* Value
22:5n-3	0.84 ± 0.05	0.77 ± 0.04	0.83 ± 0.03	0.456
22:6n-3	10.41 ± 0.19	10.14 ± 0.24	9.58 ± 0.23	0.092
n-3 PUFA	23.38 ± 0.07 ^{ab}	23.48 ± 0.16 ^{b}	21.73 ± 0.64 ^{a}	0.031
n-3/n-6	0.99 ± 0.01 ^{b}	0.97 ± 0.02 ^{b}	0.87 ± 0.02 ^{a}	0.004

Data in the same row not sharing a superscript letter were significantly ($p < 0.05$) different. SFA: saturated fatty acid; MUFA: mono-unsaturated fatty acids; PUFA: polyunsaturated fatty acid.

3.6. Biochemical Parameters in Serum and Hepatopancreas

There was no significant ($p > 0.05$) difference in serum TG, TC, LDL-C, MDA, LPL, and HL contents among all groups (Figure 5). However, the TG, TC, and MDA contents tended to be lower in the AST + LPC group compared to the AST group. The HDL-C content in the AST + LPC group was significantly ($p < 0.05$) higher than that in the control group. The LPL content showed a similar trend, but no significant difference was observed among groups.

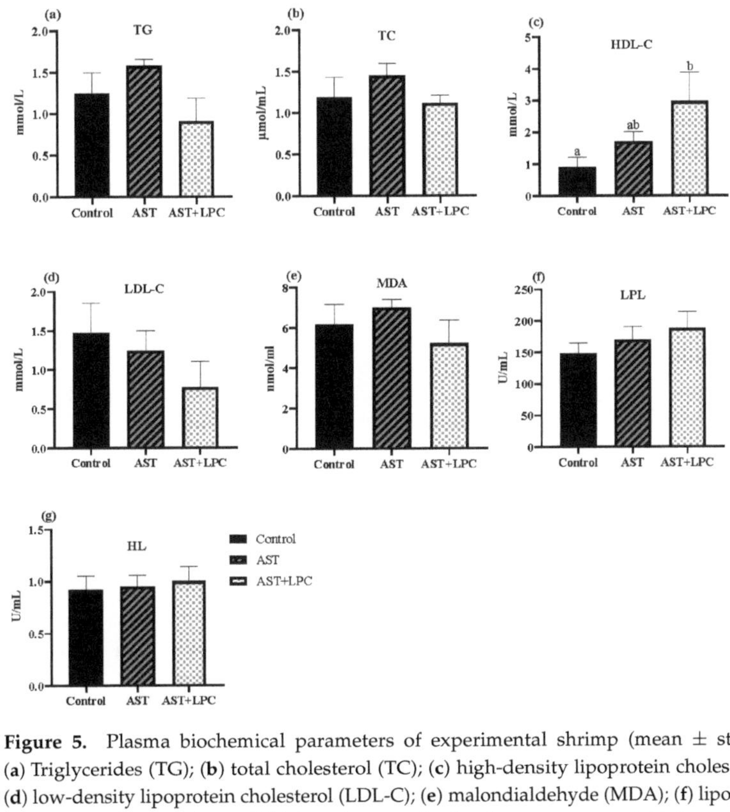

Figure 5. Plasma biochemical parameters of experimental shrimp (mean ± standard error). (**a**) Triglycerides (TG); (**b**) total cholesterol (TC); (**c**) high-density lipoprotein cholesterol (HDL-C); (**d**) low-density lipoprotein cholesterol (LDL-C); (**e**) malondialdehyde (MDA); (**f**) lipoproteins lipase (LPL); (**g**); and hepatic lipase (HL). Data bars not sharing the same superscript letter were significantly different ($p < 0.05$).

There was no significant ($p > 0.05$) difference in hepatopancreas TG, TC, and MDA contents among the groups (Figure 6). The lowest TG content was observed in the AST group, and the lowest MDA content was observed in the AST + LPC group.

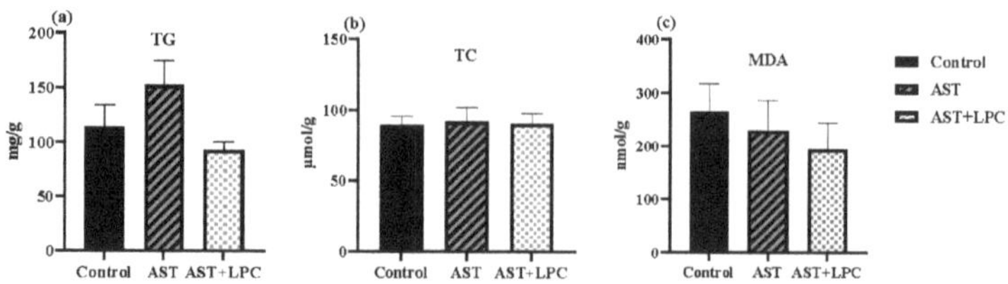

Figure 6. Hepatopancreas biochemical parameters of experimental shrimp (mean ± standard error). (**a**) Triglycerides (TG); (**b**) total cholesterol (TC); (**c**) malondialdehyde (MDA).

3.7. mRNA Expression of Genes Related to Lipid and Cholesterol Transport

In the hepatopancreas of shrimp (Figure 7), the experimental diets significantly ($p < 0.05$) regulated the expression of only *npc2* and *mttp*, both of which had the highest expression levels in the AST group. Most other genes, except *abcg5* and *apod*, showed similar trends to *npc2* and *mttp* in response to diets, although no significant ($p > 0.05$) difference was observed among groups.

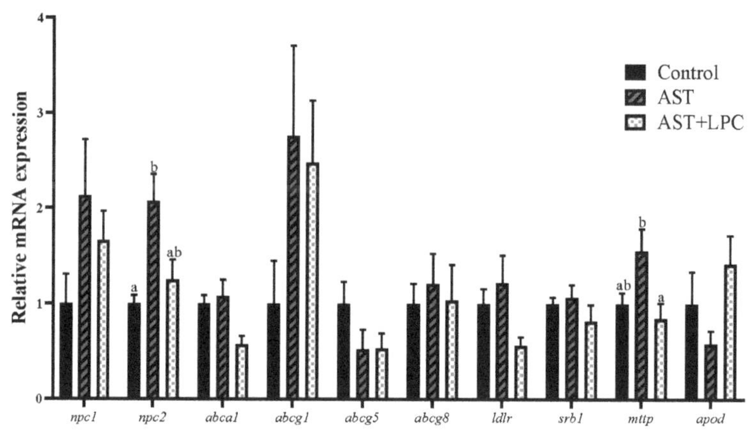

Figure 7. Relative mRNA expression of genes related to lipid and cholesterol transport in the hepatopancreas of experimental shrimp (mean ± standard error). Data bars not sharing a same lowercase letter were significantly different ($p < 0.05$).

4. Discussion

The beneficial effects of astaxanthin have been widely demonstrated in aquatic animals in terms of growth, survival, immunity, and anti-oxidation capacity [18–23]. These beneficial effects were also observed in this study, which clearly showed the growth-stimulating activity of astaxanthin (0.02%) in Pacific white shrimp. This result was consistent with the results of other studies on white shrimp [4,24,25]. Considering that the anti-oxidation capacity of astaxanthin has been a widely demonstrated and even undeniable fact, this anti-oxidation capacity was not further investigated and discussed in this study. We preferred to rather focus on effects of LPC on AST efficacy. In the present study, the growth was further improved by the addition of AST + LPC. This improvement could be, at least partly, attributed to the modulation of anti-oxidative status. Previous studies on crayfish *Procambarus clarkii* have evidenced that LPC could improve shrimp growth independently, probably by improving lipids absorption [9]. Unfortunately, in this study, there was no independent LPC treatment group, since the main purpose of this study was to evaluate the AST-stimulating effects of LPC. There was no significant difference in survival among experimental groups. The survival in all groups was higher than 95%, also indicating the accuracy of the feeding trial. Nevertheless, the improvement of survival by LPC has been observed in other fish and shrimp studies [4,8,9].

Regarding the body proximate composition, there were no significant difference in content of moisture, crude protein, and crude lipid of whole shrimp and muscle among groups. In studies on rainbow trout *Oncorhynchus mykiss* [10,26], channel catfish *Ictalurus punctatus* [11], and tiger puffer *Takifugu rubripes* [27], it has been observed that astaxanthin and LPC could promote lipid digestion and absorption and may thus increase the body lipid content. However, in this study, only an increasing trend in lipid content was observed from the control group to the AST + LPC group, but no significant differences were observed. The large intra-group data deviations could mask the significant differences between groups.

The most important aim of this study was to investigate the AST deposition-stimulating effects of LPC, as well as the subsequent effects on shrimp body color. These effects were indeed observed in this study, in particular regarding the muscle astaxanthin content. The elevation of muscle astaxanthin content by LPC was expectable considering that astaxanthin is lipid-soluble and LPC is a high-quality lipid emulsifier. However, this was the first evidence of this stimulating effect, and the stimulating efficiency was quite high. The muscle astaxanthin content in the AST + LPC group was even two times higher than that of the AST group. Considering that the astaxanthin, whether natural or artificial, is still expensive, sometime constituting 1/4–1/3 of the salmon feed cost, this result is of significance to the aquafeed industry.

Subsequently, this AST deposition-stimulating effect of LPC was reflected by the shrimp body color. Although the difference in body color between the AST and AST + LPC groups can barely be observed by naked eyes, this difference can be clearly detectable by machine [28]. In this study, for both raw and cooked shrimp, the dietary supplementation of astaxanthin significantly increased the redness but decreased the brightness. This result was similar to what was observed in other crustaceans and fish species [29–34]. More importantly, compared to the AST group, the supplementation of LPC significantly further increased the yellowness but did not affect the redness and brightness. This was quite a novel finding.

Regarding the muscle fatty acid composition, the supplementation of astaxanthin and LPC resulted in only minor changes. The increase of 18:2n-6 and n-6 PUFA contents by AST + LPC was similar to what was observed in stellate sturgeon *Acipenser stellatus* [35]. This result was probably due to the fact that the LPC used in this study was derived from soybean, which is rich in 18:2n-6 and n-6 PUFA. Some other minor changes such as those in 16:1n-7 and 18:1n-9 were difficult to explain based on the current information. A recent study on turbot also found that LPC led to minor changes in body fatty acid composition [8].

In both terrestrial animals and aquatic animals, it has been found that the effects of LPC, although mainly acting as an emulsifier, were not restricted to lipid digestion and absorption [36,37]. In fish and shrimp such as turbot *Scophthalmus maximus* [8,38], rainbow trout [10,26], channel catfish [11], and tiger shrimp *Penaeus monodon* [39], LPC has also been used to overcome or mitigate the lipid metabolism disorders caused by high dietary lipid levels or fish oil replacement. In previous studies, it was observed that dietary supplementation of LPC reduced the systemic lipid levels or serum levels of TC and TG [8,9,11,26], although some other studies did not observe these effects [39,40]. In this study, dietary AST and LPC had marginal effects on serum TG and TC levels. Species, feeding duration, and dietary formulation may largely influence the lipid-regulating effects of LPC.

However, it was clear that the astaxanthin supplementation increased the serum HDL-C concentration and decreased the LDL-C content. Moreover, LPC amplified this effect. LPL and HL also showed similar trends. Lipoprotein lipase (LPL) and hepatic lipase (HL), synthesized by liver cells, are key enzymes that regulate lipid metabolism and catalyze the hydrolysis of triglycerides. Increased serum levels or activities of LPL and HL can lead to increased lipid clearance, resulting in reduced TG and metabolic disorder mitigation [41]. Considering the roles of the parameters mentioned above in lipid and cholesterol transport, it was speculated that LPC may be involved in the transport of lipid and cholesterol. During blood circulation, LDL carries cholesterol from the liver to peripheral tissues for utilization, while HDL carries cholesterol from peripheral tissues to the liver for cholesterol removal [42,43]. Therefore, the LDL-C/HDL-C ratio is often used as a marker of cholesterol transport and an indicator of cholesterol accumulation in peripheral vasculature [44]. Results of this study suggested that the AST and LPC facilitated reverse cholesterol transport, i.e., the transport of cholesterol from extrahepatic tissue towards the liver for eventual excretion [45–47]. In particular, LPC facilitated the effects of AST in this process.

Nevertheless, from the aspect of gene expression, only the mRNA expressions of *npc2* and *mttp* were significantly affected by the diets. Npc1 and Npc2 are key proteins involved in the intracellular lipid transport. Cholesterol is firstly bound to Npc2 and then transferred to Npc1, thus increasing the rate of cholesterol transport from Npc1 to intestinal epithelial cells [48]. Mttp is not a protein specifically associated with cholesterol metabolism but a protein widely involved in the formation and secretion of ApoB-rich lipoproteins, especially in the pathways associated with chylomicrons and VLDL [49]. In general, the down-regulation of *npc2* and *mttp* mRNA expression by LPC compared to the AST group may be related to the inhibition of cholesterol absorption in the intestine.

In addition to cholesterol metabolism, the MDA content in the hepatopancreas tended to be reduced by both AST and AST + LPC. MDA is a main end-product of lipid peroxidation [50]. This result was not surprising considering the facts that astaxanthin is a strong antioxidant and that LPC promoted astaxanthin deposition. The anti-oxidative capacity of astaxanthin has been widely observed in other aquatic species [30–33,51,52] and thus was not discussed in detail in this study.

5. Conclusions

In conclusion, the dietary supplementation of lysophosphatidylcholine (LPC) improved the astaxanthin deposition in Pacific white shrimp and amplified its beneficial effects on growth performance and body color. The dietary LPC significantly increased the shell yellowness of both raw and cooked shrimp. LPC may also regulate the lipid and cholesterol transport in white shrimp. This was the first investigation of the promoting effects of LPC on dietary astaxanthin efficiency.

Author Contributions: Conceptualization and funding acquisition were assigned to H.X. and M.L. Formal analysis, data curation, and methodology were assigned to Z.S., Y.L., H.L. and Z.Y. Formal analysis and software were assigned to Q.M., L.X. and Y.W. Writing of original draft was assigned to Z.S. Writing—review and editing and supervision were assigned to H.X. All authors have read and agreed to the published version of the manuscript.

Funding: This research was funded by the Central Public-Interest Scientific Institution Basal Research Fund, CAFS/YSFRI (2024CG01 and 2023TD52), the Natural Science Foundation of Shandong Province Outstanding Youth Foundation (ZR2021YQ24), and the China Agriculture Research System (CARS-47).

Institutional Review Board Statement: All experimental protocols were approved by the Animal Care and Use Committee of Yellow Sea Fisheries Research Institute (protocol code ACUC202306125274; date of approval, 12 June 2023).

Informed Consent Statement: Not applicable.

Data Availability Statement: The raw data supporting the conclusions of this article will be made available by the authors on request.

Conflicts of Interest: Author Lindong Xiao was employed by the company Weifang Kenon Biotechnology Co., Ltd. The remaining authors declare that the research was conducted in the absence of any commercial or financial relationships that could be construed as a potential conflict of interest.

References

1. Chen, Y.K.; Mitra, A.; Rahimnejad, S.; Chi, S.Y.; Kumar, V.; Tan, B.P.; Niu, J.; Xie, S.W. Retrospect of fish meal substitution in Pacific white shrimp (*Litopenaeus vannamei*) feed: Alternatives, limitations and future prospects. *Rev. Aquac.* **2023**, *16*, 382–409. [CrossRef]
2. Nishida, Y.; Berg, P.C.; Shakersain, B.; Hecht, K.; Takikawa, A.; Tao, R.; Kakuta, Y.; Uragami, C.; Hashimoto, H.; Misawa, N. Astaxanthin: Past, present, and future. *Mar. Drugs* **2023**, *21*, 514. [CrossRef] [PubMed]
3. Matsuno, T. Aquatic animal carotenoids. *Fish. Sci.* **2001**, *67*, 5. [CrossRef]
4. Niu, J.; Tian, L.X.; Liu, Y.J.; Yang, H.J.; Ye, C.X.; Gao, W.; Mai, K.S. Effect of dietary astaxanthin on growth, survival, and stress tolerance of postlarval shrimp, *Litopenaeus vannamei*. *J. World Aquac. Soc.* **2009**, *40*, 795–802. [CrossRef]
5. Zhang, J.; Liu, Y.J.; Tian, L.X.; Yang, H.J.; Liang, G.Y.; Yue, Y.R.; Xu, D.H. Effects of dietary astaxanthin on growth, antioxidant capacity and gene expression in Pacific white shrimp *Litopenaeus vannamei*. *Aquac. Nutr.* **2013**, *19*, 917–927. [CrossRef]
6. Wang, W.L.; Ishikawa, M.; Koshio, S.; Yokoyama, S.; Dawood, M.A.O.; Hossain, M.S.; Zaineldin, A.I. Interactive effects of dietary astaxanthin and cholesterol on the growth, pigmentation, fatty acid analysis, immune response and stress resistance of kuruma shrimp (*Marsupenaeus japonicus*). *Aquac. Nutr.* **2019**, *25*, 946–958. [CrossRef]
7. Nguyen, K.D. *Astaxanthin: A Comparative Case of Synthetic vs. Natural Production*; University of Tennessee: Knoxville, TN, USA, 2013.
8. Xu, H.G.; Luo, X.; Bi, Q.Z.; Wang, Z.D.; Meng, X.X.; Liu, J.S.; Duan, M.; Wei, Y.L.; Liang, M.Q. Effects of dietary lysophosphatidylcholine on growth performance and lipid metabolism of juvenile turbot. *Aquac. Nutr.* **2022**, *2022*, 3515101. [CrossRef]
9. Cai, M.L. Effects of Dietary Lysolecithin on Growth, Lipid Metabolism and Muscle Quality of Red Swamp Crayfish (*Procambarus clarkii*). Master's Thesis, Hunan Agricultural University, Changsha, China, 2021. [CrossRef]
10. Taghavizadeh, M.; Shekarabi, H.P.S.; Mehrgan, S.M.; Islami, H.R. Efficacy of dietary lysophospholipids (Lipidol™) on growth performance, serum immuno-biochemical parameters, and the expression of immune and antioxidant-related genes in rainbow trout (*Oncorhynchus mykiss*). *Aquaculture* **2020**, *525*, 735315. [CrossRef]
11. Liu, G.X.; Ma, S.L.; Chen, F.Y.; Zhang, W.B.; Mai, K.S. Effects of dietary lysolecithin on growth performance, feed utilization, intestinal morphology and metabolic responses of channel catfish (*Ictalurus punctatus*). *Aquac. Nutr.* **2020**, *26*, 456–465. [CrossRef]
12. Eldessouki, E.A.A.; Diab, A.M.; Selema, T.A.M.A.; Sabry, N.M.; Abotaleb, M.M.; Khalil, R.H.; Abdel, T.M. Dietary astaxanthin modulated the performance, gastrointestinal histology, and antioxidant and immune responses and enhanced the resistance of *Litopenaeus vannamei* against *Vibrio harveyi* infection. *Aquac. Int.* **2022**, *30*, 1869–1887. [CrossRef]
13. Samia, F.; Wang, W.L.; Zhou, Y.; Xue, Y.C.; Yi, G.F.; Wu, M.Q.; Huang, X.X. Can dietary β-carotene supplementation provide an alternative to astaxanthin on the performance of growth, pigmentation, biochemical, and immuno-physiological parameters of *Litopenaeus vannamei*? *Aquac. Rep.* **2022**, *23*, 101054. [CrossRef]
14. Folch, J.; Lee, M.; Sloane-Stanley, G.H. A simple method for the isolation and purification of total lipids from animal tissues. *J. Biol. Chem.* **1957**, *226*, 497–509. [CrossRef] [PubMed]
15. Meng, X.X.; Bi, Q.Z.; Cao, L.; Ma, Q.; Wei, Y.L.; Duan, M.; Liang, M.Q.; Xu, H.G. Evaluation of necessity of cholesterol supplementation in diets of two marine teleosts, turbot (*Scophthalmus maximus*) and tiger puffer (*Takifugu rubripes*): Effects on growth and lipid metabolism. *Aquac. Nutr.* **2022**, *2022*, 4160991. [CrossRef]
16. Livak, K.J.; Schmittgen, T.D. Analysis of relative gene expression data using real–time quantitative PCR and the $2^{-\Delta\Delta CT}$ method. *Methods* **2001**, *25*, 402–408. [CrossRef] [PubMed]
17. Liao, Z.B.; Sun, Z.Y.; Bi, Q.Z.; Gong, Q.L.; Sun, B.; Wei, Y.L.; Liang, M.Q.; Xu, H.G. Screening of reference genes in tiger puffer (*Takifugu rubripes*) across tissues and under different nutritional conditions. *Fish. Physiol. Biochem.* **2021**, *47*, 1739–1758. [CrossRef] [PubMed]
18. Liu, L.; Li, J.; Cai, X.N.; Ai, Y.; Long, H.; Ren, W.; Huang, A.Y.; Zhang, X.; Xie, Z.Y. Dietary supplementation of astaxanthin is superior to its combination with Lactococcus lactis in improving the growth performance, antioxidant capacity, immunity and disease resistance of white shrimp (*Litopenaeus vannamei*). *Aquac. Rep.* **2022**, *24*, 101124. [CrossRef]
19. Keng, C.L.; Yusoff, M.F.; Shariff, M.; Kamarudin, M.S.; Nagao, N. Dietary supplementation of astaxanthin enhances hemato-biochemistry and innate immunity of Asian seabass, *Lates calcarifer* (Bloch, 1790). *Aquaculture* **2019**, *512*, 734339. [CrossRef]
20. Wang, W.L.; Ishikawam, M.; Koshio, S.; Yokoyama, S.; Dawood, M.A.; Zhang, Y.K. Effects of dietary astaxanthin supplementation on survival, growth and stress resistance in larval and post-larval kuruma shrimp, *Marsupenaeus japonicus*. *Aquac. Res.* **2018**, *49*, 2225–2232. [CrossRef]
21. Song, X.L.; Wang, L.; Li, X.Q.; Chen, Z.Z.; Liang, G.Y.; Leng, X.J. Dietary astaxanthin improved the body pigmentation and antioxidant function, but not the growth of discus fish (*Symphysodon* spp.). *Aquac. Res.* **2017**, *48*, 1359–1367. [CrossRef]
22. Xie, S.; Yin, P.; Tian, L.; Yu, Y.; Liu, Y.; Niu, J. Dietary supplementation of astaxanthin improved the growth performance, antioxidant ability and immune response of juvenile largemouth bass (*Micropterus salmoides*) fed high-fat diet. *Mar. Drugs* **2020**, *18*, 642. [CrossRef]
23. Xie, J.J.; Chen, X.; Liu, Y.J.; Tian, L.X.; Xie, S.W.; Niu, J. Effects of dietary astaxanthin on growth performance, hepatic antioxidative activity, hsp70, and HIF-1α gene expression of juvenile golden pompano (*Trachinotus ovatus*). *Isr. J. Aquac.-Bamidgeh* **2017**, *1430*, 12. [CrossRef]
24. Fang, H.H.; He, X.S.; Zeng, H.L.; Liu, Y.J.; Tian, L.X.; Niu, J. Replacement of astaxanthin with lutein in diets of juvenile *Litopenaeus vannamei*: Effects on growth performance, antioxidant capacity, and immune response. *Front. Mar. Sci.* **2021**, *8*, 803748. [CrossRef]

25. Tageldein, A.M.; Mohamed, A.; Eman, M.A.; Ahmed, S.A.; Mahmoud, S.K.; Mohamed, A.E.; Zaki, Z.S. Growth performance, immune-related and antioxidant genes expression, and gut bacterial abundance of pacific white leg shrimp, *Litopenaeus vannamei*, dietary supplemented with natural astaxanthin. *Front. Physiol.* **2022**, *13*, 874172. [CrossRef] [PubMed]
26. Batoul, A.; Keramat, A.A.; Hosein, O.; Mohamad, K.; Soleiman, M. Effects of lysophospholipid on rainbow trout (*Oncorhynchus mykiss*) growth, biochemical indices, nutrient digestibility and liver histomorphometry when fed fat powder diet. *Aquac. Nutr.* **2021**, *27*, 1779–1788. [CrossRef]
27. Liao, Z.B.; Xu, H.G.; Wei, Y.L.; Zhang, Q.G.; Liang, M.Q. Dietary astaxanthin differentially affected the lipid accumulation in the liver and muscle of the marine teleost, tiger puffer *Takifugu rubripes*. *Aquac. Res.* **2018**, *10*, 3421–3433. [CrossRef]
28. Smith, B.E.; Hardy, R.W.; Torrissen, O.J. Synthetic astaxanthin deposition in pan-size coho salmon (*Oncorhynchus kisutch*). *Aquaculture* **1992**, *104*, 105–119. [CrossRef]
29. Supamattaya, K.; Kiriratnikom, S.; Boonyaratpalin, M.; Borowitzka, L. Effect of a dunaliella extract on growth performance, health condition, immune response and disease resistance in black tiger shrimp (*Penaeus monodon*). *Aquaculture* **2005**, *248*, 207–216. [CrossRef]
30. Zhang, J.J.; Li, X.Q.; Leng, X.J.; Zhang, C.L.; Han, Z.Y.; Zhang, F.G. Effects of dietary astaxanthins on pigmentation of flesh and tissue antioxidation of rainbow trout (*Oncorhynchus mykiss*). *Aquac. Int.* **2013**, *21*, 579–589. [CrossRef]
31. Chen, Q.; Huang, S.; Dai, J.Y.; Wang, C.C.; Chen, S.M.; Qian, Y.X.; Han, T. Effects of synthetic astaxanthin on the growth performance, pigmentation, antioxidant capacity, and immune response in black tiger prawn (*Penaeus monodon*). *Aquac. Nutr.* **2023**, *2023*, 6632067. [CrossRef]
32. Huang, S.; Chen, Q.; Zhang, M.M.; Chen, S.M.; Dai, J.Y.; Qian, Y.X.; Gong, Y.Y.; Han, T. Synthetic astaxanthin has better effects than natural astaxanthins on growth performance, body color and n-3 PUFA deposition in black tiger prawn (*Penaeus monodon*). *Aquac. Rep.* **2023**, *33*, 101816. [CrossRef]
33. Zhao, X.P.; Wang, G.P.; Liu, X.G.; Guo, D.L.; Chen, X.L.; Liu, S.; Li, G.F. Dietary supplementation of astaxanthin increased growth, colouration, the capacity of hypoxia and ammonia tolerance of Pacific white shrimp (*Litopenaeus vannamei*). *Aquac. Rep.* **2022**, *23*, 101093. [CrossRef]
34. Lin, Y.J.; Chang, J.J.; Huang, H.T.; Lee, C.P.; Hu, Y.F.; Wu, M.L.; Huang, C.Y.; Nan, F.H. Improving red-colorperformance, immune response and resistance to *Vibrio parahaemolyticus* on white shrimp *Penaeus vannamei* by an engineered astaxanthin yeast. *Sci. Rep.* **2023**, *13*, 2248. [CrossRef]
35. Fatemeh, J.; Naser, A.; Farzaneh, N.; Enric, G.; Mansour, T.M. Supplementing lysolecithin in corn-oil based diet enhanced growth and improved body biochemical composition in juvenile stellate sturgeon (*Acipenser stellatus*). *Anim. Feed. Sci. Technol.* **2024**, *310*, 115945. [CrossRef]
36. Juntanapum, W.; Bunchasak, C.; Poeikhampha, T.; Rakangthong, C.; Poungpong, K. Effects of supplementation of lysophosphatidylcholine (LPC) to lying hens on production performance, fat digestibility, blood lipid profile, and gene expression related to nutrients transport in small intestine. *J. Anim. Feed Sci.* **2020**, *29*, 258–265. [CrossRef]
37. Nutautaitė, M.; Racevičiūtė, S.A.; Andalibizadeh, L.; Šašytė, V.; Bliznikas, S.; Pockevičius, A.; Vilienė, V. Improving broiler chickens' health by using lecithin and lysophosphatidylcholine emulsifiers: A comparative analysis of physiological indicators. *Iran. J. Vet. Res.* **2021**, *22*, 33–39. [CrossRef]
38. Li, S.H.; Luo, X.; Liao, Z.B.; Liang, M.Q.; Xu, H.G.; Mai, K.S.; Zhang, Y.J. Effects of lysophosphatidylcholine on intestinal health of turbot fed high-lipid diets. *Nutrients* **2022**, *14*, 4398. [CrossRef]
39. Imran, H.K.; Syama, J.D.; Kondusamy, A.; Purdhvi, E.M.; Rajabdeen, J.; Vanjiappan, R. Enhancing the dietary value of palm oil in the presence of lysolecithin in tiger shrimp, *Penaeus monodon*. *Aquac. Int.* **2018**, *26*, 509–522. [CrossRef]
40. Li, H.T.; Tian, L.X.; Wang, Y.D.; Hu, Y.H. Effects of lysolecithin on growth performance, body composition and hematological indices of hybrid tilapia (*Oreochromis aureus* ♂ × *Oreochromis niloticus* ♀). *J. Dalian Fish. Univ.* **2010**, *25*, 143–146. [CrossRef]
41. Wang, M.; Wu, X.H.; Li, Z. Leptin accelerates lipid metabolism by increasing lipoprotein lipase and hepatic lipase expression in insulin-resistant liver cell model. *J. Third Mil. Med. Univ.* **2014**, *36*, 1059–1063. [CrossRef]
42. Chen, J.Y.; Chen, J.C.; Wu, J.L. Molecular cloning and functional analysis of zebrafish high-density lipoprotein-binding protein Comp. *Comp. Biochem. Physiol. Part B Biochem. Mol. Biol.* **2003**, *136B*, 117–130. [CrossRef]
43. Deng, J.; Mai, K.; Ai, Q.; Zhang, W.; Wang, X.; Tan, B.; Xu, W.; Liu, Z.; Ma, H. Interactive effects of dietary cholesterol and protein sources on growth performance and cholesterol metabolism of Japanese flounder (*Paralichthys olivaceus*). *Aquac. Nutr.* **2010**, *16*, 419–429. [CrossRef]
44. Yun, B.; Mai, K.S.; Zhang, W.B.; Xu, W. Effects of dietary cholesterol on growth performance, feed intake and cholesterol metabolism in juvenile turbot (*Scophthalmus maximus* L.) fed high plant protein diets. *Aquaculture* **2011**, *319*, 105–110. [CrossRef]
45. Mardones, P.; Quiñones, V.; Amigo, L.; Moreno, M.; Miquel, J.F.; Schwarz, M.; Miettinen, H.E.; Trigatti, B.; Krieger, M.; VanPatten, S.; et al. Hepatic cholesterol and bile acid metabolism and intestinal cholesterol absorption in scavenger receptor class B type I-deficient mice. *J. Lipid Res.* **2001**, *42*, 170–180. [CrossRef]
46. Stanley, S.L.; Stephen, G.W. Implications of reverse cholesterol transport: Recent studies. *Clin. Chim. Acta* **2015**, *439*, 154–161. [CrossRef] [PubMed]
47. Brufau, G.; Groen, A.K.; Kuipers, F. Reverse cholesterol transport revisited: Contribution of biliary versus intestinal cholesterol excretion. *Arterioscler. Thromb. Vasc. Biol.* **2011**, *31*, 1726–1733. [CrossRef]

48. Wang, L.M.; Motamed, M.; Infante, E.R.; Abi-Mosleh, L.; Kwon, H.J.; Brown, M.S.; Goldstein, J.L. Identification of Surface Residues on Niemann-Pick C2 Essential for Hydrophobic Handoff of Cholesterol to NPC1 in Lysosomes. *Cell Metab.* **2010**, *12*, 166–173. [CrossRef]
49. Wei, F.; Chen, Q.C.; Cui, K.; Chen, Q.; Li, X.S.; Xu, N.; Mai, K.S.; Ai, Q.H. Lipid overload impairs hepatic VLDL secretion via oxidative stress-mediated PKCδ-HNF4α-MTP pathway in large yellow croaker (*Larimichthys crocea*). *Free Radic. Biol. Med.* **2021**, *20*, 213–225. [CrossRef]
50. Zhang, D.G.; Zhao, T.; Hogstrand, C.; Ye, H.M.; Xu, X.J.; Luo, Z. Oxidized fish oils increased lipid deposition via oxidative stress-mediated mitochondrial dysfunction and the CREB1-Bcl2-Beclin1 pathway in the liver tissues and hepatocytes of yellow catfish. *Food Chem.* **2021**, *360*, 129814. [CrossRef] [PubMed]
51. Brambilla, F.; Forchino, A.; Antonini, M.; Rimoldi, S.; Terova, G.; Saroglia, M. Effect of dietary astaxanthin sources supplementation on muscle pigmentation and lipid peroxidation in rainbow trout (*Oncorhynchus mykiss*). *Ital. J. Anim. Sci.* **2009**, *8* (Suppl. S2), 845–847. [CrossRef]
52. Xu, W.X.; Liu, Y.T.; Huang, W.X.; Yao, C.W.; Yin, Z.Y.; Mai, K.S.; Ai, Q.H. Effects of dietary supplementation of astaxanthin (Ast) on growth performance, activities of digestive enzymes, antioxidant capacity and lipid metabolism of large yellow croaker (*Larimichthys crocea*) larvae. *Aquac. Res.* **2022**, *53*, 4605–4615. [CrossRef]

antioxidants

MDPI

Article

Effects of Taurine and Vitamin C on the Improvement of Antioxidant Capacity, Immunity and Hypoxia Tolerance in Gibel Carp (*Carrassius auratus gibeilo*)

Leimin Zhang [1], Lu Zhang [2], Hualiang Liang [1,3], Dongyu Huang [3] and Mingchun Ren [1,3,*]

1 Wuxi Fisheries College, Nanjing Agricultural University, Wuxi 214081, China
2 Tongwei Agricultural Development Co., Ltd., Key Laboratory of Nutrition and Healthy Culture of Aquatic, Livestock and Poultry, Ministry of Agriculture and Rural Affairs, Healthy Aquaculture Key Laboratory of Sichuan Province, Chengdu 610093, China
3 Key Laboratory of Integrated Rice-Fish Farming Ecology, Ministry of Agriculture and Rural Affairs, Freshwater Fisheries Research Center, Chinese Academy of Fishery Sciences, Wuxi 214081, China
* Correspondence: renmc@ffrc.cn

Abstract: To investigate the effects of taurine and vitamin C on gibel carp (*Carrassius auratus gibeilo*), fish (41.85 ± 0.03 g) were fed three diets with 0% taurine + 0% vitamin C (D0), 0.1% taurine + 0% vitamin C (D1), and 0.1% taurine + 0.1% vitamin C (D2) for 8 weeks. Then 12-hour hypoxic stress test was conducted. The results showed that weight gain rate (WGR), specific growth rate (SGR), and sustained swimming time (SST) were significantly increased in the D2. CAT, SOD, T-AOC, and GSH were increased. GSH-Px and *il-6* were decreased in D1 and D2. In hypoxia, CAT and T-AOC were decreased, while GSH, *sod*, and *nrf2* were the highest in D1. Compared to normoxia, GSH-Px was increased, while SOD and MDA were decreased. *Il-10* and *nf-κb* were increased. *Vegf*, *epo*, and *ho-1* were increased and they all were higher than that in normoxia. The number of gill cell mitochondria and survival rate (SR) of gibel carp had an increasing trend but no significant difference among groups. In conclusion, taurine with vitamin C improved the growth and SST of gibel carp, and taurine and taurine with vitamin C improved antioxidant capacity, immunity, and hypoxia tolerance.

Keywords: gibel carp (*Carrassius auratus gibeilo*); hypoxia; taurine; vitamin C; survival and body condition; antioxidant capacity; dietary supplementation

Citation: Zhang, L.; Zhang, L.; Liang, H.; Huang, D.; Ren, M. Effects of Taurine and Vitamin C on the Improvement of Antioxidant Capacity, Immunity and Hypoxia Tolerance in Gibel Carp (*Carrassius auratus gibeilo*). *Antioxidants* **2024**, *13*, 1169. https://doi.org/10.3390/antiox13101169

Academic Editor: Erchao Li

Received: 22 July 2024
Revised: 20 September 2024
Accepted: 22 September 2024
Published: 26 September 2024

1. Introduction

Hypoxia is one of the main environmental factors limiting the healthy development of aquaculture [1–3]. Dissolved oxygen (DO) is necessary for the survival of aquatic animals. When the DO concentration is less than 2 mg/L, the aquatic environment is hypoxic. This can lead to various negative effects on aquatic animals, such as reduced food intake, abnormal movement, and even death [4,5]. Various attempts have been made to mitigate the effects of hypoxia on the health of fish. Studies have shown that the use of feed additives to improve fish tolerance is effective. For example, the appropriate addition of arginine to the feed can be helpful to improve the survival of the Indian major carp (*Cirrhinus mrigala*) under hypoxia [6]. The appropriate levels of γ-aminobutyric acid (GABA) in the feed can modulate the HIF-1 signaling pathway and blood biochemical levels of the Indian major carp to improve its survival under hypoxia [7]. The appropriate addition of mulberry leaf extract can reduce excess reactive oxygen species (ROS) in gibel carp (*Carrassius auratus gibeilo*), reduce oxidative damage, and improve their ability to tolerate hypoxia [8]. Therefore, food supplements are a viable method to alleviate the negative effects of hypoxic stress with the help of food.

Taurine is an organic acid that was originally isolated from the bile of bulls and is also known as sulphonic acid or sulphonic amino acid [9]. As a non-protein amino acid, taurine is not involved in protein synthesis and does not provide energy; however, it has been found in large amounts in tissues, accounting for 30–50% of total amino acids, and is an important free amino acid in the tissues of the body [10,11]. In recent years, the nutritional and physiological functions of taurine in aquatic animals and its use in feed have attracted much attention. The results of related studies have shown that the addition of taurine to feed can improve fish growth [12–15]. Other studies have shown that taurine can improve the ability of animals to tolerate hypoxia, prevent hypoxia-induced death of hepatocytes in rats [16], and increase the hypoxia tolerance of juvenile grass carp (*Ctenopharyngodon idellus*) [17]. Therefore, taurine is a beneficial additive for aquatic animals.

Vitamin C, also known as ascorbic acid, is a water-soluble antioxidant that can reduce ROS and protect cells from oxidative damage. It is an essential nutrient for animal growth and metabolism and can also improve fish growth, boost immunity and antioxidant capacity, and relieve stress [18,19]. Related studies have shown that the adequate addition of vitamin C to feed can improve the growth performance of coho salmon (*Oncorhynchus kisutch*) [20], silver carp (*Hypophthalmichthys molitrix*) [21], and tiger pufferfish (*Takifugu rubripes*) [19], while adequate levels of vitamin C can also boost the immunity of fish [22–24]. Using antioxidant enzymes as indicators, an appropriate amount of vitamin C can be used to increase the antioxidant capacity of fish [21,25,26]. An adequate supply of vitamin C as a coenzyme is beneficial to the growth and physical health of fish.

The gibel carp, *Carrassius auratus gibeilo*, is an important freshwater aquaculture species that plays a prominent role in meeting the human need for proteins and aquatic products [27]. However, climatic factors and intensive aquaculture may also expose gibel carp to the dangers of hypoxic stress, affecting its growth and production. Previous studies have shown that taurine can be used as a feed supplement to effectively improve the growth and hypoxia tolerance of some fish; however, a corresponding study on gibel carp has not yet been conducted. In addition, it is worth exploring whether hypoxic tolerance can be further improved by the combined effects of taurine and vitamin C. Therefore, this study was designed to investigate the effect of taurine and the combined effect of taurine and vitamin C on the growth performance, antioxidant capacity, immunity, and hypoxia tolerance of gibel carp in order to improve the growth and stress resistance of fish and promote the stable development of aquaculture.

2. Materials and Methods

2.1. Diet Preparation

Based on previous studies [13,28], we designed three experimental diets by adding different amounts of taurine and vitamin C to the normal diet: 0% taurine + 0% vitamin C (D0, control group), 0.1% taurine + 0% vitamin C (D1), and 0.1% taurine + 0.1% vitamin C (D2). Table 1 lists the normal feed composition. The amino acid balancing method was based on our previous study [29]. Fish, chicken, soybean, cottonseed, and rapeseed meals were used as the main protein sources. Soybean oil was the main lipid source. The feed production process includes crushing, sieving, weighing, mixing, and pelleting [30]. The ingredients were crushed, sieved, proportioned, mixed with water and oil according to the feed formulation, pelleted in a pelletizing machine (F-26(II), South China University of Technology, Guangzhou, China), and dried. The feed was frozen at −20 °C for further use.

2.2. Experimental Procedure

Fish were obtained from the Freshwater Fisheries Research Center. At the beginning of the feeding trial, the fish were fasted for 24 h and then weighed. A total of 180 healthy gibel carp (41.85 ± 0.03 g) were randomly divided into three experimental groups with three replicates each, resulting in a total of nine cages. Fish were fed twice daily based on apparent satiety (6:30 and 16:30). Fish mortality was recorded daily. Feeding trials were conducted in 1 m × 1 m × 1 m floating cages and maintained for 8 weeks. Water quality was continuously monitored, and the water temperature was 28 ± 2 °C. DO was maintained at ≥6.5 mg/L, and total ammonia nitrogen was below 0.1 mg/L throughout the feeding process. Aquaculture experiments were conducted at the Nanquan Aquaculture Basement of the Freshwater Fisheries Research Center (Wuxi, China).

Table 1. Ingredient and nutrient composition of the normal diet (% dry matter).

Ingredients	
Fish meal [1]	14.00
Chicken meal	4.00
Soybean meal [1]	22.00
Cottonseed meal	5.00
Rapeseed meal [1]	22.00
Wheat flour [1]	14.15
Rice bran	10.00
Soybean oil	4.00
Monocalcium phosphate	2.00
Vitamin premix [2]	0.20
Mineral premix [3]	2.00
Lysine	0.30
Methionine	0.10
Vc phosphate	0.05
Choline chloride	0.20
Analyzed proximate composition	
Crude protein (%)	39.43 ± 0.43
Crude lipid (%)	7.08 ± 0.33
Crude ash (%)	9.98 ± 0.25

[1] Fish meal, obtained from Wuxi Tongwei Feedstuffs Co., Ltd., Wuxi, China, crude protein 65.6%, crude lipid 9.5%; soybean meal obtained from Wuxi Tongwei Feedstuffs Co., Ltd., Wuxi, China, crude protein 46.0%, crude lipid 4.3%; rapeseed meal obtained from Wuxi Tongwei Feedstuffs Co., Ltd., Wuxi, China, crude protein 39.2%, crude lipid 6.1%; wheat flour obtained from Wuxi Tongwei Feedstuffs Co., Ltd., Wuxi, China, crude protein 13.1%, crude lipid 4.0%. [2] Vitamin premix (IU or mg/kg of premix, purchased by HANOVE Biotechnology Co., Ltd., Wuxi, China): vitamin A, 800,000 IU; vitamin D3, 250,000 IU; vitamin E, 4500 IU; vitamin K3, 600 mg; thiamin, 800 mg; riboflavin, 800 mg; calcium pantothenate, 2000 mg; pyridoxine HCl, 2500 mg; cyanocobalamin, 8 mg; biotin, 16 mg; folic acid, 400 mg; niacin, 2800 mg; inositol, 10,000 mg; vitamin C, 10,000 mg. [3] Mineral premix (g/kg of premix, purchased by HANOVE Biotechnology Co., Ltd. Wuxi, China): magnesium sulfate, 1.5%; ferrous sulfate, 30 g; zinc sulfate, 13.5 g; cupric sulfate, 0.8 g; manganese sulfate, 6 g; zeolite was used as a carrier.

2.3. Sample Collection

After the feeding trial, fish were fasted for 24 h before sampling. Fish in all cages were weighed to obtain data on growth performance. Two randomly selected fish from each cage were used to collect liver tissue samples. Liver samples were collected and temporarily stored in a tank of liquid nitrogen at −196 °C. After sampling, all liver samples were stored at −80 °C for subsequent gene expression analysis and determination of antioxidant enzyme activity. Two additional fish were randomly selected from each cage and stored at −20 °C for whole-body composition analysis. A SY28060 swimming flume (Loligo Systems, Viborg, Denmark) was used to test the swimming ability of gibel carp. Specific measurements were based on previous studies [31]. One fish was placed in the flume test area. The flow velocity was set at 5 cm/s to acclimatize the fish, and the flow velocity of the water was gradually increased to 30 cm/s. At the same time, the time between the fish swimming against the current and the point at which the fish swam with the current to the wall of the flume and stopped moving was recorded as the sustained swimming time (SST).

The remaining fish were safely transferred to a recirculating water culture system (300 L per tank) for hypoxia experiments. Before starting hypoxic stress, the water surface in the tanks was set high enough to allow the fish to swim freely, and the water surface was covered with a transparent suspension film to prevent oxygen exchange between air and water. The DO content of the water was monitored during the experiment using a portable dissolved oxygen meter. Stress was triggered when the DO content in the water dropped to 0.8 ± 0.1 mg/L. The stress period was 12 h, during which the fish in each tank were observed and recorded. Twelve hours later, the SR of the fish was noted. During the process, the water temperature was approximately 19.5 °C. Liver tissue was collected from six of the remaining fish for analysis of gene expression and antioxidant enzyme activity, and gill tissue was collected for transmission electron microscopy.

2.4. Chemical Analysis

The moisture, crude protein, crude lipid, and ash contents (%) of the experimental feed and whole fish body were measured using standard methods with three replicates per group [32]. The samples were first dried in an oven at 105 °C to determine the moisture content. Subsequently, crude protein content was measured using the Kjeldahl nitrogen determination method on an automatic instrument (Haineng K1100, Jinan Haineng Instrument Co., Ltd., Jinan, China). Crude lipid content was analyzed by the Soxhlet extraction method in an automatic fat analyzer (Haineng SOX606, Jinan Haineng Instrument Co., Ltd., China) and samples were calcined at 560 °C for 6 h in a Muffle furnace (XL-2A, Hangzhou Zhuochi Instrument Co., Ltd., Hangzhou, China) to obtain ash content data. Activities of antioxidant enzymes such as catalase (CAT) (U/mgprot), superoxide dismutase (SOD) (U/mgprot), glutathione peroxidase (GSH-Px) (U/mgprot), total antioxidant capacity (T-AOC) (mmol/gprot), glutathione (GSH) (μmol/gprot), and malondialdehyde (MDA) (nmol/mgprot) were measured using corresponding kits provided by Nanjing Jiancheng Bioengineering Institute (Nanjing, China) following previous studies. The expertise of the measurement methods is widely recognized by universities and research institutes [33–36]. The main instrument used was the spectrophotometer (Thermo Fisher Multiskan GO, Shanghai, China). The succinct methods of determining antioxidant enzyme activities as well as MDA are shown in Table 2. The specific methods for determination were as follows: All operations were performed in strict accordance with the instructions of the kits. (1) CAT was measured using the kit model A007-1-1 by the ammonium molybdenum acid method. Each milligram of protein in the sample corresponds to 1 μmol of H_2O_2 decomposed by CAT per second as one unit of activity (U). (2) T-AOC was determined using the kit A015-2-1 by the ABTS method. The principle is that ABTS is oxidized to green ABTS+ in the presence of a suitable oxidant. Antioxidants inhibit the production of ABTS+. The total antioxidant capacity of a sample can be determined and calculated by measuring the absorbance of ABTS+ at 405 nm or 734 nm, and Trolox is used as a reference for the total antioxidant capacity of antioxidants. (3) The activity of SOD is determined using kit model A001-3-2 with the WST-1 method, and the meaning of the result is that one unit of SOD activity (U) is the amount of enzyme required to inhibit 50% of the oxidation rate per reactive solution and per milligram of protein in 1 mL of reactive solution. (4) The TBA method was applied to determine MDA using the kit model A003-1-2. The principle is that MDA in the degradation products of lipid peroxidation can condense with thiobarbituric acid (TBA) to form a red product with a maximum absorption peak at 532 nm. Therefore, the results of the determination at this wavelength can reflect the content of MDA. (5) The kit model A006-2-1 using the microplate method was used for the determination of GSH. The principle of the determination is that GSH can react with dithiobisnitrobenzoic acid to produce a yellow compound, which can be used for colorimetric quantitative determination of the content of GSH under the condition of 405 nm wavelength. (6) The activity of GSH-Px is measured using the kit A005-1-2 by the colorimetric method. A unit of enzyme activity (U) is defined as one unit of enzyme activity (U) per milligram of protein per minute at 37 °C, minus the effect of the nonenzymatic reaction, which reduces the concentration of GSH in the reaction system by 1 μmol/L. The calculation of all the above indexes needs the protein content in liver tissue, so we use the kit model A045-2-2, applying the Caulmers Brilliant Blue method to determine the protein content. The principle is that the protein molecule has the $-NH^{3+}$ group. When the brownish-red Caulmers Brilliant Blue colorant is added to the protein standard solution or the sample, the anion on the Caulmers Brilliant Blue dye combines with the protein -NH3+, making the solution turn blue. The solution turns blue, and the protein content can be calculated by measuring the absorbance. Gill cell sections were prepared by Wuhan Servicebio Technology Co., Ltd. (Wuhan, China).

Table 2. Liver parameters related to antioxidant capacity.

Index	Measurement Methods	Note
CAT	Ammonium molybdenum acid method	Assay kits purchased from Jian Cheng Bioengineering Institute (Nanjing, China); Spectrophotometer (Thermo Fisher Multiskan GO, Shanghai, China).
T-AOC	ABTS method	
SOD	WST-1 method	
MDA	TBA method	
GSH	Microplate method	
GSH-Px	Colorimetric method	

2.5. RNA Extraction and Real-Time PCR Analysis

RNA was extracted from the liver tissue of gibel carp using an RNA extraction reagent (Vazyme, Nanjing, China). The concentration and quality of RNA were determined using a NanoDrop 2000 spectrophotometer. The concentration of the RNA samples was adjusted to 60 ng/μL, with A260/A280 ranging between 1.8 and 2.0. qPCR was performed using the One-Step SYBR Prime Script TM PLUS RT-PCR kit (Takara, Dalian, China) on a CFX96 Touch (Bio-Rad, Hercules, CA, USA). The qPCR reactions were programmed as follows: reverse transcription at 42 °C for 5 min, pre-denaturation at 95 °C for 10 s, then denaturation at 95 °C for 5 s and 60 °C (annealing temperature) for 30 s for 40 cycles. The melting curves were analyzed by increasing the temperature from 65 °C to 95 °C while observing fluorescence. Beta-actin (*β-actin*), which has high and stable expression, was chosen as the internal reference gene, and no significant difference was detected. The primer sequences used are listed in Table 3. Some of the primer sequences were obtained from previous research, and the remaining primers were designed online using Primer Premier 6.0. Sangon Biotech (Shanghai) Co., Ltd. (Shanghai, China) helped to synthesize the primers. The standard curve method was used to calculate the relative expression of genes [33].

Table 3. Primer sequences for RT-qPCR analysis.

Genes	Forward (5′–3′)	Reverse (5′–3′)	Primer Source
il-10	AGTGAGACTGAAGGAGCTCCG	TGGCAGAATGGTGTCCAAGTA	[37]
tgf-β	GTTGGCGTAATAACCAGAAGG	AACAGAACAAGTTTGTACCGATAAG	[38]
il-1β	GCGCTGCTCAACTTCATCTTG	GTGACACATTAAGCGGCTTCA C	[38]
il-6	CGGAGGGGCTTAACAGGATG	GCTGGCTCAGGAATGGGTAT	DQ861993.1
il-8	ATTGGTGAAGGAATGAGTCT	CCACAGATGACCTTGACAT	KC184490.1
tnf-α	CATTCCTACGGATGGCATTTACTT	CCTCAGGAATGTCAGTCTTGCAT	[38]
nf-κb	GCTCTGACTGCGGTCTTATAC	GCGCTTCATCGAGGATAGTT	[39]
cat	TGAAGTTCTACACCGATGAG	CTGAGAGTGGACGAAGGA	XM_026238665.1
sod	TCGGAGACCTTGGTAATGT	CGCCTTCTCATGGATCAC	JQ776518.1
gpx	GAAGTGAACGGTGTGAACGC	GATCCCCCATCAAGGACACG	DQ983598.1
keap1	CTCCGCTGAATGCTACAA	GGTCATAACACTCCACACT	XM_026245355.1
nrf2	TACCAAAGACAAGCAGAAGAAACG	GCCTCGTTGAGCTGGTGTTTGG	[40]
hif-1α	CTGCCGATCAGTCTGTCTCC	TTTGTGGAGTCTGGACCACG	DQ306727.1
vegf	ATCGAGCACACGTACATCCC	CCTTTGGCCTGCATTCACAC	NM_131408.3
epo	CGAAGTGTCAGCATACCGGA	GCAGATGACGCACTTTTCCC	KC460317.1
ho-1	GCAAACCAAGAGAAGCCACC	GGAAGTAGACGGGCTGAACC	KC758864
angpt1	CCAAACCTCACCAAGCAAGC	GGATTACAGTCCAGCCTCCG	XM_059556208.1
et1	TAAAGCAGCGTCAGACAGGG	CTGCCAGCTTGTGTTTGCAT	NM_131519.1
nos	GGGGACCCTCCTGAAAATGG	TTCTGTCCTCAACGCTGGTG	AY644726.1
tf	CCGAGAAGATGCACGCAAAG	TGTGCATGCCTTGACCAGAT	AF518747.1
tfr1	CTTTGTCAACGAAGTGGCTGAAT	TACCAAAGAAAATGTGGCGGAAC	XM_052542523.1
β-actin	TCCATTGTTGGACGACCCAG	TGGGCCTCATCTCCCACATA	LC382464.1

Note: *il-10*, interleukin-10; *tgf-β*, transforming growth factor-β; *il-1β*, interleukin-1β; *il-6*, inter leukin-6; *il-8*, interleukin-8; *tnf-α*, tumor necrosis factor-α; *nf-κb*, nuclear factor kappa-β; *cat*, catalase; *sod*, superoxide dismutase; *gpx*, glutathione peroxidase; *keap1*, recombinant kelch-like ECH-associated protein 1; *nrf2*, nuclear factor erythroid 2-related factor 2; *hif-1α*, hypoxia-inducible factor 1-α; *vegf*, vascular endothelial growth factor; *epo*, erythropoietin; *ho-1*, heme oxygenase-1; *angpt1*, angiopoietin-1; *et1*, endothelin 1; *nos*, nitric oxide synthase; *tf*, transferrin; *tfr1*, transferrin receptor protein 1; *β-actin*, beta-actin.

2.6. Statistical Analysis

The data were subjected to normality and homogeneity tests before any statistical analysis. The statistical software SPSS 25.0 was used to analyze all data. A one-way ANOVA (Duncan's test) was used to evaluate the effects of the addition of taurine and vitamin C on growth performance, whole-body composition, SST in normoxia, SR, and the number of cellular mitochondria in hypoxia. The effects of DO and the addition of taurine and vitamin C, their interaction on the liver genes, and the antioxidant enzyme activities of gibel carp were evaluated using a two-way ANOVA. When a significant interaction between the main effects of the variables was observed, the main effects of taurine and vitamin C were evaluated using a one-way ANOVA (Duncan's test), and those of DO were evaluated using an independent samples t-test. The results are presented as means with standard deviations. Statistical significance was set at $p < 0.05$.

3. Results

3.1. Growth Performance

The growth performance-related indices and feed conversion ratio (FCR) are listed in Table 4. With the addition of taurine and vitamin C, final body weight (FBW), specific growth rate (SGR), and weight gain rate (WGR) showed an increasing trend. Among all groups, WGR and SGR were highest in the D2 group, and both were higher than those in the D0 group ($p < 0.05$), and the survival rate (SR) of all groups was 100%. The D2 group showed an increasing trend and the highest FCR, but no significant difference was observed ($p > 0.05$).

Table 4. Growth performance, whole-body composition, and swimming ability of gibel carp fed graded dietary of taurine and vitamin C levels for 8 weeks.

Parameters	Dietary Taurine and Vitamin C Levels (%)		
	0 + 0	0.1 + 0	0.1 + 0.1
Growth performance			
IBW (g) [1]	41.92 ± 0.08	41.82 ± 0.04	41.78 ± 0.11
FBW (g) [2]	101.57 ± 0.58	102.97 ± 0.84	104.20 ± 0.77
FCR [3]	1.37 ± 0.02	1.39 ± 0.02	1.43 ± 0.01
SGR (% day-1) [4]	0.95 ± 0.01 [b]	0.97 ± 0.01 [ab]	0.98 ± 0.01 [a]
WGR (%) [5]	142 ± 0.02 [b]	146 ± 0.02 [ab]	149 ± 0.02 [a]
SR (%) [6]	100.0 ± 0.00	100.0 ± 0.00	100.0 ± 0.00
Whole-body composition (%)			
Moisture	75.21 ± 1.02	74.88 ± 0.95	74.99 ± 0.62
Crude protein	15.92 ± 0.38	15.80 ± 0.42	16.42 ± 0.36
Crude lipid	2.62 ± 0.93	2.37 ± 0.75	1.82 ± 0.26
Ash	4.56 ± 0.16	4.75 ± 0.06	4.76 ± 0.14
Swimming ability			
SST (sec) [7]	37.17 ± 8.63 [b]	43.00 ± 8.79 [b]	142.20 ± 8.63 [a]

Note: Data are mean value ± SEM. Means in the same row with different super scripts are significantly different ($p < 0.05$). [1] IBM, Initial body weight. [2] FBW, Final body weight. [3] FCR, Feed conversion ratio = dry feed fed (g)/wet weight gain (g). [4] SGR, Specific growth rate (%/d) = 100 × [(Ln (final body weight (g)) − Ln (initial body weight (g)))/days]. [5] WGR, Weight gain rate (%) = 100 × (final weight (g) − initial weight (g))/initial weight (g). [6] SR, Survival rate (%) = 100 × (survival fish number/total fish). [7] SST, Sustained swimming time (s).

3.2. Whole-Body Composition

The whole-body composition of gibel carp is shown in Table 4. Moisture content was lower in both the D1 and D2 groups than in the D0 group. Crude protein content was lower in the D1 group than in the D0 group, whereas it was highest in the D2 group. The crude lipid content showed a decreasing trend and reached the lowest value in the D2 group, whereas the ash content showed an increasing trend and reached the highest value in the

D2 group. None of the above body composition indices showed a significant difference between the groups ($p > 0.05$).

3.3. Swimming Ability

Table 4 shows the swimming ability of the different groups of gibel carp. No significant difference in SST was found between groups D1 and D0, while the SST of group D2 was significantly longer than those of the other two groups ($p > 0.05$).

3.4. Antioxidant Capacity of the Liver (Nrf2 Signaling Pathway)

Results of the antioxidant enzyme activities in gibel carp liver under normoxic and hypoxic environments are shown in Table 5. CAT, T-AOC, SOD, and GSH levels tended to increase with the addition of taurine and taurine with vitamin C in a normoxic environment, and the differences were significant compared with the D0 group ($p < 0.05$). GSH-Px tended to decrease ($p < 0.05$), and no significant difference was observed in MDA. In a hypoxic environment, CAT and T-AOC showed a decreasing trend, and GSH showed an increasing, and then decreasing trend, and the maximum value was observed in the D1 group ($p < 0.05$). The differences were not significant in SOD, MDA, and GSH-Px levels between the groups. The results of the two-way ANOVA showed that DO has a significant effect on SOD, MDA, and GSH-Px. Compared with normoxia, SOD and MDA were significantly decreased, and GSH-Px was significantly increased in hypoxia ($p < 0.05$), while no significant difference was found in the other indices. The addition of taurine and taurine with vitamin C significantly affected the T-AOC and GSH levels. The interaction of DO and the addition of taurine or taurine with vitamin C had significant effects on CAT, T-AOC, SOD, and GSH-Px ($p < 0.05$).

Table 5. Antioxidant enzyme activities and MDA of gibel carp fed graded dietary of taurine and vitamin C levels in normoxia for 8 weeks and hypoxia for 12 h.

Experimental Groups	Parameters					
Dissolved Oxygen (mg/L)/Taurine and Vitamin C Levels (%)	**CAT (U/mgprot)** [1]	**T-AOC (mmol/gprot)** [2]	**SOD (U/mgprot)** [3]	**MDA (nmol/mgprot)** [4]	**GSH (μmol/gprot)** [5]	**GSH-Px (U/mgprot)** [6]
6.5/(0 + 0)	235.26 ± 6.40^{c}	0.20 ± 0.03^{b}	9.72 ± 0.41^{b}	8.12 ± 0.43	341.36 ± 39.41^{b}	389.08 ± 24.74^{a}
6.5/(0.1 + 0)	256.69 ± 4.69^{b}	0.47 ± 0.06^{a}	11.29 ± 0.28^{a}	6.95 ± 0.24	492.17 ± 16.72^{a}	344.99 ± 29.15^{ab}
6.5/(0.1 + 0.1)	287.47 ± 2.43^{a}	0.34 ± 0.05^{ab}	11.07 ± 0.30^{a}	7.13 ± 0.67	416.84 ± 39.82^{ab}	299.65 ± 8.92^{b}
0.8/(0 + 0)	307.08 ± 10.93^{A}	0.39 ± 0.04^{A}	7.90 ± 0.53	3.17 ± 0.41	355.38 ± 26.57^{B}	448.10 ± 31.79
0.8/(0.1 + 0)	258.22 ± 14.23^{B}	0.46 ± 0.06^{A}	7.42 ± 0.53	2.70 ± 0.43	476.97 ± 41.14^{A}	529.97 ± 28.06
0.8/(0.1 + 0.1)	255.88 ± 13.15^{B}	0.19 ± 0.02^{B}	6.76 ± 0.19	3.48 ± 0.26	377.81 ± 21.70^{B}	483.13 ± 15.06
Dissolved oxygen (mg/L)						
6.5	261.54 ± 6.14	0.34 ± 0.04	10.75 ± 0.24^{y}	7.34 ± 0.27^{y}	416.79 ± 25.22	344.54 ± 15.55^{x}
0.8	272.53 ± 9.63	0.35 ± 0.04	7.33 ± 0.26^{x}	3.10 ± 0.22^{x}	401.79 ± 20.87	487.34 ± 17.47^{y}
Taurine and vitamin C levels (%)						
0 + 0	267.18 ± 13.80	0.32 ± 0.04^{b}	8.82 ± 0.44	5.15 ± 0.86	349.15 ± 21.47^{b}	418.59 ± 21.39
0.1 + 0	257.45 ± 7.07	0.46 ± 0.04^{a}	9.35 ± 0.65	4.83 ± 0.68	483.72 ± 22.89^{a}	447.75 ± 37.64
0.1 + 0.1	274.83 ± 7.19	0.26 ± 0.03^{b}	8.92 ± 0.67	5.11 ± 0.71	395.55 ± 21.32^{b}	381.20 ± 33.15
Two-way ANOVA						
Dissolved oxygen	0.073	0.725	0.000	0.000	0.619	0.009
Taurine and vitamin C levels	0.219	0.000	0.341	0.138	0.002	0.210
Interaction	0.000	0.006	0.009	0.331	0.714	0.025

Note: Data are mean value ± SEM. Means with different superscript letters in the same column are significantly different ($p < 0.05$), normoxic data are labeled with a lower case "abc" and hypoxic data are labeled with an upper case "AB", and the superscript letter "x" and "y" indicates a significant difference ($p < 0.05$) between values of 0.8 mg/L and 6.5 mg/L dissolved oxygen as determined by *t*-test. Means with the same letters or no letters indicate no significant difference among treatments. [1] CAT, catalase; [2] T-AOC, total antioxidant capacity; [3] SOD, superoxide dismutase; [4] MDA, malondialdehyde; [5] GSH, glutathione; [6] GSH-Px, glutathione peroxidase.

Figure 1 shows the expression levels of antioxidant-related genes in gibel carp liver under normoxic and hypoxic environments. From Figure 1, we observe that in the normoxic environment, after adding taurine and taurine with vitamin C, *sod*, *gpx*, and *nrf2* had an increasing trend, but none of them had significant differences ($p > 0.05$). *cat* and *keap1* had no significant trend. In the hypoxic environment, the expression of *sod* and *nrf2* increased and then decreased after the addition of taurine and taurine with vitamin C ($p < 0.05$). The expression levels of *cat*, *gpx*, and *keap1* were positively associated with the addition of taurine and taurine with vitamin C, but none were significantly different ($p > 0.05$). The results of the two-way ANOVA showed that DO significantly affected the expression of *cat*, *sod*, and *keap1*, while the addition of taurine and taurine with vitamin C significantly affected *sod* and *nrf2*. The expression of *sod* was significantly affected by the interaction between DO and the addition of taurine and taurine with vitamin C.

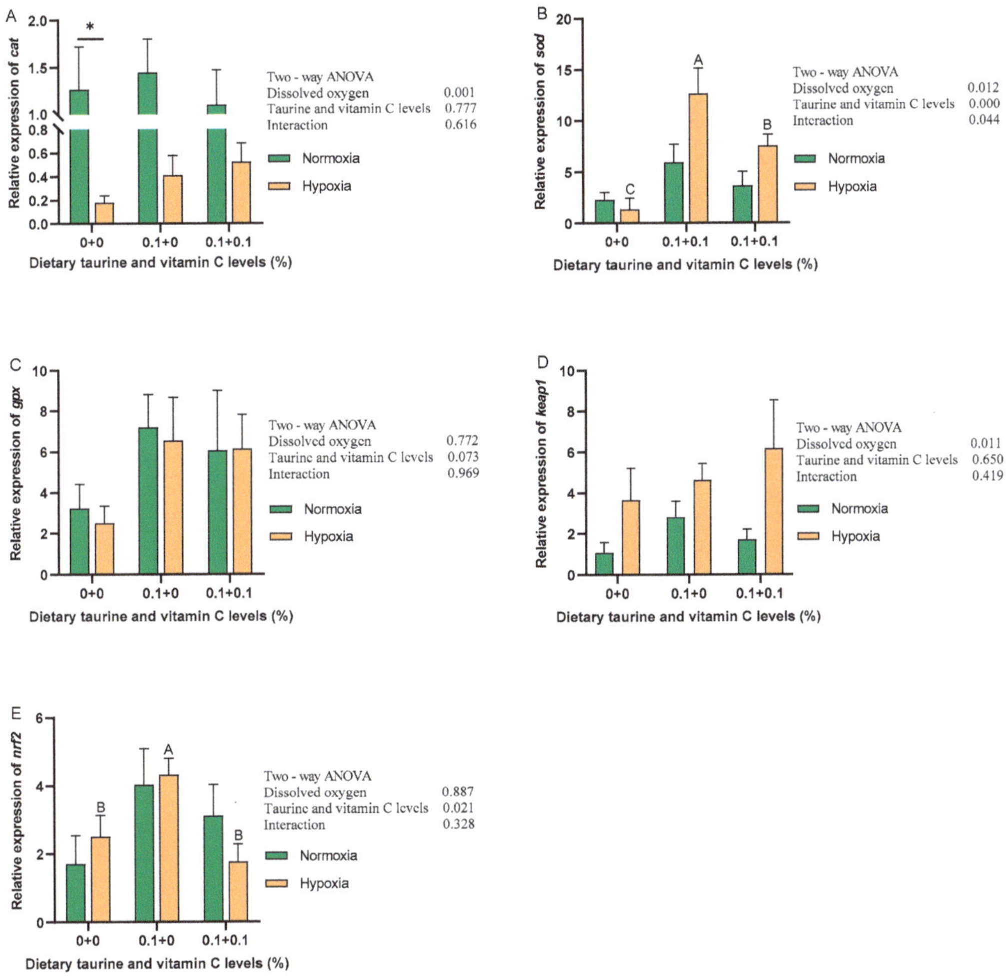

Figure 1. Relative expression levels of mRNA of antioxidant-related genes of gibel carp fed graded dietary of taurine and vitamin C in normoxia for 8 weeks and hypoxia for 12 h. Data with different superscript letters are significantly different ($p < 0.05$); the hypoxic data were labeled with an upper case "ABC", "*" indicates a significant difference ($p < 0.05$) between values of 0.8 mg/L and 6.5 mg/L dissolved oxygen as determined by *t*-test. (**A**) *cat*, catalase; (**B**) *sod*, superoxide dismutase; (**C**) *gpx*, glutathione peroxidase; (**D**) *keap1*, recombinant kelch-like ECH-associated protein 1; (**E**) *nrf2*, nuclear factor erythroid 2-related factor 2.

3.5. Immune Response of the Liver (NF-κB Signaling Pathway)

The expression levels of immune-related genes in gibel carp liver under normoxic and hypoxic environments are shown in Figure 2. In the normoxic environment, all inflammatory factors showed a decreasing trend, and the expression of *il-6* in the D1 and D2 groups was significantly lower compared to the D0 group ($p < 0.05$); the expressions of the rest of the genes were not significantly different among groups. In hypoxia, as the additions of taurine and taurine with vitamin C, an increasing trend in the expression of *il-10*, *tgf-β*, and *nf-κb* was observed in gibel carp liver, among which the changes in *il-10* and *nf-κb* were significant ($p < 0.05$). There were decreasing trends in the expressions of *il-1β* and *il-8*, but the differences were not significant. The remaining genes showed no significant change in trend. The results of a two-way ANOVA showed that DO could significantly affect the expression of *il-10*, *il-6*, *il-8*, and *tnf-α* ($p < 0.05$). The additions of taurine and taurine with vitamin C did not significantly affect the expression of each gene; *il-10* and *nf-κb* were significantly affected by the interaction of DO and the additions of taurine and taurine with vitamin C.

3.6. Hypoxia Signaling Pathway (HIF-1 Signaling Pathway)

Figure 3 shows the expression levels of hypoxia-related genes in gibel carp liver from each group in normoxic and hypoxic environments. In a normoxic environment, the addition of taurine and taurine with vitamin C to the diet had no significant effect on each gene. In a hypoxic environment, the expression of *vegf*, *epo*, *ho-1*, *angpt1*, *nos*, and *tfr1* was increased by the addition of taurine and taurine with vitamin C. Among them, *epo*, *ho-1*, and *tfr1* were significantly increased ($p < 0.05$), while *hif-1α* showed a decreasing trend in expression levels, but there was no significant difference ($p > 0.05$). There was no significant trend in the expression levels of *et1* and *tf*. The results of the two-way ANOVA showed that *ho-1* was significantly affected by DO, the addition of taurine and taurine with vitamin C, and the interaction of DO and the addition of taurine and taurine with vitamin C. DO significantly affected the expression levels of *hif-1α*, *vegf*, *epo*, *ho-1*, and *et1* ($p < 0.05$). None of the other genes was significantly affected by these three factors.

3.7. Survival Rate of Fish and Mitochondrial Number in Gill Cells in Hypoxia

The results of SR and the number of cellular mitochondria (NCM) of the gibel carp in each group under hypoxia are shown in Table 6. With the addition of taurine and taurine with vitamin C, the SR and NCM increased, and the SR of the D2 group was much better than that of the other groups, although the difference was not significant. The statistical results of the NCM for each group are presented in Table 6 and Figure 4. The NCM of the D1 and D2 groups was more than the D0 group; however, the difference was not significant.

Table 6. Survival rate of fish and the number of cellular mitochondria of gibel carp in hypoxia.

Parameters	Dietary Taurine and Vitamin C levels (%)		
	0 + 0	0.1 + 0	0.1 + 0.1
SR (%) [1]	12.50 ± 12.50^a	8.33 ± 4.17^a	33.33 ± 4.17^a
NCM (pcs) [2]	2.00 ± 1.00^a	6.00 ± 1.53^a	5.67 ± 2.91^a

Note: Data are mean value ± SEM. Means in the same row with different superscripts are significantly different ($p < 0.05$). [1] SR, survival rate (%) = 100 × (survival fish number/total fish). [2] NCM (pcs), The number of cellular mitochondria.

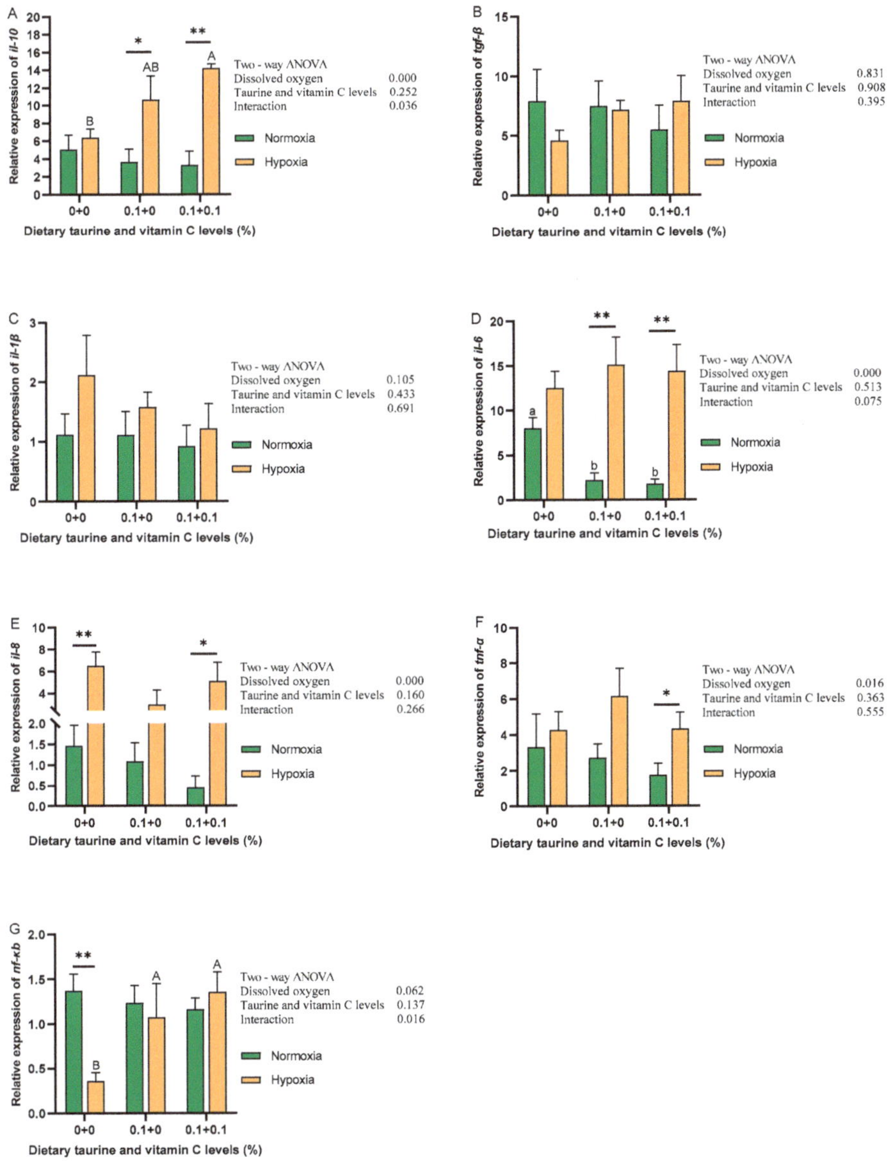

Figure 2. Relative expression levels of mRNA of hepatic inflammation-associated genes of gibel carp fed graded dietary of taurine and vitamin C in normoxia for 8 weeks and hypoxia for 12 h. Data with different superscript letters are significantly different ($p < 0.05$); the normoxic data were labeled with a lower case "ab" and the hypoxic data were labeled with an upper case "AB", "*" indicates a significant difference ($p < 0.05$) and "**" indicates a significant difference ($p < 0.01$) between values of 0.8 mg/L and 6.5 mg/L dissolved oxygen as determined by *t*-test. (**A**) *il-10*, interleukin-10; (**B**) *tgf-β*, transforming growth factor-β; (**C**) *il-1β*, interleukin-1β; (**D**) *il-6*, interleukin-6; (**E**) *il-8*, interleukin-8; (**F**) *tnf-α*, tumor necrosis factor-α; (**G**) *nf-κb*, nuclear factor kappa-β.

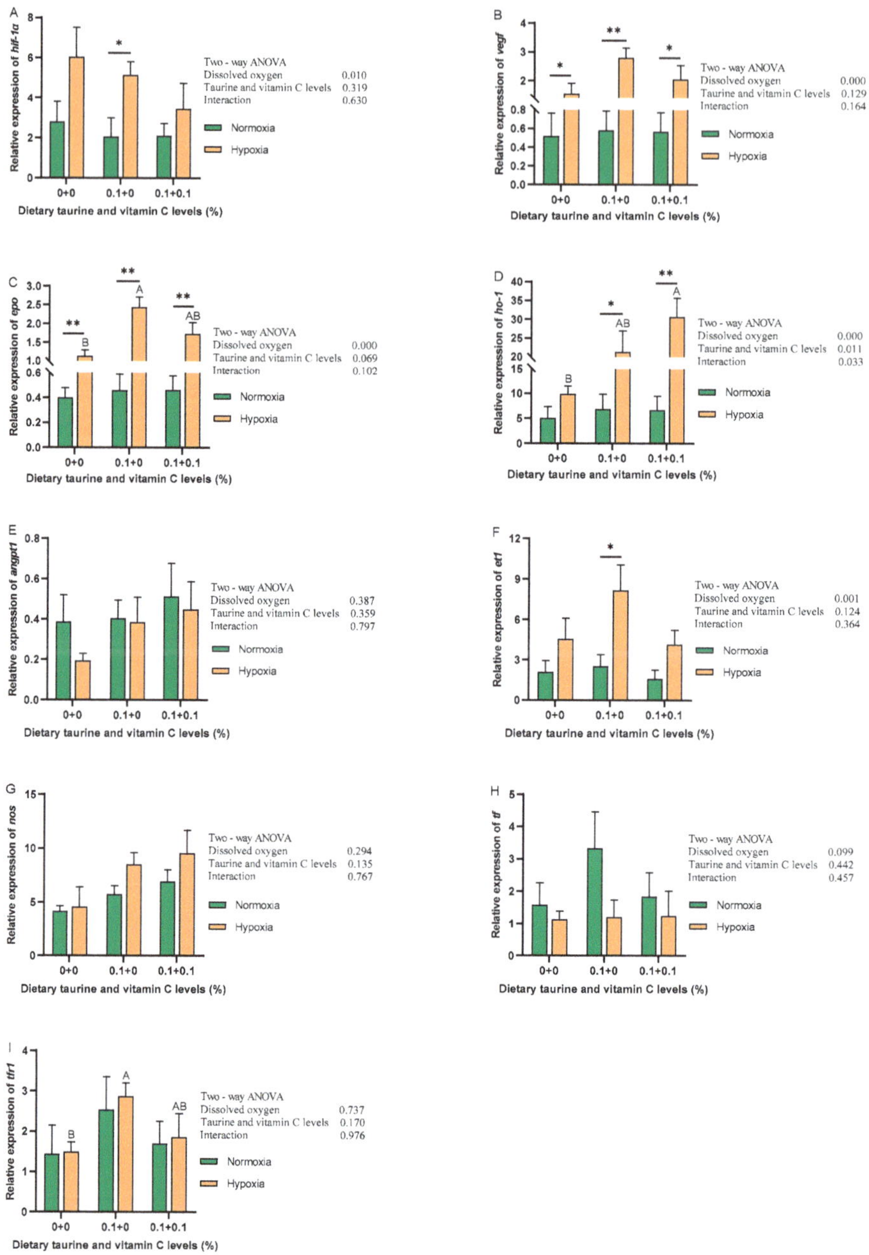

Figure 3. Relative expression levels of mRNA of hypoxia genes of gibel carp fed graded dietary of taurine and vitamin C in normoxia for 8 weeks and hypoxia for 12 h. Data with different superscript letters are significantly different ($p < 0.05$); the hypoxic data were labeled with an upper case "AB", "*" indicates a significant difference ($p < 0.05$) and "**" indicates a significant difference ($p < 0.01$) between values of 0.8 mg/L and 6.5 mg/L dissolved oxygen as determined by *t*-test. (**A**) *hif-1α*, hypoxia-inducible factor 1-α; (**B**) *vegf*, vascular endothelial growth factor; (**C**) *epo*, erythropoietin; (**D**) *ho-1*, heme oxygenase-1; (**E**) *angpt1*, angiopoietin-1; (**F**) *et1*, endothelin 1; (**G**) *nos*, nitric oxide synthase; (**H**) *tf*, transferrin; (**I**) *tfr1*, transferrin receptor protein 1.

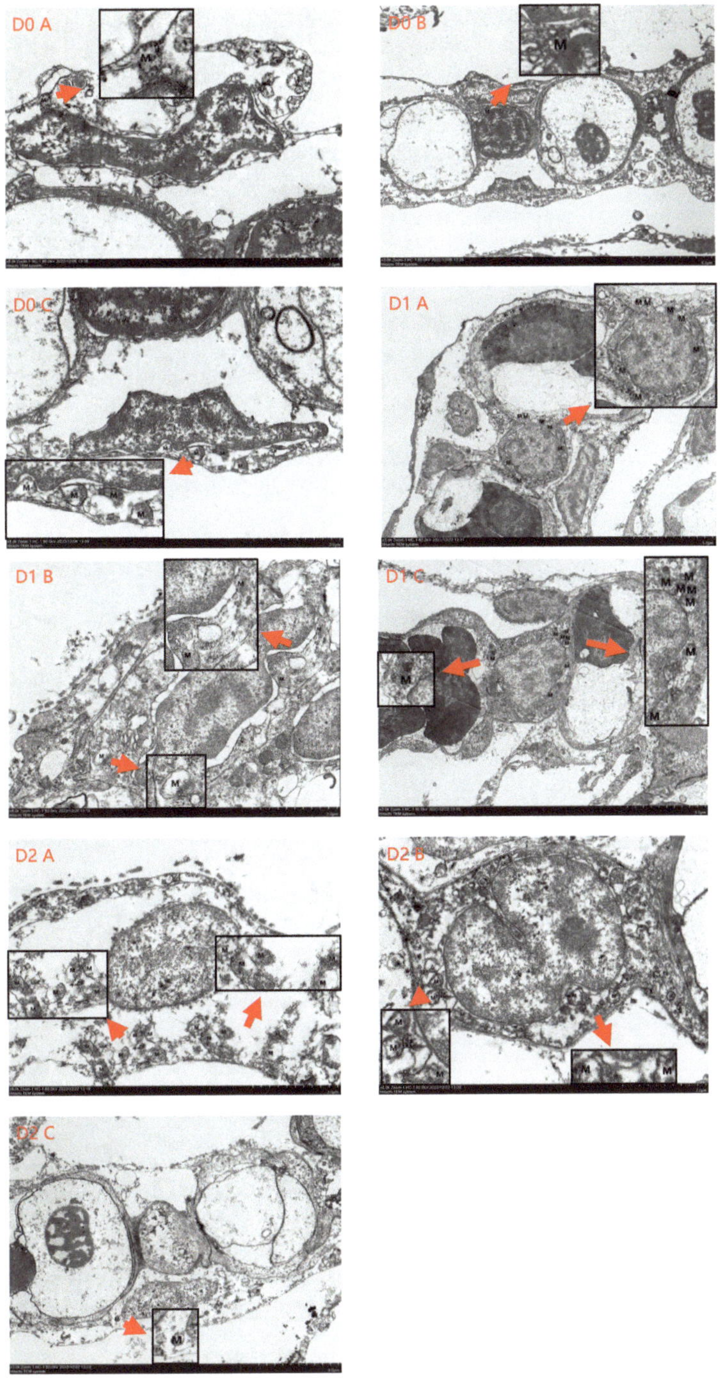

Figure 4. Transmission electron microscopy results of gibel carp gill cells in each group in hypoxia. Note: "D0/D1/D2" denote the three treatment groups, and "A/B/C" represent three samples from the same treatment group. "M" indicates the location and number of mitochondria.

4. Discussion

4.1. Growth Performance and Whole-Body Composition

The results of the present study showed that the addition of taurine and taurine with vitamin C was beneficial to the growth performance of gibel carp, suggesting that taurine and vitamin C play an important role in regulating the growth performance of fish. Numerous studies on other fish species have also reported similar results. For example, the addition of adequate amounts of taurine in the diet has a positive effect on the growth of yellowtail (*Seriola quinqueradiata*) [41], Japanese flounder (*Paralichthys olivaceus*) [42], and turbot [43] by increasing their FBW and SGR and decreasing FCR. Adequate amounts of vitamin C supplements can improve various growth indices, such as WGR and SGR, of Lemeduk fish larvae (*Barbonymus schwanenfeldii*) [44] and yellow catfish [25]. One reason is that vitamin C is an essential nutrient for fish survival. It can act as an enzyme cofactor and help fish maintain optimal health and normal metabolic function, which in turn promotes their growth [45,46]. Taurine is a neuronal trophic factor that has a stimulating effect on the olfactory and gustatory senses of aquatic animals in terms of nitrogen content, water solubility, and acidity and can, therefore, be used as an attractant to promote feeding by aquatic animals [47,48]. Due to the effect of taurine to promote fish feeding and vitamin C to promote fish growth, the addition of both taurine and vitamin C to the feed significantly improved the growth performance of gibel carp.

In terms of whole-body composition, the results showed that the addition of taurine with vitamin C to the feed showed a tendency to increase the crude protein content and decrease the crude lipid content of the gibel carp, but the degree of increase and decrease was not significant compared to the D0 group. Studies have shown that the addition of taurine to feed can increase crude protein content and decrease the crude lipid content of juvenile Atlantic salmon (*Salmo salar*) [49]. The addition of adequate amounts of vitamin C can lead to a significant increase in crude protein and crude lipid content of the fish body [24,50]. However, vitamin C did not significantly affect the body composition of sabaki tilapia (*Oreochromis spilurus*) [51]; therefore, the effect of vitamin C on body composition may depend to some extent on the fish species. The results of this study indicate that exogenous taurine promotes the involvement of sulfur-containing amino acids in protein anabolism by reducing taurine synthesis in fish [52]. At the same time, taurine may be involved in the synthesis of bile acids, promote lipid metabolism [12], increase the activity of the rate-limiting enzyme, and lead to lipid-lowering effects [53]. This effect can be further enhanced by using vitamin C as a coenzyme [45]. The fact that the increase in crude protein and the decrease in crude lipid were not significant may be due to the small amount of added taurine and vitamin C.

4.2. Antioxidant Capacity of the Liver (Nrf2 Signaling Pathway)

In the normoxic experiment, the addition of taurine and taurine with vitamin C to the feed significantly increased the enzyme activities of CAT, SOD, T-AOC, and GSH. This was consistent with the results of studies on juvenile black carp [13] and yellowfin bream (*Acanthopagrus latus*) [54], in which the addition of moderate amounts of taurine to the feeds significantly increased the activities of antioxidant enzymes, such as SOD and CAT in fish. Similarly, the addition of vitamin C to feed could also increase the activities of related antioxidant enzymes in yellow catfish [25] and heterozygous gibel carp [28]. It has been suggested that the addition of taurine and taurine with vitamin C could improve the antioxidant ability of gibel carp by promoting the activities of related enzymes [55,56]. Interestingly, the GSH-Px results showed a significant decrease. A rational explanation for this phenomenon is that SOD can decompose the superoxide anion into H_2O_2; whereas, both CAT and GSH-Px are important detoxification enzymes for H_2O_2 [57]. When the H_2O_2 concentration is low, organic peroxides are the preferred substrates for GSH-Px; however, at higher H_2O_2 concentrations, they are metabolized by CAT [58]. The addition of taurine and taurine with vitamin C to the diet increased the expression of *nrf2*, *sod*, and the activity of GSH, while the activity of CAT and T-AOC decreased under hypoxic conditions.

Compared with the normoxic results, GSH-Px activity was significantly higher in the group with 0.1% taurine + 0% vitamin C (D1) and in the group with 0.1% taurine + 0.1% vitamin C (D2) under hypoxia, which is also consistent with the above explanation. Moreover, the MDA levels of all groups were significantly lower under hypoxia than under normoxia, suggesting that the addition of taurine and vitamin C could lower the MDA levels and alleviate the hypoxic stress of gibel carp by increasing the antioxidant enzyme activities in the hypoxic environment. It can be concluded that the addition of taurine and taurine with vitamin C to the diet can increase the antioxidant capacity of gibel carp in both normoxic and hypoxic environments and alleviate hypoxic stress.

4.3. Immune Response of the Liver (NF-κB Signaling Pathway)

Studies have shown that the nuclear transcription factor κB (NF-κB) signaling pathway regulates various physiological activities and that activation of this pathway may promote inflammatory responses and is involved in immune responses [59,60]. Taurine and vitamin C have also been shown to be effective modulators of the organismal inflammatory response and are involved in physiological activities, such as immunomodulation [61,62]. For example, taurine and vitamin C can be used as effective additives to improve the immunity of the olive flounder (*Paralichthys olivaceus*) [63] and rainbow trout [22] by modulating inflammatory cytokines. In this experiment, the relative expression levels of the *il-6* gene in the 0.1% taurine + 0% vitamin C group (D1) and the 0.1% taurine + 0.1% vitamin C group (D2) were significantly lower than those in the 0% taurine + 0% vitamin C group (D0) in normoxia. *Il-6* is a pleiotropic immunocytokine with both pro-inflammatory and anti-inflammatory effects. Studies have shown that prolonged activation of the il-6 signaling pathway can cause major damage to the liver [64]. The levels of other pro-inflammatory cytokines, including *il-8* and *tnf-α*, also showed a decreasing trend. Therefore, the addition of taurine and taurine with vitamin C to the diet can improve the immunity of gibel carp in normoxia by downregulating the expression of pro-inflammatory cytokines. Hypoxic stress can inhibit and impair immune functions in fish [65,66]. In this study, the relative expression levels of *il-10*, *il-6*, and *il-8* were significantly higher under hypoxic conditions than under normoxia, suggesting that hypoxic stress exacerbates the inflammatory response in the liver. The relative expression of *nf-κb* in the 0.1% taurine + 0% vitamin C group (D1), and the 0.1% taurine + 0.1% vitamin C group (D2) was significantly higher than that in the 0% taurine + 0% vitamin C group (D0). *Il-10* also increased, and the level in the D2 group was significantly higher than that in the D0 group. The phenomenon that both anti-inflammatory and pro-inflammatory cytokines show the same expression trend has also been observed in other studies [33,67]. It is speculated that there is a negative feedback mechanism in the NF-κB signaling pathway and that the hypoxic environment promotes the inflammatory response in the liver of gibel carp, while taurine and vitamin C activate the expression of *il-10* by upregulating the expression of *nf-κb* to alleviate hypoxic stress.

4.4. Hypoxia Signaling Pathway (HIF-1 Signaling Pathway)

It is well known that hypoxia can be extremely detrimental to the survival and growth of fish [4,5]. Several biological mechanisms resist hypoxic stress in fish. The HIF-1 signaling pathway, which is mainly regulated by hypoxia-inducible factor 1-α (*hif-1α*), plays a crucial role in fish adaptation to hypoxia [68]. In this signaling pathway, *hif-1α*, as an upstream gene, regulates the expression of a variety of genes that can respond to the hypoxic environment and thus control the adaptation of fish to hypoxia [69–72]. In this experiment, the addition of taurine and taurine with vitamin C to the feed did not significantly affect the relative expression levels of genes in the HIF-1 signaling pathway when the gibel carp were in normoxia. However, compared to normoxia, this pathway was activated during hypoxia, and the expression of the genes *hif-1α*, erythropoietin (*epo*), vascular endothelial growth factor (*vegf*), and heme oxygenase-1 (*ho-1*), which are associated with the production and degradation of erythrocytes, vessels, and hemoglobin [69–71], was significantly increased, indicating that the gibel carp was affected by hypoxic stress and activated the corresponding defense mechanisms. At the same time, the addition of taurine and taurine with vitamin C significantly increased the relative expression levels of *epo*, *ho-1*, and *tfr1* under hypoxia. *Tfr1* is the major regulatory receptor in the process of iron uptake in cells and has a crucial function in the development of many diseases [72], suggesting that the defenses of gibel carp against hypoxia were further enhanced. This finding is consistent with previous research, showing that taurine and vitamin C can inhibit the transcription and expression of *hif-1* and, thus, have a therapeutic effect on hypoxia-induced vascular remodeling [73,74]. Under hypoxia, *hif-1α* had a certain decreasing trend between the groups of gibel carp, but the difference was not significant, which was probably the reason for the low dose of added taurine and vitamin C. However, due to the sensitive regulatory effect of *hif-1α* on downstream genes, some genes showed significant differences. The above results suggest that the addition of taurine and taurine with vitamin C may further increase the tolerance of gibel carp to hypoxia by modulating the HIF-1 signaling pathway.

4.5. Survival Rate and Mitochondrial Number in Gill Cells of Gibel Carp in Hypoxia

The survival of fish in a hypoxic environment can be used as a visual indicator to assess the damage caused by hypoxia and their ability to tolerate it. In this experiment, although the difference in SR between the groups was not significant, the SR of the 0.1% taurine + 0.1% vitamin C group (D2) was much better than that of the D0 group and the 0.1% taurine + 0% vitamin C group (D1), which is consistent with previous studies showing that taurine can significantly prolong the survival time of hypoxic mice [75] and that vitamin C can improve the hypoxia tolerance of channel catfish (*Ictalurus punctatus*) to a certain extent [76]. At the same time, the results of this experiment indicate that feeding taurine and taurine with vitamin C can improve the antioxidant capacity, immune response, and cardiovascular oxygen transport capacity of gibel carp by modulating the Nrf2, NF-κB, and HIF-1 signaling pathways, thus intuitively showing stronger hypoxia tolerance ability and improved SR. In addition, the results of the swimming ability test conducted at the end of the feeding experiment showed that the simultaneous addition of taurine and vitamin C to the diet could significantly increase the SST of gibel carp in a certain water current environment. The results of previous studies suggested that swimming ability can be used as an important indicator to evaluate the ability of fish to adapt to the environment and that fish with strong swimming ability tend to have better metabolism and a stronger body [77–79]. This result can, therefore, be taken as evidence that the addition of taurine and vitamin C to the feed can improve the SR of gibel carp in hypoxic environments by increasing their physical fitness.

Mitochondria are sensitive organelles in cells under stress and are one of the most important organelles that serve as sources of energy for cellular and oxygen metabolism and play important roles in physiological and pathological processes such as cell growth, apoptosis, and senescence [80,81]. Gills are an important organ for oxygen respiration in fish, and their physicochemical state may better reflect the ability of fish to cope with hypoxia [82]. In this experiment, the number of mitochondria in the gill cells of gibel carp was higher in both the group with 0.1% taurine + 0% vitamin C (D1) and the group with 0.1% taurine + 0.1% vitamin C (D2) than in the group with 0% taurine + 0% vitamin C (D0) under hypoxic condition, although the difference was not significant. This result is generally consistent with previous studies showing that both taurine and vitamin C play a crucial role in regulating the functional expression of mitochondria [83,84]. It has been shown that hypoxia leads to high ROS production, resulting in a reduced mitochondrial inner membrane area and decreased oxidative capacity, while mitochondria themselves produce ROS through the respiratory chain, such that large amounts of ROS trigger mitochondrial autophagy in response to a hypoxic environment [85]. In conjunction with other results from this study and other studies, taurine and vitamin C may increase the antioxidant capacity of fish and reduce the damage to organisms by ROS, thereby reducing mitochondrial autophagy and improving mitochondrial function, which in turn may improve the ability of gibel carp to tolerate hypoxia.

5. Conclusions

In summary, what we have demonstrated by conducting culture trials as well as hypoxic stress experiments is that the addition of 0.1% taurine + 0.1% vitamin C to the diet could improve the growth performance and swimming ability of gibel carp, while the addition of 0.1% taurine and 0.1% taurine + 0.1% vitamin C improved the antioxidant capacity, immune response, and hypoxia tolerance of gibel carp by modulating the Nrf2, NF-κB, and HIF-1 signaling pathways.

The results of this research confirm that taurine and vitamin C are beneficial feed additives and provide referable nutritional approaches to improve the efficiency of fish farming such as gibel carp and to reduce the damage caused by hypoxic environmental stress in aquaculture. The effects of hypoxia on aquaculture are objective and long term. In the future, more relevant research needs to be invested in alleviating the problem of hypoxia in fish, such as researching more fish species, selecting more suitable feed additives, and determining the optimal amount of additives by setting different additivity levels, in order to ensure the healthy development of the aquaculture industry.

Author Contributions: M.R., H.L. and D.H. designed the study. L.Z. (Leimin Zhang) carried out the experiments and wrote the manuscript. L.Z. (Lu Zhang) provided technical guidance. All authors have read and agreed to the published version of the manuscript.

Funding: This study was financially supported by the National Key R & D Program of China (2023YFD2400601), the earmarked fund for CARS (CARS-46), the National Natural Science Foundation of China (32102806) and Central Public-interest Scientific Institution Basal Research Fund, Freshwater Fisheries Research Center, CAFS (NO. 2024JBFR01).

Institutional Review Board Statement: The study was conducted according to the Management Rule of Laboratory Animals (Chinese Order No. 676 of the State Council, revised 1 March 2017). The study was approved by the Laboratory Animal Ethics Committee of the Freshwater Fisheries Research Center (LAECFFRC-2023-03-03).

Informed Consent Statement: Not applicable.

Data Availability Statement: Data are contained within the article.

Conflicts of Interest: Lu Zhang is employed by Tongwei Agricultural Development Co., Ltd. The remaining authors declare that the research was conducted in the absence of any commercial or financial relationships that could be construed as potential conflicts of interest. Lu Zhang made important contributions to the experimental technique.

References

1. Qin, F.; Shi, M.; Yuan, H.; Yuan, L.; Lu, W.; Zhang, J.; Tong, J.; Song, X. Dietary nano-selenium relieves hypoxia stress and, improves immunity and disease resistance in the Chinese mitten crab (*Eriocheir sinensis*). *Fish Shellfish Immunol.* **2016**, *54*, 481–488. [CrossRef] [PubMed]
2. Li, F.; Sun, Z.; Qi, H.; Zhou, X.; Xu, C.; Wu, D.; Fang, F.; Feng, J.; Zhang, N. Effects of Rice-Fish Co-culture on Oxygen Consumption in Intensive Aquaculture Pond. *Rice Sci.* **2019**, *26*, 50–59. [CrossRef]
3. Tanner, C.A.; Burnett, L.E.; Burnett, K.G. The effects of hypoxia and pH on phenoloxidase activity in the Atlantic blue crab, Callinectes sapidus. *Comp. Biochem. Physiol. Part A Mol. Integr. Physiol.* **2006**, *144*, 218–223. [CrossRef] [PubMed]
4. Wu, R. Hypoxia: From molecular responses to ecosystem responses. *Mar. Pollut. Bull.* **2002**, *45*, 35–45. [CrossRef] [PubMed]
5. Wu, R.S.S.; Zhou, B.S.; Randall, D.J.; Woo, N.Y.S.; Lam, P.K.S. Aquatic Hypoxia Is an Endocrine Disruptor and Impairs Fish Reproduction. *Environ. Sci. Technol.* **2003**, *37*, 1137–1141. [CrossRef]
6. Varghese, T.; Rejish Kumar, V.; Gopan, A.; Valappil, R.K.; Sajina, K.A.; Mishal, P.; Pal, A.K. Dietary arginine modulates nonspecific immune responses in Indian Major Carp, *Cirrhinus mrigala* exposed to hypoxia. *Aquaculture* **2020**, *529*, 735613. [CrossRef]
7. Varghese, T.; Rejish Kumar, V.J.; Anand, G.; Dasgupta, S.; Pal, A. Dietary GABA enhances hypoxia tolerance of a bottom-dwelling carp, *Cirrhinus mrigala* by modulating HIF-1α, thyroid hormones and metabolic responses. *Fish Physiol. Biochem.* **2020**, *46*, 199–212. [CrossRef]
8. Li, H.; Lu, L.; Wu, M.; Xiong, X.; Luo, L.; Ma, Y.; Liu, Y. The effects of dietary extract of mulberry leaf on growth performance, hypoxia-reoxygenation stress and biochemical parameters in various organs of fish. *Aquac. Rep.* **2020**, *18*, 100494. [CrossRef]
9. Tiedemann, F.; Gmelin, L. Einige neue Bestandtheile der Galle des Ochsen. *Ann. Der Phys.* **2006**, *85*, 326–337. [CrossRef]
10. Kuzmina, V.; Gavrovskaya, L.; Ryzhova, O. Taurine. Effect on Exotrophia and Metabolism in Mammals and Fish. *J. Evol. Biochem. Physiol.* **2010**, *46*, 19–27. [CrossRef]

11. Jacobsen, J.G.; Smith, L.H. Biochemistry and physiology of taurine and taurine derivatives. *Physiol. Rev.* **1968**, *48*, 424–511. [CrossRef] [PubMed]
12. Kim, S.K.; Kim, K.G.; Kim, K.D.; Kim, K.W.; Son, M.H.; Rust, M.; Johnson, R. Effect of dietary taurine levels on the conjugated bile acid composition and growth of juvenile Korean rockfish Sebastes schlegeli (*Hilgendorf*). *Aquac. Res.* **2015**, *46*, 2768–2775. [CrossRef]
13. Zhang, J.; Hu, Y.; Ai, Q.; Mao, P.; Tian, Q.; Zhong, L.; Xiao, T.; Chu, W. Effect of dietary taurine supplementation on growth performance, digestive enzyme activities and antioxidant status of juvenile black carp (*Mylopharyngodon piceus*) fed with low fish meal diet. *Aquac. Res.* **2018**, *49*, 3187–3195. [CrossRef]
14. Zhang, Y.; Wei, Z.; Liu, G.; Deng, K.; Yang, M.; Pan, M.; Gu, Z.; Liu, D.; Zhang, W.; Mai, K. Synergistic effects of dietary carbohydrate and taurine on growth performance, digestive enzyme activities and glucose metabolism in juvenile turbot *Scophthalmus maximus* L. *Aquaculture* **2019**, *499*, 32–41. [CrossRef]
15. Salze, G.; McLean, E.; Craig, S.R. Dietary taurine enhances growth and digestive enzyme activities in larval cobia. *Aquaculture* **2012**, *362–363*, 44–49. [CrossRef]
16. Nakashima, T.; Seto, Y.; Nakajima, T.; Shima, T.; Sakamoto, Y.; Cho, N.; Sano, A.; Iwai, M.; Kagawa, K.; Okanoue, T.; et al. Calcium-associated cytoprotective effect of taurine on the calcium and oxygen paradoxes in isolated rat hepatocytes. *Liver* **1990**, *10*, 167–172. [CrossRef]
17. Yang, H.; Tian, L.; Huang, J.; Liang, G.; Liu, Y. Dietary taurine can improve the hypoxia-tolerance but not the growth performance in juvenile grass carp *Ctenopharyngodon idellus*. *Fish Physiol. Biochem.* **2013**, *39*, 1071–1078. [CrossRef]
18. Lim, C.; Lovell, R.T. Pathology of the vitamin C deficiency syndrome in channel catfish (*Ictalurus punctatus*). *J. Nutr.* **1978**, *108*, 1137. [CrossRef]
19. Eo, J.; Lee, K. Effect of dietary ascorbic acid on growth and non-specific immune responses of tiger puffer, *Takifugu rubripes*. *Fish Shellfish Immunol.* **2008**, *25*, 611–616. [CrossRef]
20. Xu, C.; Yu, H.; Li, L.; Li, M.; Qiu, X.; Fan, X.; Fan, Y.; Shan, L. Effects of Dietary Vitamin C on the Growth Performance, Biochemical Parameters, and Antioxidant Activity of Coho Salmon *Oncorhynchus kisutch* (Walbaum, 1792) Postsmolts. *Aquac. Nutr.* **2022**, *2022*, 6866578. [CrossRef]
21. Khan, M.; Fatima, M.; Shah, S.Z.H.; Khan, N.; Khizar, A.; Nadeem, H.; Khan, F. Evaluation of dietary Vitamin C requirement of Hypophthalmichthys molitrix fingerlings and its effects on growth, haematology and serum enzyme activities. *Aquac. Res.* **2022**, *53*, 5582–5593. [CrossRef]
22. Leal, E.; Zarza, C.; Tafalla, C. Effect of vitamin C on innate immune responses of rainbow trout (*Oncorhynchus mykiss*) leukocytes. *Fish Shellfish Immunol.* **2017**, *67*, 179–188. [CrossRef] [PubMed]
23. Yusuf, A.; Huang, X.; Chen, N.; Apraku, A.; Wang, W.; Cornel, A.; Rahman, M.M. Impact of dietary vitamin c on plasma metabolites, antioxidant capacity and innate immunocompetence in juvenile largemouth bass, *Micropterus salmoides*. *Aquac. Rep.* **2020**, *17*, 100383. [CrossRef]
24. Zou, W.; Lin, Z.; Huang, Y.; Limbu, S.M.; Rong, H.; Yu, C.; Lin, F.; Wen, X. Effect of dietary vitamin C on growth performance, body composition and biochemical parameters of juvenile Chu's croaker (*Nibea coibor*). *Aquac. Nutr.* **2020**, *26*, 60–73. [CrossRef]
25. Liang, X.; Li, Y.; Hou, Y.; Qiu, H.; Zhou, Q. Effect of dietary vitamin C on the growth performance, antioxidant ability and innate immunity of juvenile yellow catfish (*Pelteobagrus fulvidraco Richardson*). *Aquac. Res.* **2017**, *48*, 149–160. [CrossRef]
26. Imanpoor, M.; Imanpoor, M.R.; Roohi, Z. Effects of dietary vitamin C on skeleton abnormalities, blood biochemical factors, haematocrit, growth, survival and stress response of Cyprinus carpio fry. *Aquac. Int.* **2017**, *25*, 793–803. [CrossRef]
27. Henchion, M.; Hayes, M.; Mullen, A.; Fenelon, M.; Brijesh Kumar, T. Future Protein Supply and Demand: Strategies and Factors Influencing a Sustainable Equilibrium. *Foods* **2017**, *6*, 53. [CrossRef]
28. Wu, L.; Xu, W.; Li, H.; Dong, B.; Geng, H.; Jin, J.; Han, D.; Liu, H.; Zhu, X.; Yang, Y.; et al. Vitamin C Attenuates Oxidative Stress, Inflammation, and Apoptosis Induced by Acute Hypoxia through the Nrf2/Keap1 Signaling Pathway in Gibel Carp (*Carassius gibelio*). *Antioxidants* **2022**, *11*, 935. [CrossRef]
29. Huang, D.; Liang, H.; Ge, X.; Zhu, J.; Li, S.; Wang, Y.; Ren, M.; Chen, X. Effects of Dietary Lysine Levels on Growth Performance and Glycolipid Metabolism via the AKT/FoxO1 Pathway in Juvenile Largemouth Bass, *Micropterus salmoides*. *Aquac. Nutr.* **2022**, *2022*, 1372819. [CrossRef]
30. Ren, M.; Liao, Y.; Xie, J.; Liu, B.; Zhou, Q.; Ge, X.; Cui, H.; Pan, L.; Chen, R. Dietary arginine requirement of juvenile blunt snout bream, *Megalobrama amblycephala*. *Aquaculture* **2013**, *414–415*, 229–234. [CrossRef]
31. Tang, B.; Chen, X.; Hu, C.; Dai, J.; Wang, X.; Zhang, J. Flow velocity effects on swimming behavior and exercise physiology of juvenile Acanthopagrus Schlegel II. *Acta Hydrobiol. Sin.* **2023**, *47*, 1993–2002. (In Chinese)
32. AOAC. *Official Methods of Analysis of the Association of Official Analytical Chemists*, 15th ed.; Association of Official Analytical Chemists, Inc.: Arlington, VA, USA, 2003.
33. Zhao, F.; Xu, P.; Xu, G.; Huang, D.; Zhang, L.; Ren, M.; Liang, H. Dietary valine affects growth performance, intestinal immune and antioxidant capacity in juvenile largemouth bass (*Micropterus salmoides*). *Anim. Feed Sci. Technol.* **2023**, *295*, 115541. [CrossRef]
34. Sun, X.; Yu, H.; Xing, K.; Tian, Y.; Chen, C.; Guo, Y.; Shi, H.; Yang, S.; Chen, S.; Wang, Q. Effects of taurine levels in feed on blood biochemical parameters and antioxidant indexes of Cynoglossus semilaevis and their responses to fishing stress. *Isr. J. Aquac. Bamidgeh* **2021**, *73*, 1–16. [CrossRef]

35. Zhao, W.; Liu, Z.; Niu, J. Growth performance, intestinal histomorphology, body composition, hematological and antioxidant parameters of Oncorhynchus mykiss were not detrimentally affected by replacement of fish meal with concentrated dephenolization cottonseed protein. *Aquac. Rep.* **2021**, *19*, 100557. [CrossRef]
36. Zhang, C.; Yao, W.; Li, X.; Duan, Z.; Dang, J.; Cao, K.; Leng, X. Dietary emulsifier and antioxidant improved astaxanthin utilization and antioxidant capacity of rainbow trout (*Oncorhynchus mykiss*). *Aquac. Nutr.* **2021**, *27*, 2416–2426. [CrossRef]
37. Khieokhajonkhet, A.; Suwannalers, P.; Aeksiri, N.; Ratanasut, K.; Chitmanat, C.; Inyawilert, W.; Phromkunthong, W.; Kaneko, G. Effects of dietary red pepper extracts on growth, hematology, pigmentation, disease resistance, and growth- and immune-related gene expressions of goldfish (*Carassius auratus*). *Anim. Feed Sci. Technol.* **2023**, *301*, 115658. [CrossRef]
38. Yang, K.; Qi, X.; He, M.; Song, K.; Luo, F.; Qu, X.; Wang, G.; Ling, F. Dietary supplementation of salidroside increases immune response and disease resistance of crucian carp (*Carassius auratus*) against Aeromonas hydrophila. *Fish Shellfish Immunol.* **2020**, *106*, 1–7. [CrossRef] [PubMed]
39. Gu, Y.; Chen, K.; Xi, B.; Xie, J.; Bing, X. Protective effects of paeonol against lipopolysaccharide-induced liver oxidative stress and inflammation in gibel carp (*Carassius auratus gibelio*). *Comp. Biochem. Physiol. Part C Toxicol. Pharmacol.* **2022**, *257*, 109339. [CrossRef] [PubMed]
40. Sun, L.; Wang, Q.; Wang, R.; Sun, K.; Li, S.; Lin, G.; Lei, P.; Xu, H. Effect of dietary poly-γ-glutamic acid on growth, digestive enzyme activity, antioxidant capacity, and TOR pathway gene expression of gibel carp (*Carassius auratus gibelio*). *Aquac. Rep.* **2022**, *27*, 101412. [CrossRef]
41. Matsunari, H.; Takeuchi, T.; Takahashi, M.; Mushiake, K. Effect of dietary taurine supplementation on growth performance of yellowtail juveniles *Seriola quinqueradiata*. *Fish. Sci.* **2005**, *71*, 1131–1135. [CrossRef]
42. Kim, S.; Matsunari, H.; Takeuchi, T.; Yokoyama, M.; Murata, Y.; Ishihara, K. Effect of different dietary taurine levels on the conjugated bile acid composition and growth performance of juvenile and fingerling Japanese flounder Paralichthys olivaceus. *Aquaculture* **2007**, *273*, 595–601. [CrossRef]
43. Qi, G.; Ai, Q.; Mai, K.; Xu, W.; Liufu, Z.; Yun, B.; Zhou, H. Effects of dietary taurine supplementation to a casein-based diet on growth performance and taurine distribution in two sizes of juvenile turbot (*Scophthalmus maximus* L.). *Aquaculture* **2012**, *358–359*, 122–128. [CrossRef]
44. Mellisa, S.; Fitria Hasri, I.; Nurfadillah, N.; Arisa, I.I. The effectiveness of feeding artemia enriched with vitamin c on the growth performance and survival of Lemeduk fish larvae (*Barbonymus schwanenfeldii*). *Iop Conf. Ser. Earth Environ. Sci.* **2021**, *674*, 12061. [CrossRef]
45. Gasco, L.; Gai, F.; Maricchiolo, G.; Genovese, L.; Ragonese, S.; Bottari, T.; Caruso, G. Supplementation of Vitamins, Minerals, Enzymes and Antioxidants in Fish Feeds. In *Feeds for the Aquaculture Sector*; Springer International Publishing AG: Cham, Switzerland, 2018; pp. 63–103. [CrossRef]
46. Kandeel, M.M.A.; Magouz, F.I.; Omar, A.A.; Amer, A.A.; Zaineldin, A.I.; Ashry, A.M.; Dawood, M.A.O. Combined effects of butyl hydroxytoluene and vitamin C on the growth performance, blood biochemistry, and antioxidative status of common carp (*Cyprinus carpio*). *Ann. Anim. Sci.* **2024**, *24*, 881–888. [CrossRef]
47. Carr, W.E.S.; Netherton, I.J.C.; Gleeson, R.A.; Derby, C.D. Stimulants of Feeding Behavior in Fish: Analyses of Tissues of Diverse Marine Organisms. *Biol. Bull.* **1996**, *190*, 149–160. [CrossRef]
48. Martinez, J.; Chatzifotis, S.; Divanach, P.; Takeuchi, T. Effect of dietary taurine supplementation on growth performance and feed selection of sea bass *Dicentrarchus labrax* fry fed with demand-feeders. *Fish. Sci.* **2004**, *70*, 74–79. [CrossRef]
49. Espe, M.; Ruohonen, K.; El-Mowafi, A. Effect of taurine supplementation on metabolism and body lipid to protein ratio in juvenile Atlantic salmon (*Salmo salar*). *Aquac. Res.* **2011**, *43*, 349–360. [CrossRef]
50. Nasar, M.F.; Shah, S.Z.H.; Aftab, K.; Fatima, M.; Bilal, M.; Hussain, M. Dietary vitamin C requirement of juvenile grass carp (*Ctenopharyngodon idella*) and its effects on growth attributes, organ indices, whole-body composition and biochemical parameters. *Aquac. Nutr.* **2021**, *27*, 1903–1911. [CrossRef]
51. Al-Amoudi, M.M.; El-Nakkadi, A.M.N.; El-Nouman, B.M. Evaluation of optimum dietary requirement of vitamin C for the growth of *Oreochromis spilurus* fingerlings in water from the Red Sea. *Aquaculture* **1992**, *105*, 165–173. [CrossRef]
52. Li, P.; Mai, K.; Trushenski, J.; Wu, G. New developments in fish amino acid nutrition: Towards functional and environmentally oriented aquafeeds. *Amino Acids* **2008**, *37*, 43–53. [CrossRef]
53. Yun, B.; Ai, Q.; Mai, K.; Xu, W.; Qi, G.; Luo, Y. Synergistic effects of dietary cholesterol and taurine on growth performance and cholesterol metabolism in juvenile turbot (*Scophthalmus maximus L.*) fed high plant protein diets. *Aquaculture* **2012**, *324–325*, 85–91. [CrossRef]
54. Dehghani, R.; Oujifard, A.; Torfi Mozanzadeh, M.; Morshedi, V.; Bagheri, D. Effects of dietary taurine on growth performance, antioxidant status, digestive enzymes activities and skin mucosal immune responses in yellowfin seabream, *Acanthopagrus latus*. *Aquaculture* **2019**, *517*, 734795. [CrossRef]
55. Salze, G.P.; Davis, D.A. Taurine: A critical nutrient for future fish feeds. *Aquaculture* **2015**, *437*, 215–229. [CrossRef]
56. Ming, J.; Xie, J.; Xu, P.; Ge, X.; Liu, W.; Ye, J. Effects of emodin and vitamin C on growth performance, biochemical parameters and two HSP70s mRNA expression of Wuchang bream (*Megalobrama amblycephala Yih*) under high temperature stress. *Fish Shellfish Immunol.* **2012**, *32*, 651–661. [CrossRef] [PubMed]

57. Borković, S.S.; Pavlović, S.Z.; Kovačević, T.B.; Štajn, A.Š.; Petrović, V.M.; Saičić, Z.S. Antioxidant defence enzyme activities in hepatopancreas, gills and muscle of Spiny cheek crayfish (*Orconectes limosus*) from the River Danube. *Comp. Biochem. Physiol. Part C Toxicol. Pharmacol.* **2008**, *147*, 122–128. [CrossRef]
58. Yu, B. Cellular Defenses Against Damage From Reactive Oxygen Species. *Physiol. Rev.* **1995**, *75*, 236. [CrossRef]
59. Xie, X.; Xu, Z.; Xu, K.; Xiao, Y. DUSP19 mediates spinal cord injury-induced apoptosis and inflammation in mouse primary microglia cells via the NF-kB signaling pathway. *Neurol. Res.* **2019**, *42*, 31–38. [CrossRef]
60. Qiu, W.; Hu, J.; Magnuson, J.; Greer, J.; Yang, M.; Chen, Q.; Fang, M.; Zheng, C.; Schlenk, D. Evidence linking exposure of fish primary macrophages to antibiotics activates the NF-kB pathway. *Environ. Int.* **2020**, *138*, 105624. [CrossRef]
61. Zhang, F.; Mao, Y.; Qiao, H.; Jiang, H.; Zhao, H.; Chen, X.; Tong, L.; Sun, X. Protective effects of taurine against endotoxin-induced acute liver injury after hepatic ischemia reperfusion. *Amino Acids* **2009**, *38*, 237–245. [CrossRef]
62. Zhao, Y.; Wang, L.; Gao, J. Effects of dietary highly unsaturated fatty acid levels on growth, fatty acid profiles, antioxidant activities, mucus immune responses and hepatic lipid metabolism related gene expressions in loach (*Misgurnus anguillicaudatus*) juveniles. *Aquac. Res.* **2019**, *50*, 2486–2495. [CrossRef]
63. Liu, J.; Pan, M.; Liu, Y.; Huang, D.; Luo, K.; Wu, Z.; Zhang, W.; Mai, K. Taurine alleviates endoplasmic reticulum stress, inflammatory cytokine expression and mitochondrial oxidative stress induced by high glucose in the muscle cells of olive flounder (*Paralichthys olivaceus*). *Fish Shellfish Immunol.* **2022**, *123*, 358–368. [CrossRef] [PubMed]
64. Schmidt-Arras, D.; Rose-John, S. IL-6 pathway in the liver: From physiopathology to therapy. *J. Hepatol.* **2016**, *64*, 1403–1415. [CrossRef] [PubMed]
65. Nam, S.; Haque, M.N.; Shin, Y.K.; Park, H.S.; Rhee, J. Constant and intermittent hypoxia modulates immunity, oxidative status, and blood components of red seabream and increases its susceptibility to the acute toxicity of red tide dinoflagellate. *Fish Shellfish Immunol.* **2020**, *105*, 286–296. [CrossRef] [PubMed]
66. Abdel-Tawwab, M.; Hagras, A.E.; Elbaghdady, H.A.M.; Monier, M.N. Effects of dissolved oxygen and fish size on Nile tilapia, *Oreochromis niloticus* (L.): Growth performance, whole-body composition, and innate immunity. *Aquac. Int.* **2015**, *23*, 1261–1274. [CrossRef]
67. Gu, J.; Liang, H.; Ge, X.; Xia, D.; Pan, L.; Mi, H.; Ren, M. A study of the potential effect of yellow mealworm (*Tenebrio molitor*) substitution for fish meal on growth, immune and antioxidant capacity in juvenile largemouth bass (*Micropterus salmoides*). *Fish Shellfish Immunol.* **2022**, *120*, 214–221. [CrossRef]
68. Xiao, W. The hypoxia signaling pathway and hypoxic adaptation in fishes. Science China. *Life Sci.* **2015**, *58*, 148–155. [CrossRef]
69. Wanner, R.M.; Spielmann, P.; Stroka, D.M.; Camenisch, G.; Wenger, R.H. Epolones induce erythropoietin expression via hypoxia-inducible factor-1α activation. *Blood* **2000**, *96*, 1558–1565. [CrossRef]
70. Shweiki, D.; Itin, A.; Soffer, D.; Keshet, E. Vascular endothelial growth factor induced by hypoxia may mediate hypoxia-initiated angiogenesis. *Nature* **1992**, *359*, 843–845. [CrossRef]
71. Lee, P.; Jiang, B.; Chin, B.Y.; Iyer, N.; Alam, J.; Semenza, G.; Choi, A. Hypoxia-inducible Factor-1 Mediates Transcriptional Activation of the Heme Oxygenase-1 Gene in Response to Hypoxia. *J. Biol. Chem.* **1997**, *272*, 5375–5381. [CrossRef]
72. Tacchini, L.; Bianchi, L.; Zazzera-A, B.; Cairo, G. Transferrin receptor induction by hypoxia. HIF-1-mediated transcriptional activation and cell-specific post-transcriptional regulation. *J. Biol. Chem.* **1999**, *274*, 24142–24146. [CrossRef]
73. Amano, H.; Maruyama, K.; Naka, M.; Tanaka, T. Target validation in hypoxia-induced vascular remodeling using transcriptome/metabolome analysis. *Pharmacogenom. J.* **2003**, *3*, 183–188. [CrossRef] [PubMed]
74. Pan, Y.; Mansfield, K.; Bertozzi, C.; Rudenko, V.; Chan, D.; Giaccia, A.; Simon, M. Multiple Factors Affecting Cellular Redox Status and Energy Metabolism Modulate Hypoxia-Inducible Factor Prolyl Hydroxylase Activity In Vivo and In Vitro. *Mol. Cell. Biol.* **2007**, *27*, 912–925. [CrossRef] [PubMed]
75. Malcangio, M.; Bartolini, A.; Ghelardini, C.; Bennardini, F.; Malmberg-Aiello, P.; Franconi, F.; Giotti, A. Effect of ICV taurine on the impairment of learning, convulsions and death caused by hypoxia. *Psychopharmacology* **1989**, *98*, 316–320. [CrossRef]
76. Xiao, K.; Wang, X.; Wang, M.; Guo, H.; Liu, W.; Guang-zhen, J. Metabolism, antioxidant and immunity in acute and chronic hypoxic stress and the improving effect of vitamin C in the channel catfish (*Ictalurus punctatus*). *Fish Physiol. Biochem.* **2023**, *50*, 183–196. [CrossRef] [PubMed]
77. Roche, D.G.; Binning, S.A.; Bosiger, Y.; Johansen, J.L.; Rummer, J.L. Finding the best estimates of metabolic rates in a coral reef fish. *J. Exp. Biol.* **2013**, *216*, 2103–2110. [CrossRef]
78. John, J.S.; Thometz, N.M.; Boerner, K.; Denum, L.; Kendall, T.L.; Richter, B.P.; Gaspard, J.C.; Williams, T.M. Metabolic trade-offs in tropical and subtropical marine mammals: Unique maintenance and locomotion costs in West Indian manatees and Hawaiian monk seals. *J. Exp. Biol.* **2021**, *224*, jeb237628. [CrossRef]
79. Killen, S.S.; Glazier, D.S.; Rezende, E.L.; Clark, T.D.; Atkinson, D.; Willener, A.S.T.; Halsey, L.G. Ecological Influences and Morphological Correlates of Resting and Maximal Metabolic Rates across Teleost Fish Species. *Am. Nat.* **2016**, *187*, 592–606. [CrossRef]
80. Zhu, X.; Xiao, Z.; Xu, Y.; Zhao, X.; Cheng, P.; Cui, N.; Cui, M.; Li, J.; Zhu, X. Differential Impacts of Soybean and Fish Oils on Hepatocyte Lipid Droplet Accumulation and Endoplasmic Reticulum Stress in Primary Rabbit Hepatocytes. *Gastroenterol. Res. Pract.* **2016**, *2016*, 9717014. [CrossRef]
81. Kriváková, P.; Cervinkova, Z.; Lotkova, H.; Kucera, O.; Rousar, T. Mitochondria and their role in cell metabolism. *Acta Med.* **2005**, *48*, 57–67.

82. Nilsson, G.E.; Dymowska, A.; Stecyk, J.A.W. New insights into the plasticity of gill structure. *Respir. Physiol. Neurobiol.* **2012**, *184*, 214–222. [CrossRef]
83. Jong, C.J.; Sandal, P.; Schaffer, S.W. The Role of Taurine in Mitochondria Health: More Than Just an Antioxidant. *Molecules* **2021**, *26*, 4913. [CrossRef] [PubMed]
84. Nishinaka, Y.; Sugiyama, S.; Yokota, M.; Saito, H.; Ozawa, T. The effects of a high dose of ascorbate on ischemia-reperfusion-induced mitochondrial dysfunction in canine hearts. *Heart Vessel.* **1992**, *7*, 18–23. [CrossRef] [PubMed]
85. Scherz-Shouval, R.; Shvets, E.; Fass, E.; Shorer, H.; Gil, L.; Elazar, Z. Reactive oxygen species are essential for autophagy and specifically regulate the activity of Atg4. *EMBO J.* **2007**, *26*, 1749–1760. [CrossRef] [PubMed]

antioxidants

MDPI

Article

Effects of Berberine on Lipid Metabolism, Antioxidant Status, and Immune Response in Liver of Tilapia (*Oreochromis niloticus*) under a High-Fat Diet Feeding

Rui Jia [1,2], Yiran Hou [1,2], Liqiang Zhang [1,2], Bing Li [1,2,*] and Jian Zhu [1,2,*]

1 Key Laboratory of Integrated Rice-Fish Farming Ecology, Ministry of Agriculture and Rural Affairs, Freshwater Fisheries Research Center, Chinese Academy of Fishery Sciences, Wuxi 214081, China; jiar@ffrc.cn (R.J.); houyr@ffrc.cn (Y.H.); zhangliqiang@ffrc.cn (L.Z.)

2 Wuxi Fisheries College, Nanjing Agricultural University, Wuxi 214081, China

* Correspondence: lib@ffrc.cn (B.L.); zhuj@ffrc.cn (J.Z.)

Abstract: Berberine, a natural alkaloid found abundantly in various medicinal plants, exhibits antioxidative, anti-inflammatory, and lipid metabolism-regulatory properties. Nonetheless, its protective effects and the molecular mechanisms underlying liver injury in fish have not been fully elucidated. The aims of this study were to investigate the antioxidative, anti-inflammatory, and lipid metabolism-regulating effects of berberine against high-fat diet (HFD)-induced liver damage and to clarify the underlying molecular mechanisms. Tilapia were fed diets containing two doses of berberine (50 and 100 mg/kg diet) alongside high fat for 60 days. The results showed that berberine treatments (50 and/or 100 mg/kg) significantly reduced elevated aminotransferases, triglycerides (TG), total cholesterol (TC), and low-density lipoprotein cholesterol (LDL-c) in the plasma. In the liver, berberine treatments significantly increased the expression of peroxisome proliferator-activated receptor α (*pparα*) and carnitine palmitoyltransferase 1 (*cpt-1*) genes, leading to a reduction in lipid accumulation. Meanwhile, berberine treatment suppressed lipid peroxidation formation and enhanced antioxidant capacity. Berberine upregulated the mRNA levels of erythroid 2-related factor 2 (*nrf2*) and its downstream genes including heme oxygenase 1 (*ho-1*) and glutathione-S-transferase (*gstα*). Additionally, berberine attenuated the inflammation by inhibiting the expression of toll-like receptor 2 (*tlr2*), myeloid differential protein-88 (*myd88*), *relb*, and inflammatory cytokines such as interleukin-1β (*il-1β*), tumor necrosis factor-α (*tnf-α*), and *il-8*. In summary, this study suggested that berberine offers protection against HFD-induced liver damage in tilapia via regulating lipid metabolism, antioxidant status, and immune response. This protective effect may be attributed to the modulation of the Nrf2, TLR2/MyD88/NF-κB, and PPARα signaling pathways.

Keywords: berberine; high-fat diet; lipid metabolism; liver damage; *Oreochromis niloticus*

check for updates

Citation: Jia, R.; Hou, Y.; Zhang, L.; Li, B.; Zhu, J. Effects of Berberine on Lipid Metabolism, Antioxidant Status, and Immune Response in Liver of Tilapia (*Oreochromis niloticus*) under a High-Fat Diet Feeding. *Antioxidants* **2024**, *13*, 548. https://doi.org/10.3390/antiox13050548

Academic Editors: Evangelos Zoidis and Erchao Li

Received: 14 April 2024
Revised: 28 April 2024
Accepted: 28 April 2024
Published: 29 April 2024

1. Introduction

The liver is crucial in fish, serving as the central organ for metabolism of substances and energy and providing vital barrier functions through detoxification and phagocytosis. It is susceptible to damage due to a variety of factors, including exposure to heavy metals, misuse of chemical medications or antibiotics, and changes in environments [1,2]. Abnormal liver function can suppress growth, disrupt normal metabolism, reduce immunity and stress tolerance, and may even lead to death. In aquaculture, fatty liver is a common metabolic dysfunction disease of fish liver [3]. There are numerous inducing factors, such as nutritional imbalances, environmental stress, and abnormalities in physiological functions, all of which can lead to excessive lipid accumulation or lipid metabolic disorders in the liver of fish, thereby causing liver lesions [4]. The mechanism underlying fatty liver injury has been extensively reported in fish, implicating lipid accumulation, oxidative stress, and inflammatory responses [5,6]. Diets rich in fat have been found to disrupt lipid metabolism, exacerbate lipid peroxidation, and impair immune function in the liver of tilapia (*Oreochromis niloticus*) [7,8]. Similarly, a high-fat diet (HFD) has been observed to cause lipid deposition, oxidative stress, and chronic inflammation in the liver of blunt snout bream (*Megalobrama amblycephala*) [9,10].

Given the adverse impact of fatty liver on aquaculture fish, researchers have been exploring preventive and therapeutic measures. Du et al., (2013) have recommended employing nutritionally balanced diets, preventing feed deterioration, enhancing aquaculture technology, and mitigating stress in fish as viable strategies [11]. Alongside, the exploration of pharmacological agents, particularly herbal extracts with hepatoprotective and antioxidant properties, has opened new avenues for treating fatty liver disease in fish. Notably, dietary *Eucommia ulmoides* leaf extract alleviated liver steatosis and improved liver function in *Ictalurus punctatus* [12]. Similarly, resveratrol has been found to regulate lipid synthesis and metabolism in the liver of *O. niloticus*, thus reducing liver damage [13]. Saikosaponin d, by acting on the AMPK/PPARα pathway, has been effective in countering hepatic steatosis induced by an HFD in hybrid grouper (*Epinephelus lanceolatusd*♂ × *Epinephelus fuscoguttatus*♀) [14]. Additionally, dietary betaine has been shown to effectively mitigate hepatic inflammation induced by an HFD in *Acanthopagrus schlegelii* [15]. These findings clearly demonstrate the substantial potential of herbal extracts in improving liver health and combating lipid accumulation in fish. Therefore, integrating these natural compounds into aquaculture practices could serve as an effective strategy for mitigating fatty liver disease, ultimately enhancing the welfare and productivity of cultured fish.

Berberine, a natural alkaloid, is prevalent in numerous medicinal plants, especially in traditional Chinese medicinal species like *Coptis chinensis*, *Berberis vulgaris*, and *Hydrastis Canadensis* [16]. Recent studies highlighted berberine's extensive potential in medicine, especially as a treatment option for diabetes, cardiovascular disease, fatty liver, and specific chronic inflammatory disorders [17]. Its capacity to regulate metabolic pathways, decrease blood glucose and cholesterol levels, alongside its antioxidant capabilities, further validate berberine's therapeutic promise [18]. In aquaculture, berberine has been investigated as a dietary supplement, demonstrating beneficial effects in various fish species [19]. Dietary berberine was reported to promote growth and decrease the mortality of *M. amblycephala* induced by *Aeromonas hydrophila* exposure [20]. Yu et al. (2020) found that berberine improved intestinal barrier function by modulating the intestinal microbiota in *M. amblycephala* [21]. Furthermore, it has been demonstrated that berberine could promote lipid metabolism and enhance antioxidant capacity in *Mylopharyngodon piceus* [22]. In studies focusing on liver health, berberine has shown its effectiveness by reducing hepatocyte apoptosis induced by an HFD in *M. amblycephala* [23] and by alleviating chronic liver injury induced by copper exposure in *Acrossocheilus fasciatus* [24]. These findings highlight the necessity for further investigations into the impact of berberine on liver functions across a broader spectrum of fish species. Conducting such extensive research is crucial to deepen our understanding of berberine's therapeutic potential and its possible application in aquaculture.

Tilapia (*O. niloticus*) is extensively cultured in regions including China, Asia, and Africa. In 2022, the production of tilapia in China reached 1,738,947 tons. In intensive tilapia aquaculture, the common practice of overfeeding or providing diets high in fats and sugars to accelerate growth frequently leads to the emergence of fatty liver disease. HFD has been confirmed to induce oxidative stress, disrupt lipid metabolism, reduce immune capacity, and damage liver tissue in tilapia [25–27]. Several extracts from traditional Chinese herbs, such as resveratrol [13], total flavanones from *Sedum sarmentosum Bunge* [28], and total flavones from *Glycyrrhiza* [29], have been found to regulate lipid metabolism, suppress oxidative stress, alleviate inflammation, and consequently ameliorate HFD-induced liver damage in tilapia. However, there is a notable absence of research on the protective effects and the underlying molecular mechanisms of berberine against fatty liver damage in tilapia. Therefore, it is interesting to investigate the protective effects of berberine using HFD-induced liver damage model in tilapia, focusing on its impacts on lipid metabolism, oxidative stress, and immune responses.

2. Materials and Methods

2.1. Tilapia, Experimental Design, and Sampling

Juvenile tilapia, weighing 52 ± 2.2 g, were obtained from the Freshwater Fish Research Center of the Chinese Academy of Fishery Sciences (Wuxi, China) and underwent a two-week acclimatization to laboratory conditions in a recirculation system, which maintained a temperature of 29 ± 2 °C, dissolved oxygen levels above 6 mg/L, and a pH range of 7.4–7.9. Prior to initiating the experiment, these fish were fed a control diet twice daily.

Post-acclimatization, the tilapia were weighed and systematically allocated into four distinct groups: a normal control group (NC), a high-fat diet group (HFD), and two groups receiving 50 mg/kg and 100 mg/kg of berberine, respectively. The NC group received a control diet consisting of 6% fat, while the HFD group was fed a high-fat diet containing 21% fat. The berberine-supplemented groups were fed diets containing either 50 mg/kg or 100 mg/kg of berberine, complemented by 21% fat. The formulation of the high-fat diet was based on methodologies validated in previous studies [30,31]. The inclusion rate of berberine in the diets was selected according to previous studies [32,33]. Each group consisted of 60 fish, tested across three replicates. The fish were fed approximately 4% of their body weight twice daily (09:00 and 16:00), over a period of 60 days.

After 60 days of feeding, the tilapia were weighed, and nine fish from each group were randomly selected for the collection of liver and blood tissues, conducted under anesthesia using 100 mg/L MS-222 (Sigma-Aldrich, Shanghai, China). The plasma was isolated from the blood via centrifugation at 5000 rpm, 4 °C, for 10 min, facilitating the analysis of blood biochemical parameters. Liver samples were immediately flash-frozen in liquid nitrogen to preserve them for subsequent assessments of enzymatic activity and gene expression.

2.2. Biochemical Parameter Analysis

Plasma biochemical parameters, including total triacylglycerol (TG), total cholesterol (TC), low-density lipoprotein cholesterol (LDL-c), high-density lipoprotein cholesterol (HDL-c), glutamate pyruvate transaminase (GPT), glutamate oxaloacetate transaminase (GOT), total protein (TP), albumin (Alb), alkaline phosphatase (AKP), and acid phosphatase (ACP), were quantified using commercial assay kits. These measurements followed the protocols provided by the Nanjing Jiancheng Bioengineering Institute (Nanjing, China).

Antioxidative parameters, including superoxide dismutase (SOD), total antioxidant capacity (T-AOC), glutathione (GSH), and malondialdehyde (MDA), were measured in liver samples utilizing commercial assay kits, following the protocols specified by the manufacturer (Nanjing Jiancheng Bioengineering Institute, Nanjing, China).

2.3. Measurement of Target Gene Expression

RNA was extracted from 100 mg of tilapia liver tissue employing RNAiso Plus (Takara, Beijing, China), chloroform, isopropanol, and ethanol. Spectrophotometric analysis measured the $OD_{260/280}$ values to assess RNA quality and concentration. The PrimeScript™ RT Reagent Kit with gDNA Eraser (Takara) was utilized to convert RNA into first-strand cDNA via a two-step reverse transcription process. This cDNA was then used as a template for quantitative real-time PCR (qPCR). The qPCR reactions utilized TB Green™ Premix EX Taq™ II (Takara) in a total volume of 25 μL, comprising 12.5 μL TB Green™ Premix EX Taq™ II, 1 μL each of forward and reverse primers, 8.5 μL of ddH2O, and 2 μL of cDNA. For normalization of gene expression, the ubiquitin-conjugating enzyme (*ucbe*) was utilized as a reference gene. The relative mRNA levels were determined using the $2^{-\Delta\Delta Cq}$ method [34]. The specific primers used for qPCR in tilapia are detailed in Table 1.

2.4. Statistical Analysis

In this study, data were processed using SPSS 24.0 for analysis and GraphPad Prism 5 software for visualization. Analyses for normal distribution and homogeneity of variance were performed on all data. Differences between groups were determined using one-way analysis of variance (ANOVA), followed by the LSD test for instances of equal variances and Tamhane's T2 test for cases of unequal variances, with significance established at $p < 0.05$. Results are expressed as mean ± standard error of the mean (mean ± SEM).

Table 1. The primer sequences used in the present study.

Gene	Primer Sequence (5′-3′)	GenBank Number/References
Nuclear factor erythroid 2-related factor 2 (*nrf2*)	F: CTGCCGTAAACGCAAGATGG R: ATCCGTTGACTGCTGAAGGG	XM_003447296.5
NAD(P)H dehydrogenase 1 (*nqo1*)	F: TGGATTTCAGGTTCTGGCTCC R: TCCTGTGGAGATGCCGAGA	XM_013273094.3
Glutathione S-transferase (*gsta*)	F: TAATGGGAGAGGGAAGATGG R: CTCTGCGATGTAATTCAGGA	NM_001279635.1
Heme oxygenase (*ho-1*)	F: CTTGCCCGTGTGGAATCACT R: AGATCACCGAGGTAGCGAGT	XM_013270165.3
Peroxisome proliferator activated receptor alpha (*pparα*)	F: CTGATAAAGCTTCGGGCTTCCA R: CGCTCACACTTATCATACTCCAGCT	NM_001290066.1
Carnitine O-palmitoyltransferase 1 (*cpt-1*)	F: TTTCCAGGCCTCCTTACCCA R: TGTACTGCTCATTGTCCAGCAGA	XM_013268638.3
Acyl-CoA oxidase 1 (*acox-1*)	F: GGTCAAAGGCAACAATCAGGAG R: GACTCTGCCAAAGGCAACCA	NM_001290199.1
Toll-like receptor 2 (*tlr2*)	F: AAAAGCATAGATGAGTTCCACATCC R: GTAAGACAAGGCATCACAAACACC	JQ809459.1
Myloid differentiation factor 88 (*myd88*)	F: CAGGTTCCTGAGGTCGACAG R: CATTTCGTGGACGAACGCAA	KJ130039.1
NF-kB subunit (*relb*)	F: TCACTGCCTCCACCTTTGCT R: ATCCTCATAGTTCCTCTTCCGTTTT	XM_005459330.4
Tumor necrosis factor-alpha (*tnf-α*)	F: AAGCCAAGGCAGCCATCCAT R: TTGACCATTCCTCCACTCCAGA	[35]
Interleukin-1 beta (*il-1β*)	F: TCAGTTCACCAGCAGGGATG R: GACAGATAGAGGTTTGTGCC	[36]
Interleukin-8 (*il-8*)	F: CTGTGAAGGCATGGGTGTGGAG R: TCGCAGTGGGAGTTGGGAAGAA	[35]
Glutamine synthase a (*gs*)	F: AGCTACCACATTCGTGCCTAC R: TACGAGGAATGCGAATGCTGG	NM_001279668.1
UDP-glucuronosyltransferase 2A2 (*ugt2a2*)	F: GGTGCTGTGTCAGGAAAGGAA R: ATCAAACTAGCCACCTTTGGCA	XM_025896953.1
NADH-cytochrome b5 reductase 2 (*cbr2*)	F: ATCGCTGGTGGAACAGGTATC R:TGTGGAGGTTTGTCCAGTGT	XM_003439423.3
Lysozyme C (*lzm*)	F: AAGGGAAGCAGCAGCAGTTGTG R: CGTCCATGCCGTTAGCCTTGAG	XM_019361339.1
Immunoglobulin (*igm*)	F: ACCGAATCGAAAAATGCGGC R: AACACAACCAGGACATTGGTTC	KJ676389.1
Complement C3 (*c3*)	F: GGTGTGGATGCACCTGAGAA R: GGGAAATCGGTACTTGGCCT	XM_013274267.3
Hepcidin (*hep*)	F: GACACAAGCGTGGCATCAAG R: GTTGAGGCAGTAACTGAGGACA	XM_019365122.2
Ubiquitin-conjugating enzyme (*ubce*)	F: CTCTCAAATCAATGCCACTTCC R: CCCTGGTGGAGGTTCCTTGT	[37]

3. Results

3.1. Changes in Hepatic Damage Parameters in Plasma

In the plasma, HFD feeding alone significantly elevated the levels of GPT and GOT after 60 days. However, these alterations were significantly mitigated by treatment with berberine at the dose of 50 mg/kg (Figure 1A,B). Similarly, the increased GOT was markedly decreased in the group treated with 100 mg/kg berberine (Figure 1B). Moreover, the AKP level was elevated in the HFD group, whereas berberine treatment failed to mitigate this increase (Figure 1E). Additionally, the levels of TP, Alb, and ACP were unchanged by either the HFD or berberine treatment (Figure 1C,D,F).

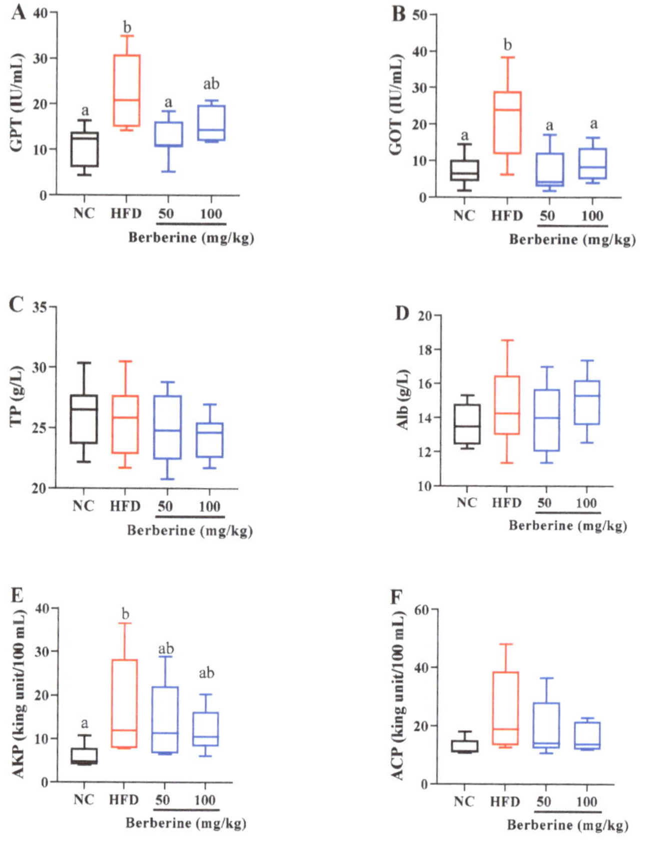

Figure 1. Changes in plasma hepatic damage parameters in tilapia fed berberine-inclusive high-fat diet. Different lowercase letters indicate significant differences between groups. (**A**) Glutamate pyruvate transaminase (GPT). (**B**) Glutamate oxaloacetate transaminase (GOT). (**C**) Total protein (TP) (**D**) Albumin (Alb). (**E**) Alkaline phosphatase (AKP). (**F**) Acid phosphatase (ACP).

3.2. Change in Lipid Metabolism in Plasma

As shown in Figure 2, the plasma parameters' analysis displayed that an HFD-alone treatment significantly raised the levels of TG, TC, LDL-c, and HDL-c after 60 days. Notably, the elevation in TG and TC levels was significantly ameliorated in the group receiving 50 mg/kg of berberine. Furthermore, berberine treatment at dosages of both 50 and 100 mg/kg markedly reduced the increases in LDL-c ($p < 0.05$).

3.3. Changes in the Expression of Genes Related to Metabolism Function

To explore berberine's role in regulating lipid metabolism, we examined the mRNA levels of fatty acid β-oxidation-related genes, including *pparα*, *acox1*, and *cpt-1* in the liver (Figure 3). The results highlighted a significant decrease in the expression of *pparα* in the group treated with an HFD in comparison to that in the NC group ($p < 0.05$; Figure 3A). However, this decrease was significantly reversed by berberine treatment with 50 and 100 mg/kg when compared to the HFD-only treatment ($p < 0.05$). The *cpt-1* mRNA level was significantly reduced in the HFD group; however, this downregulation was mitigated by treatments with 100 mg/kg of berberine ($p < 0.05$; Figure 3C). In addition, *acox1* mRNA level was notably lower in the HFD group, but the downregulation was not significantly counteracted by berberine treatments (Figure 3B).

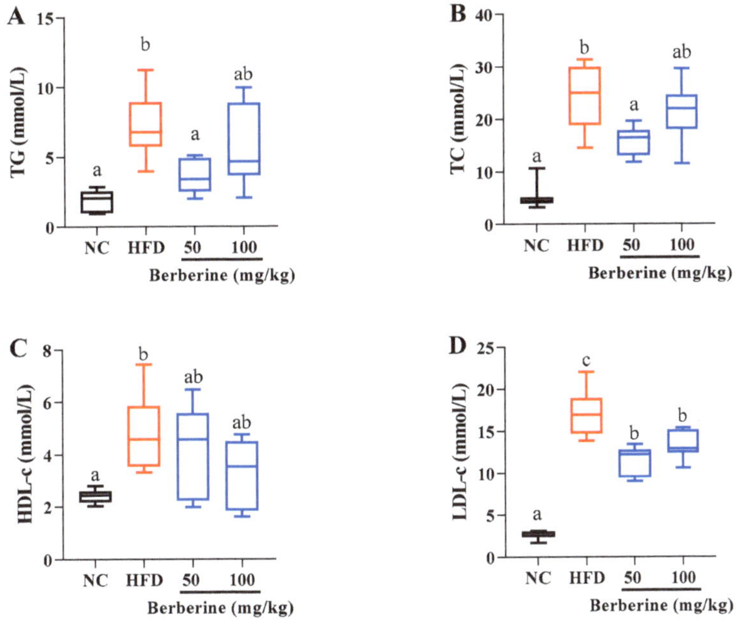

Figure 2. Changes in plasma lipid metabolism parameters in tilapia fed berberine-inclusive high-fat diet. Different lowercase letters indicate significant differences between groups. (**A**) Total triacylglycerol (TG). (**B**) Total cholesterol (TC). (**C**) Low-density lipoprotein cholesterol (LDL-c). (**D**) High-density lipoprotein cholesterol (HDL-c).

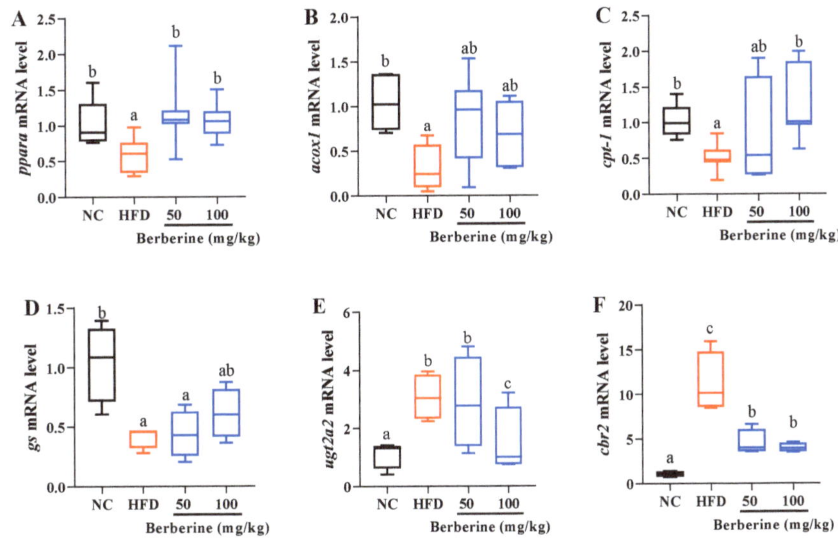

Figure 3. Relative expression of genes related to metabolism function in liver of tilapia fed berberine-inclusive high-fat diet. Different lowercase letters indicate significant differences between groups. (**A**) Peroxisome proliferator activated receptor alpha (*pparα*). (**B**) Acyl-CoA oxidase 1 (*acox1*). (**C**) Carnitine O-palmitoyltransferase 1 (*cpt-1*). (**D**) Glutamine synthase a (*gs*). (**E**) UDP-glucuronosyltransferase 2A2 (*ugt2a2*). (**F**) NADH-cytochrome b5 reductase 2 (*cbr2*).

To investigate the relationship between other metabolism function and the hepatoprotective effects of berberine against HFD-induced liver damage, we measured the mRNA levels of *gs*, *ugt2a2*, and *cbr2*. Compared with the NC group, HFD treatment led to an increase in the mRNA levels of *ugt2a2* and *cbr2*, while causing a decrease in *gs* mRNA level ($p < 0.05$; Figure 3D–F). The mRNA levels of *ugt2a2* and *cbr2* were significantly reduced following berberine treatment with 100 mg/kg compared with the HFD group ($p < 0.05$; Figure 3D,E). However, berberine treatment did not change the expression of *gs* compared with the HFD group ($p > 0.05$; Figure 3F).

3.4. Changes in Antioxidation Status in Liver

In the liver, there was a marked reduction in the levels of SOD, GSH, and T-AOC, accompanied by an increase in the MDA level, in tilapia treated with an HFD alone, compared with those in the NC group ($p < 0.05$; Figure 4). The decrease in SOD level was markedly improved by berberine treatment at both 50 and 100 mg/kg ($p < 0.05$). Similarly, the reduction in GSH level was significantly ameliorated with berberine treatment at a dose of 100 mg/kg ($p < 0.05$). Furthermore, the elevation in the MDA level was substantially reduced under berberine treatment at both 50 and 100 mg/kg doses ($p < 0.05$).

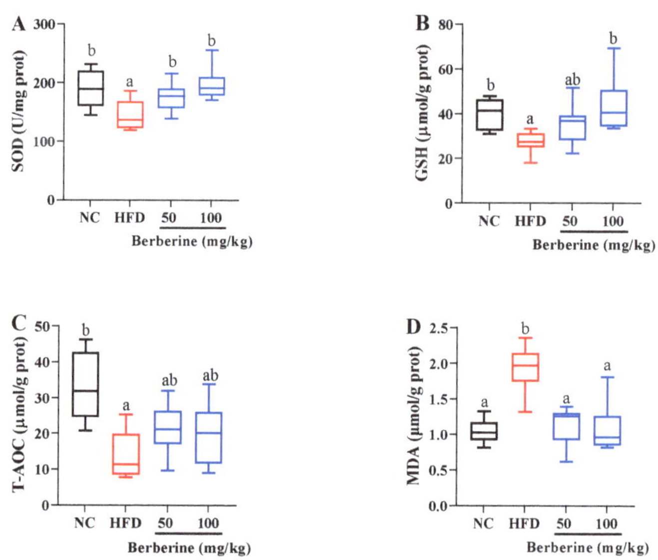

Figure 4. Changes in antioxidant status in liver of tilapia fed berberine-inclusive high-fat diet. Different lowercase letters indicate significant differences between groups. (**A**) Superoxide dismutase (SOD). (**B**) Glutathione (GSH). (**C**) Total antioxidant capacity (T-AOC). (**D**) Malondialdehyde (MDA).

3.5. Changes in the Expression of Genes Related to Antioxidant Status

After HFD feeding, the mRNA levels of *nrf2* and *gstα* were notably downregulated when compared with those in the NC group ($p < 0.05$; Figure 5A,C), while this downregulation was significantly alleviated in the groups treated with berberine at doses of 50 and 100 mg/kg ($p < 0.05$; Figure 5A,C). Similarly, the mRNA level of *ho-1* was markedly reduced after 60 days of HFD feeding, whereas this reduction was prevented by treatment with 100 mg/kg of berberine ($p < 0.05$; Figure 5B). In addition, the expression of *nqo1* was significantly reduced after HFD feeding, whereas berberine treatments did not influence the *nqo1* expression ($p > 0.05$; Figure 5D).

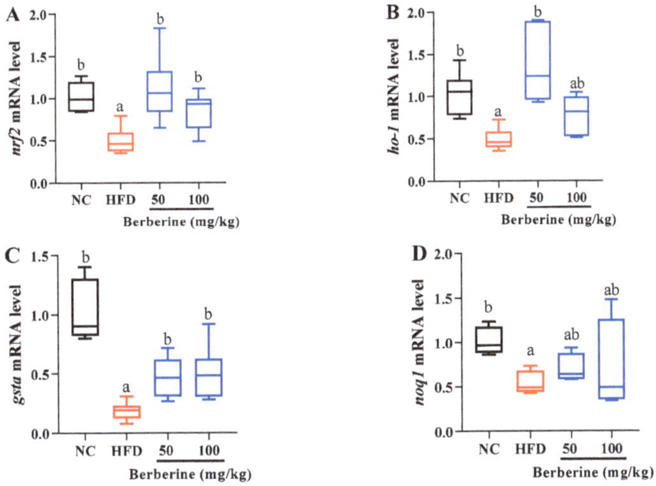

Figure 5. Relative expression of genes related to antioxidant status in liver of tilapia fed berberine-inclusive high-fat diet. Different lowercase letters indicate significant differences between groups. (**A**) Nuclear factor erythroid 2-related factor 2 (*nrf2*) (**B**) Heme oxygenase (*ho-1*). (**C**) Glutathione S-transferase (*gsta*). (**D**) NAD(P)H dehydrogenase 1 (*nqo1*).

3.6. Changes in the Expression of Genes Related to Inflammatory Response

The expression of genes associated with the inflammatory response in the liver of tilapia are depicted in Figure 6. The mRNA levels of *tlr2*, *myd88*, *relb*, *tnf-α*, *il-1β*, and *il-8* were significantly elevated in the HFD group compared to those in the NC group ($p < 0.05$). These genes exhibited a decreasing trend following treatment with berberine, showing significant differences at a dosage of 100 mg/kg of berberine ($p < 0.05$). Additionally, *tnf-α* expression was notably suppressed in the group treated with 50 mg/kg of berberine compared with that in the HFD group ($p < 0.05$).

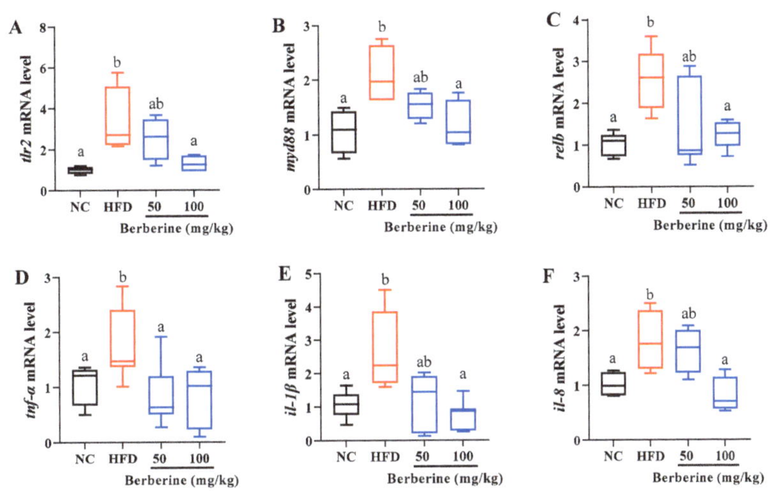

Figure 6. Relative expression of genes related to inflammatory response in liver of tilapia fed berberine-inclusive high-fat diet. Different lowercase letters indicate significant differences between groups. (**A**) Toll-like receptor 2 (*tlr2*). (**B**) Myloid differentiation factor 88 (*myd88*). (**C**) NF-kB subunit (*relb*). (**D**) Tumor necrosis factor-alpha (*tnf-α*). (**E**) Interleukin-1 beta (*il-1β*). (**F**) Interleukin-8 (*il-8*).

3.7. Changes in the Expression of Genes Related to Immune Function

The qPCR analysis revealed that in the liver of tilapia subjected solely to an HFD, there was a significant reduction in the transcript levels of *c3*, *lzm*, *igm*, and *hep*. However, this reduction was ameliorated under treatments with berberine at concentrations of 50 and 100 mg/kg ($p < 0.05$; Figure 7).

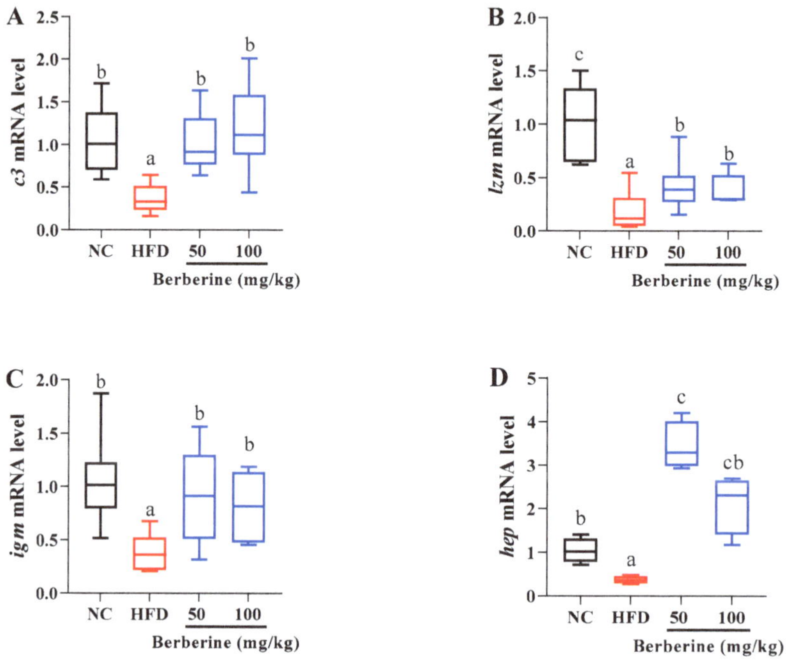

Figure 7. Relative expression of genes related to immune function in liver of tilapia fed berberine-inclusive high-fat diet. Different lowercase letters indicate significant differences between groups. (**A**) Complement C3 (*c3*). (**B**) Lysozyme C (*lzm*). (**C**) Immunoglobulin (*igm*). (**D**) Hepcidin (*hep*).

4. Discussion

Berberine has been identified as a therapeutic agent for liver diseases, including both chronic and acute hepatic damage [16]. Earlier studies have highlighted its hepatoprotective properties, demonstrating that berberine mitigates CCl_4-induced acute hepatotoxicity in rats [38]. An in vitro study has further confirmed berberine's capability to protect hepatocytes from hypoxia/reoxygenation (H/R)-induced damage [39]. In fish, berberine has been shown to protect against chronic copper-induced liver injury in *A. fasciatus*, significantly diminishing the serum levels of GPT and GOT [24]. Consistent with these findings, our study revealed a marked elevation in the activities of GPT and GOT in tilapia subjected to an HFD alone, indicative of severe liver injury. Conversely, this injurious trend was notably reversed following berberine treatment with 50 and/or 100 mg/kg, where the levels of these markers were nearly normalized, indicating berberine's efficacy in counteracting HFD-induced liver damage in fish.

4.1. Effects of Berberine on the Metabolism Function

TC and TG are principal components of fish blood lipids, serving as critical markers for pathological diagnosis. They reflect the lipid metabolism of the liver and the overall health status of the fish. The elevations in plasma TC, TG, LDL-c, and HDL-c following an HFD indicate a significant disruption in lipid metabolism [40]. Specifically, the increase in LDL-c and TC are concerning, pointing to an elevated risk of fatty liver disease development, which can further compromise fish health through the enhancement of oxidative stress and inflammatory responses [41]. Although the increase in HDL-c is often deemed positive in health due to its role in facilitating the reverse transport of cholesterol, its simultaneous elevation with LDL-c and TG after an HFD feeding may suggest an incapacity in the lipid regulatory system in fish [42]. In this study, the ability of berberine (50 and/or 100 mg/kg) to significantly reduce LDL-c, TC, and TG levels in tilapia highlighted its comprehensive effectiveness in ameliorating disruptions in lipid metabolism. Similar results were also reported in other fish species. For instance, dietary berberine decreased serum levels of LDL-c, TC, and TG in *Pelteobagrus fulvidraco* [43]; similarly, it resulted in reductions in serum levels of LDL-c, HDL-c, TC, and TG in *Ctenopharyngodon idella* [44]. These findings suggest berberine's potential beneficial impact in ameliorating lipid imbalances, highlighting its versatile role in aquaculture nutrition. Nonetheless, in *M. amblycephala*, dietary berberine failed to suppress the elevation of TG and TC induced by an HFD [23]. The observed variability in response across different species underscores the imperative for additional research to elucidate the underlying mechanisms of berberine's effects.

In the liver, lipid accumulation is closely associated with fatty acid β-oxidation. The activity of rate-limiting enzymes for β-oxidation, such as CPT-1 and AOX-1, plays a significant role in this process. PPARα, in particular, is recognized as a central regulator of lipid metabolism. It activates specific target genes (e.g., *cpt-1 and acox-1*) to facilitate fatty acid transport, oxidation, and lipogenesis [45]. Activation of PPARα not only ameliorates the metabolic syndrome but also exhibits anti-inflammatory effects [46]. In HFD-fed fish, a significant consequence was the downregulation of *cpt-1*, *acox-1*, and *pparα*, leading to reduced β-oxidation [47]. This diminished β-oxidation signifies a dysregulation in hepatic fatty acids, which in turn accelerates lipid accumulation and induces liver injury. It has been reported that berberine can enhance fatty acid β-oxidation, leading to a reduction in lipid accumulation. For instance, in *M. piceus*, the inclusion of dietary berberine mitigated the suppression of *cpt-1*, *acox-1*, and *pparα* genes provoked by an HFD feeding [32]. Likewise, in *M. amblycephala*, there was a notable reduction in the expression of *cpt-1*, *pparα*, and *acox* genes after an HFD feeding; however, supplementation with berberine effectively reversed this downregulation [32,48]. In line with these findings, our study revealed that the expression levels of *cpt-1* and *pparα* genes were significantly lower in fish fed an HFD, but this downregulation was counteracted by 100 mg/kg of berberine supplementation. Additionally, the lower *pparα* expression was also ameliorated with 50 mg/kg of berberine. This suggests that berberine may alleviate liver lipid accumulation in tilapia by modulating β-oxidation through the PPARα pathway.

Fatty liver disease is associated not only with lipid metabolism but also with other metabolic functions. Glutamine synthase (GS) is pivotal in nitrogen metabolism, regulates the homeostasis of blood ammonium ions and glutamine, and contributes to the modulation of liver functions [49]. In the model of CCl_4-induced liver damage, GS activity was observed to decrease [50]; similarly, in the case of HFD-induced metabolic disorders, the expression of hepatic GS was found to be downregulated [51]. This study also noted a reduction in *gs* expression following HFD feeding, yet this downregulation was not significantly reversed by berberine treatment, indicating that berberine may not effectively mitigate the abnormalities in glutamine metabolism caused by an HFD feeding.

UDP-glucuronosyltransferases (UGTs) are essential phase II drug-metabolizing enzymes that facilitate the detoxification of various substances, whose dysregulation can lead to metabolic disorders and improper management of xenobiotics. In rats subjected to chronic CCl_4-induced liver fibrosis, an alteration in the mRNA level of *ugt* isoforms was observed, with *ugt1a1*, *1a6*, *2b1*, and *2b2* mRNA levels being elevated, while *ugt2b3*, *2b6*, and *2b12* mRNA levels were found to be reduced [52]. Similarly, the progression of human non-alcoholic fatty liver disease (NAFLD) is accompanied by a selective upregulation of certain UGT isoforms, notably UGT1A1, UGT1A3, UGT2B10, and UGT1A9 [53]. In our study, the mRNA level of *ugt2a2* was upregulated in the HFD group; however, treatment with 100 mg/kg of berberine mitigated this upregulation. This finding suggests that berberine possesses the potential to ameliorate UGT dysregulation and consequently alleviate metabolic disturbances.

NADH-cytochrome b5 reductase (CBR), an integral membrane enzyme, plays a pivotal role in several biochemical processes linked to liver health and disease, including fatty acid metabolism, drug processing, and antioxidant function. It has reported that the activity of CBR showed a significant increase in rats with liver injury induced by CCl_4 and HFD, while this augmentation is mitigated by hepatoprotective agents such as d-limonene, silymarin, and trans-anethole [54,55]. Consistent with these findings, an increase in the mRNA level of the *cbr2* was observed in tilapia with HFD-induced liver injury, which was, however, alleviated by 50 and 100 mg/kg of berberine treatments. This suggests that berberine may have a regulatory effect on detoxification processes, contributing to the mitigation of liver damage.

4.2. Effect of Berberine on Antioxidant Status

Berberine is known for its antioxidant activity and its capacity to scavenge free radicals. It mitigates oxidative stress, as demonstrated by the modulation of antioxidant enzyme activities and the levels of oxidative stress indicators, including GSH and MDA [56]. Berberine can enhance the levels of T-AOC, SOD, and catalase (CAT), while suppressing the formation of MDA in the liver of *Micropterus salmoides* [57]. Berberine also prevents the reduction of antioxidant activity and formation of lipid peroxidation induced by acetaminophen in the liver of *Cyprinus carpio* [58]. Similarly, our findings demonstrated the remarkable antioxidative properties of berberine (50 and/or 100 mg/kg); it effectively mitigated oxidative damage induced by HFD feeding, maintained the normal levels of SOD (50 and 100 mg/kg) and GSH (100 mg/kg), and inhibited the formation of MDA (50 and 100 mg/kg) in liver tissues. This aligns with previously published findings that demonstrated the concomitant administration of berberine in HFD-fed *M. amblycephala*, *I. punctatus*, and rats significantly mitigated the reduction of antioxidant components (e.g., SOD and GSH) and the elevation of lipid peroxidation [23,59,60].

In liver injury induced by oxidative stress, a variety of critical molecules and pathways have been identified. Among these, the Nrf2 pathway is particularly notable for its extensive research focus related to oxidative stress. Nrf2, a transcription factor, plays a crucial role in protecting against oxidative damage via enhancing the expression of various cellular antioxidant defense proteins, including HO-1, GSTα, and NQO1 [61]. Studies have consistently demonstrated the therapeutic and biological effects of berberine through the activation of the Nrf2 pathway [62]. Specifically, berberine has been shown to mitigate liver injury induced by methotrexate, HFD, and CCl_4 through the activation of the Nrf2 pathway in rats [60,63,64]. In fish, it was also found that dietary berberine enhanced the antioxidant capacity by upregulating *nrf2* expression [57,65]. In this study, the *nrf2* expression was markedly downregulated in HFD-induced liver injury, but this downregulation was reversed to normal levels following treatments with 50 and 100 mg/kg of berberine. Correspondingly, the mRNA levels of its downstream antioxidant genes *ho-1* and *gstα* were significantly downregulated in the HFD-treated group. However, berberine treatment at doses of 50 and 100 mg/kg notably increased *gstα* expression, and the 50 mg/kg of berberine markedly elevated *ho-1* expression. These findings suggest that berberine's antioxidative effects may be attributed not only to its ROS scavenging activity but also to its enhancement of detoxifying/antioxidant enzyme expression through the activation of the Nrf2 signaling pathway in the liver of tilapia.

4.3. Effect of Berberine on Inflammatory and Immune Response

TLR2, a pattern recognition receptor of the innate immune system, is essential in connecting inflammation and liver injury, especially through its regulation of inflammatory responses under HFD-induced liver damage [66]. It has been reported that TLR2 activation triggers MyD88, subsequently initiating the NF-κB pathway to modulate the inflammatory response [67]. Mice lacking TLR2 gene exhibited notable reductions in inflammation, steatosis, and the development of non-alcoholic steatohepatitis [68,69]. Meanwhile,

TLR2/NF-κB pathway activation was observed in mice subjected to an HFD, contributing to vascular inflammation [70]. Our study also found similar results, as we noted significant upregulation of *tlr2*, *myd88*, and NF-κB (*relb*) mRNA levels in HFD-induced liver injury. Importantly, treatment with 100 mg/kg berberine significantly attenuated these upregulations, suggesting that berberine may mitigate inflammation through the modulation of the TLR2/Myd88/NF-kB pathway. This finding corroborates previous research results, demonstrating that the anti-inflammatory effects of berberine are associated with the TLR (including TLR2) pathway [71].

Activation of the TLR2/MyD88/NF-κB signaling pathway initiates a cascade of genes associated with inflammatory cytokines, such as IL-1β, IL-8, and TNF-α, thereby exacerbating hepatic injury [72]. Indeed, after HFD feeding, upregulation of pro-inflammatory cytokines in the livers of both mice and fish was observed [59,73]. Similarly, in this study, *il-1β*, *il-8*, and *tnf-α* exhibited high expression levels in the liver of tilapia subjected to HFD treatment. In contrast, these elevated expressions of *il-1β*, *il-8*, and *tnf-α* were significantly reduced following treatment with 100 mg/kg of berberine. These findings suggest that the protective effect of berberine on the liver may be attributed to its anti-inflammatory properties. Supporting this notion, Wang et al., (2022) found that berberine mitigated the highly expressed *il-1β*, *tnf-α*, and *nf-κB* in *A. schlegelii* treated with an HFD [33]. Moreover, berberine diminished inflammation and lowered the levels of TNF-α, IL-6, and IL-1β in rats with non-alcoholic fatty liver disease by regulating the TLR4/MyD88/NF-κB signaling pathway [74].

In addition to inflammation, a suppression of immunity emerged as another consequence of liver injury induced by an HFD [59]. In fish, several innate immune parameters, such as LZM, HEP, and C3, play a key role in protecting against pathogens and contributing to the overall defense mechanisms [75,76]. The reduced LZM activity was observed in the plasma of *C. carpio* fed an HFD [77]. Similarly, reductions in LZM and C3 were noted in the plasma of blunt snout bream fed an HFD [78]. Furthermore, an HFD led to the downregulation of these proteins in the liver of mice [79,80]. In addition, IgM, a key player in the primary immune response, was also found to be decreased in the plasma of *C. carpio* and *O. niloticus* after HFD feeding [77,81]. Our data aligned with those in these studies, demonstrating that HFD treatment led to the suppression of mRNA levels of *lzm*, *hep*, *c3*, and *igm* in the liver of tilapia. The immunostimulatory effects of berberine have been well-documented, evidenced by a significant increase in LZM activity in the plasma of *O. niloticus* [82], alongside elevated levels of IgM and C3 in the intestine of *A. schlegelii* [65]. In line with these findings, our data demonstrated that berberine treatments (50 and 100 mg/kg) markedly reversed the HFD-induced downregulation of *lzm*, *hep*, *c3*, and *igm*, suggesting that berberine exerted a positive effect on the immune function in the liver of tilapia.

5. Conclusions

In summary, our findings indicated that berberine conferred hepatoprotective effects in tilapia by enhancing antioxidative and immune capacity, reducing inflammation, and improving lipid metabolism (Figure 8). The beneficial effects of berberine may be attributed to the enhancement of Nrf2 and PPARα signaling pathways, along with the inhibition of the TLR2/MyD88/NF-κB signaling pathway. Specifically, the upregulation of the Nrf2 pathway triggers the production of phase II detoxifying/antioxidant enzymes, such as HO-1 and GSTa, while the downregulation of the TLR2/MyD88/NF-κB pathway results in a decreased production of pro-inflammatory cytokines, including IL-1β, IL-8, and TNF-α. Additionally, increased PPARα is believed to enhance fatty acid β-oxidation, which alleviates liver lipid deposition.

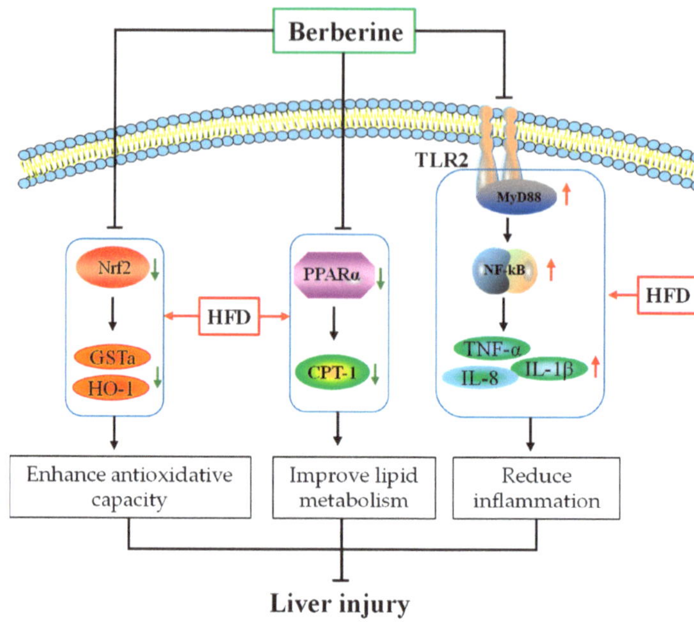

Figure 8. Possible mechanisms of berberine in ameliorating liver injury induced by HFD in tilapia. Red arrows indicate stimulatory modification, green arrows indicate inhibitory modification.

Author Contributions: Conceptualization, R.J. and J.Z.; methodology, Y.H.; software, R.J.; validation, R.J., Y.H., and L.Z.; formal analysis, B.L.; investigation, R.J.; resources, B.L.; data curation, L.Z.; writing—original draft preparation, R.J.; writing—review and editing, J.Z.; visualization, Y.H.; supervision, J.Z.; project administration, B.L.; funding acquisition, B.L. All authors have read and agreed to the published version of the manuscript.

Funding: This research was funded by Wuxi modern industry development fund project (K20221053), Central Public-Interest Scientific Institution Basal Research Fund, CAFS (2023TD64), Young Science-Technology Talents Support Project of Jiangsu Association Science and Technology (TJ-2021-076), the National Key R&D Program of China (2019YFD0900305).

Institutional Review Board Statement: All animals in this study were approved by Freshwater Fisheries Research Center (20 April 2023), and all procedures were performed according to Jiangsu Laboratory's Animal Management Guidelines (014000319/2008-00079).

Informed Consent Statement: Not applicable.

Data Availability Statement: All data are contained within the main manuscript.

Conflicts of Interest: The authors declare no conflicts of interest.

References

1. de Lima-Faria, J.M.; da Silva, V.C.; Chen, L.C.; Martinez, D.S.T.; de Sabóia-Morais, S.M.T. Co-exposure of iron oxide nanoparticles with glyphosate herbicides in Poecilia reticulata: Fish liver damages is reversible during iron accumulation and elimination period. *Chemosphere* **2023**, *328*, 138590. [CrossRef] [PubMed]
2. Wolf, J.C.; Wolfe, M.J. A brief overview of nonneoplastic hepatic toxicity in fish. *Toxicol. Pathol.* **2005**, *33*, 75–85. [CrossRef] [PubMed]
3. Wang, T.; Wei, X.; Chen, T.; Wang, W.; Xia, X.; Miao, J.; Yin, S. Studies of the mechanism of fatty liver formation in Takifugu fasciatus following copper exposure. *Ecotoxicol. Environ. Saf.* **2019**, *181*, 353–361. [CrossRef] [PubMed]
4. Wang, X.; Li, Y.; Hou, C.; Gao, Y.; Wang, Y. Physiological and molecular changes in large yellow croaker (P seudosciaena crocea R.) with high-fat diet-induced fatty liver disease. *Aquac. Res.* **2015**, *46*, 272–282. [CrossRef]

5. Schlegel, A. Studying non-alcoholic fatty liver disease with zebrafish: A confluence of optics, genetics, and physiology. *Cell. Mol. Life Sci.* **2012**, *69*, 3953–3961. [CrossRef] [PubMed]
6. Asaoka, Y.; Terai, S.; Sakaida, I.; Nishina, H. The expanding role of fish models in understandingnon-alcoholic fatty liver disease. *Dis. Models Mech.* **2013**, *6*, 905–914.
7. Tao, Y.-F.; Qiang, J.; Bao, J.-W.; Chen, D.-J.; Yin, G.-J.; Xu, P.; Zhu, H.-J. Changes in Physiological Parameters, Lipid Metabolism, and Expression of MicroRNAs in Genetically Improved Farmed Tilapia (*Oreochromis niloticus*) With Fatty Liver Induced by a High-Fat Diet. *Front. Physiol.* **2018**, *9*, 1521. [CrossRef]
8. Qiang, J.; He, J.; Yang, H.; Sun, Y.-L.; Tao, Y.-F.; Xu, P.; Zhu, Z.-X. Dietary lipid requirements of larval genetically improved farmed tilapia, *Oreochromis niloticus* (L.), and effects on growth performance, expression of digestive enzyme genes, and immune response. *Aquac. Res.* **2017**, *48*, 2827–2840. [CrossRef]
9. Dai, Y.-J.; Cao, X.-F.; Zhang, D.-D.; Li, X.-F.; Liu, W.-B.; Jiang, G.-Z. Chronic inflammation is a key to inducing liver injury in blunt snout bream (*Megalobrama amblycephala*) fed with high-fat diet. *Dev. Comp. Immunol.* **2019**, *97*, 28–37. [CrossRef] [PubMed]
10. Cao, X.-F.; Dai, Y.-J.; Liu, M.-Y.; Yuan, X.-Y.; Wang, C.-C.; Huang, Y.-Y.; Liu, W.-B.; Jiang, G.-Z. High-fat diet induces aberrant hepatic lipid secretion in blunt snout bream by activating endoplasmic reticulum stress-associated IRE1/XBP1 pathway. *Biochim. Biophys. Acta BBA Mol. Cell Biol. Lipids* **2019**, *1864*, 213–223. [CrossRef] [PubMed]
11. Du, Z. Causes of fatty liver in farmed fish: A review and new perspectives. *J. Fish. China* **2014**, *38*, 1628–1638.
12. Zhang, F.-L.; Hao, Q.; Zhang, Q.-S.; Lv, H.-Y.; Yang, Y.-L.; Chao, R.; Zhang, Z.; Zhou, Z.-G. Influences of dietary Eucommia ulmoides leaf extract on the hepatic lipid metabolism, inflammation response, intestinal antioxidant capacity, intestinal microbiota, and disease resistance of the channel catfish (*Ictalurus punctatus*). *Fish Shellfish Immunol.* **2022**, *123*, 75–84. [CrossRef] [PubMed]
13. Zheng, Y.; Shi, Y.; Yang, X.; Gao, J.; Nie, Z.; Xu, G. Effects of resveratrol on lipid metabolism in liver of red tilapia Oreochromis niloticus. *Comp. Biochem. Physiol. Part C Toxicol. Pharmacol.* **2022**, *261*, 109408. [CrossRef]
14. Zou, C.Y.; Du, L.K.; Wu, J.H.; Gan, S.Y.; Li, Q.Q.; Babu, V.S.; Wu, Y.X.; Lin, L. Saikosaponin d alleviates high-fat-diet induced hepatic steatosis in hybrid grouper (*Epinephelus lanceolatusd*♂ × *Epinephelus fuscoguttatus*♀) by targeting AMPK/PPARα pathway. *Aquaculture* **2022**, *553*, 738088. [CrossRef]
15. Jin, M.; Shen, Y.D.; Pan, T.T.; Zhu, T.T.; Li, X.J.; Xu, F.M.; Betancor, M.B.; Jiao, L.F.; Tocher, D.R.; Zhou, Q.C. Dietary Betaine Mitigates Hepatic Steatosis and Inflammation Induced by a High-Fat-Diet by Modulating the Sirt1/Srebp-1/Pparalpha Pathway in Juvenile Black Seabream (*Acanthopagrus schlegelii*). *Front. Immunol.* **2021**, *12*, 694720. [CrossRef] [PubMed]
16. Zhou, M.; Deng, Y.; Liu, M.; Liao, L.; Dai, X.; Guo, C.; Zhao, X.; He, L.; Peng, C.; Li, Y. The pharmacological activity of berberine, a review for liver protection. *Eur. J. Pharmacol.* **2021**, *890*, 173655. [CrossRef] [PubMed]
17. Kavyani, Z.; Shahhosseini, E.; Moridpour, A.H.; Falahatzadeh, M.; Vajdi, M.; Musazadeh, V.; Askari, G. The effect of berberine supplementation on lipid profile and obesity indices: An umbrella review of meta-analysis. *PharmaNutrition* **2023**, *26*, 100364. [CrossRef]
18. Gaba, S.; Saini, A.; Singh, G.; Monga, V. An insight into the medicinal attributes of berberine derivatives: A review. *Bioorg. Med. Chem.* **2021**, *38*, 116143. [CrossRef] [PubMed]
19. Wang, L.; Sagada, G.; Wang, C.; Gao, C.; Wang, B.; Shao, Q.; Yan, Y. Berberine in fish nutrition: Impact on hepatoenteric health, antioxidative and immune status. *Front. Mar. Sci.* **2022**, *9*, 967748. [CrossRef]
20. Xu, W.-N.; Chen, D.-H.; Chen, Q.-Q.; Liu, W.-B. Growth performance, innate immune responses and disease resistance of fingerling blunt snout bream, Megalobrama amblycephala adapted to different berberine-dietary feeding modes. *Fish Shellfish Immunol.* **2017**, *68*, 458–465. [CrossRef]
21. Yu, C.B.; Zhang, J.; Qin, Q.; Liu, J.; Xu, J.X.; Xu, W.N. Berberine improved intestinal barrier function by modulating the intestinal microbiota in blunt snout bream (*Megalobrama amblycephala*) under dietary high-fat and high-carbohydrate stress. *Fish Shellfish Immunol.* **2020**, *102*, 336–349. [CrossRef]
22. Ming, J.H.; Wang, T.; Wang, T.H.; Ye, J.Y.; Zhang, Y.X.; Yang, X.; Shao, X.P.; Ding, Z.Y. Effects of dietary berberine on growth performance, lipid metabolism, antioxidant capacity and lipometabolism-related genes expression of AMPK signaling pathway in juvenile black carp (*Mylopharyngodon piceus*) fed high-fat diets. *Fish Physiol. Biochem.* **2023**, *49*, 769–786. [CrossRef] [PubMed]
23. Lu, K.-L.; Wang, L.-N.; Zhang, D.-D.; Liu, W.-B.; Xu, W.-N. Berberine attenuates oxidative stress and hepatocytes apoptosis via protecting mitochondria in blunt snout bream *Megalobrama amblycephala* fed high-fat diets. *Fish Physiol. Biochem.* **2017**, *43*, 65–76. [CrossRef]
24. Wang, C.; Wang, L.; Yang, L.; Gao, C.; Wang, B.; Shu, Y.; Wang, H.; Yan, Y. Protective effects of berberine in chronic copper-induced liver and gill injury in freshwater grouper (*Acrossocheilus fasciatus*). *Ecotoxicol. Environ. Saf.* **2023**, *267*, 115672. [CrossRef] [PubMed]
25. Xu, F.; Xu, C.; Xiao, S.; Lu, M.; Limbu, S.M.; Wang, X.; Du, Z.; Qin, J.G.; Chen, L. Effects of α-lipoic acid on growth performance, body composition, antioxidant profile and lipid metabolism of the GIFT tilapia (Oreochromis niloticus) fed high-fat diets. *Aquac. Nutr.* **2019**, *25*, 585–596. [CrossRef]
26. Lin, J.-J.; Liu, Y.-C.; Chang, C.-J.; Pan, M.-H.; Lee, M.-F.; Pan, B.S. Hepatoprotective mechanism of freshwater clam extract alleviates non-alcoholic fatty liver disease: Elucidated in vitro and in vivo models. *Food Funct.* **2018**, *9*, 6315–6325. [CrossRef] [PubMed]

27. Qiang, J.; Tao, Y.F.; Bao, J.W.; Chen, D.J.; Li, H.X.; He, J.; Xu, P. High Fat Diet-Induced miR-122 Regulates Lipid Metabolism and Fat Deposition in Genetically Improved Farmed Tilapia (GIFT, *Oreochromis niloticus*) Liver. *Front. Physiol.* **2018**, *9*, 1422. [CrossRef] [PubMed]
28. Yu, K.; Huang, K.; Tang, Z.; Huang, X.; Sun, L.; Pang, L.; Mo, C. Metabolism and antioxidation regulation of total flavanones from *Sedum sarmentosum* Bunge against high-fat diet-induced fatty liver disease in Nile tilapia (*Oreochromis niloticus*). *Fish Physiol. Biochem.* **2021**, *47*, 1149–1164. [CrossRef] [PubMed]
29. Du, J.; Jia, R.; Cao, L.; Gu, Z.; He, Q.; Xu, P.; Yin, G.; Ma, Y. Regulatory effects of *Glycyrrhiza* total flavones on fatty liver injury induced by a high-fat diet in tilapia (*Oreochromis niloticus*) via the Nrf2 and TLR signaling pathways. *Aquac. Int.* **2022**, *30*, 1527–1548. [CrossRef]
30. Sheng, C.H.; Jing, J.L.; Lee, M.F.; Liu, Y.C.; Pan, B.S. Freshwater clam extracts alleviate dyslipidaemia of tilapia fed a high-fat diet as an animal model. *J. Funct. Foods* **2016**, *25*, 559–567.
31. He, A.Y.; Ning, L.J.; Chen, L.Q.; Chen, Y.L.; Xing, Q.; Li, J.M.; Qiao, F.; Li, D.L.; Zhang, M.L.; Du, Z.Y. Systemic adaptation of lipid metabolism in response to low- and high-fat diet in Nile tilapia (*Oreochromis niloticus*). *Physiol. Rep.* **2015**, *3*, e12485. [CrossRef]
32. Zhou, W.H.; Rahimnejad, S.; Lu, K.L.; Wang, L.N.; Liu, W.B. Effects of berberine on growth, liver histology, and expression of lipid-related genes in blunt snout bream (*Megalobrama amblycephala*) fed high-fat diets. *Fish Physiol. Biochem.* **2019**, *45*, 83–91. [CrossRef] [PubMed]
33. Wang, L.; Sagada, G.; Xu, B.; Zhang, J.; Shao, Q. Influence of dietary berberine on liver immune response and intestinal health of black sea bream (*Acanthopagrus schlegelii*) fed with normal and high-lipid diets. *Aquac. Nutr.* **2022**, *2022*, 6285266. [CrossRef]
34. Livak, K.; Schmittgen, T. Analysis of relative gene expression data using real-time quantitative PCR and the $2^{-\Delta\Delta CT}$ Method. *Methods-Companion Methods Enzymol.* **2001**, *25*, 402–408. [CrossRef] [PubMed]
35. Limbu, S.M.; Zhou, L.; Sun, S.X.; Zhang, M.L.; Du, Z.Y. Chronic exposure to low environmental concentrations and legal aquaculture doses of antibiotics cause systemic adverse effects in Nile tilapia and provoke differential human health risk. *Environ. Int.* **2018**, *115*, 205–219. [CrossRef] [PubMed]
36. Ken, C.F.; Chen, C.N.; Ting, C.H.; Pan, C.Y.; Chen, J.Y. Transcriptome analysis of hybrid tilapia (*Oreochromis* spp.) with Streptococcus agalactiae infection identifies Toll-like receptor pathway-mediated induction of NADPH oxidase complex and piscidins as primary immune-related responses. *Fish Shellfish Immunol.* **2017**, *70*, 106–120. [CrossRef] [PubMed]
37. Chang, G.Y.; Xian, L.W.; Tian, J.; Wei, L.; Fan, W.; Ming, J.; Hua, W. Evaluation of reference genes for quantitative real-time RT-PCR analysis of gene expression in Nile tilapia (*Oreochromis niloticus*). *Gene* **2013**, *527*, 183–192.
38. Feng, Y.; Siu, K.-Y.; Ye, X.; Wang, N.; Yuen, M.-F.; Leung, C.-H.; Tong, Y.; Kobayashi, S. Hepatoprotective effects of berberine on carbon tetrachloride-induced acute hepatotoxicity in rats. *Chin. Med.* **2010**, *5*, 33. [CrossRef]
39. Zhu, Y.; Li, J.; Zhang, P.; Peng, B.; Li, C.; Ming, Y.; Liu, H. Berberine protects hepatocyte from hypoxia/reoxygenation-induced injury through inhibiting circDNTTIP2. *PeerJ* **2023**, *11*, e16080. [CrossRef] [PubMed]
40. Gong, X.; Li, T.; Wan, R.; Sha, L. Cordycepin attenuates high-fat diet-induced non-alcoholic fatty liver disease via down-regulation of lipid metabolism and inflammatory responses. *Int. Immunopharmacol.* **2021**, *91*, 107173. [CrossRef] [PubMed]
41. Yang, B.; Shen, Y.; Monroig, Ó.; Zhao, W.; Bao, Y.; Tao, S.; Jiao, L.; Zhou, Q.; Jin, M. The ameliorative role of methionine in hepatic steatosis and stress response in juvenile black seabream (*Acanthopagrus schlegelii*) fed with a high-fat diet. *Aquaculture* **2024**, *580*, 740306. [CrossRef]
42. Barter, P. HDL-C: Role as a risk modifier. *Atheroscler. Suppl.* **2011**, *12*, 267–270. [CrossRef] [PubMed]
43. Wang, L.; Gao, C.; Yang, L.; Wang, C.; Wang, B.; Wang, H.; Shu, Y.; Yan, Y. The growth-promoting and lipid-lowering effects of berberine are associated with the regulation of intestinal bacteria and bile acid profiles in yellow catfish (*Pelteobagrus fulvidraco*). *Aquac. Rep.* **2023**, *33*, 101848. [CrossRef]
44. Tian, J.-J.; Jin, Y.-Q.; Yu, E.-M.; Sun, J.-H.; Xia, Y.; Zhang, K.; Li, Z.-F.; Gong, W.-B.; Wang, G.-J.; Xie, J. Intestinal farnesoid X receptor mediates the effect of dietary berberine on lipid accumulation in grass carp (*Ctenopharyngodon idella*). *Aquaculture* **2022**, *553*, 738055. [CrossRef]
45. Wang, Y.P.; Nakajima, T.; Gonzalez, F.J.; Tanaka, N. PPARs as Metabolic Regulators in the Liver: Lessons from Liver-Specific PPAR-Null Mice. *Int. J. Mol. Sci.* **2020**, *21*, 2061. [CrossRef]
46. Yang, Z.; Roth, K.; Agarwal, M.; Liu, W.; Petriello, M.C. The transcription factors CREBH, PPARa, and FOXO1 as critical hepatic mediators of diet-induced metabolic dysregulation. *J. Nutr. Biochem.* **2021**, *95*, 108633. [CrossRef] [PubMed]
47. Jia, R.; Cao, L.-P.; Du, J.-L.; He, Q.; Gu, Z.-Y.; Jeney, G.; Xu, P.; Yin, G.-J. Effects of high-fat diet on steatosis, endoplasmic reticulum stress and autophagy in liver of tilapia (*Oreochromis niloticus*). *Front. Mar. Sci.* **2020**, *7*, 363. [CrossRef]
48. Lu, K.-L.; Zhang, D.-D.; Wang, L.-N.; Xu, W.-N.; Liu, W.-B. Molecular characterization of carnitine palmitoyltransferase IA in Megalobrama amblycephala and effects on its expression of feeding status and dietary lipid and berberine. *Comp. Biochem. Physiol. Part B Biochem. Mol. Biol.* **2016**, *191*, 20–25. [CrossRef] [PubMed]
49. Frieg, B.; Görg, B.; Gohlke, H.; Häussinger, D. Glutamine synthetase as a central element in hepatic glutamine and ammonia metabolism: Novel aspects. *Biol. Chem.* **2021**, *402*, 1063–1072. [CrossRef] [PubMed]
50. Savilov, P.N.; Yakovlev, V.N. Effect of Liver Damage and Hyperbaric Oxygenation on Glutamine Synthetase of Hepatocytes. *Bull. Exp. Biol. Med.* **2016**, *160*, 295–297. [CrossRef] [PubMed]
51. Fan, Z.; Wang, S.; Meng, Y.; Wen, C.; Xu, M.; Li, X. Butyrate Alleviates High-Fat-Induced Metabolic Disorders Partially through Increasing Systematic Glutamine. *J. Agric. Food Chem.* **2023**, *72*, 449–460. [CrossRef] [PubMed]

52. Xu, L.; Zheng, R.; Xie, P.; Guo, Q.; Ji, H.; Li, T. Dysregulation of UDP-glucuronosyltransferases in CCl4 induced liver injury rats. *Chem. Biol. Interact.* **2020**, *325*, 109115. [CrossRef] [PubMed]
53. Hardwick, R.N.; Ferreira, D.W.; More, V.R.; Lake, A.D.; Lu, Z.; Manautou, J.E.; Slitt, A.L.; Cherrington, N.J. Altered UDP-glucuronosyltransferase and sulfotransferase expression and function during progressive stages of human nonalcoholic fatty liver disease. *Drug Metab. Dispos.* **2013**, *41*, 554–561. [CrossRef] [PubMed]
54. Pandit, K.; Kumar, A.; Kaur, S.; Kumar, V.; Jain, S.K.; Bhardwaj, R.; Kaur, S. Amelioration of oxidative stress by *trans*-Anethole via modulating phase I and phase II enzymes against hepatic damage induced by CCl_4 in male Wistar rats. *Environ. Sci. Pollut. Res. Int.* **2022**, *29*, 6317–6333. [CrossRef] [PubMed]
55. Victor Antony Santiago, J.; Jayachitra, J.; Shenbagam, M.; Nalini, N. Dietary d-limonene alleviates insulin resistance and oxidative stress-induced liver injury in high-fat diet and L-NAME-treated rats. *Eur. J. Nutr.* **2012**, *51*, 57–68. [CrossRef] [PubMed]
56. Ma, X.; Chen, Z.; Wang, L.; Wang, G.; Wang, Z.; Dong, X.; Wen, B.; Zhang, Z. The pathogenesis of diabetes mellitus by oxidative stress and inflammation: Its inhibition by berberine. *Front. Pharmacol.* **2018**, *9*, 782. [CrossRef] [PubMed]
57. Chen, S.; Jiang, X.; Liu, N.; Ren, M.; Wang, Z.; Li, M.; Chen, N.; Li, S. Effects of dietary berberine hydrochloride inclusion on growth, antioxidant capacity, glucose metabolism and intestinal microbiome of largemouth bass (*Micropterus salmoides*). *Aquaculture* **2022**, *552*, 738023. [CrossRef]
58. Grădinariu, L.; Dediu, L.; Crețu, M.; Grecu, I.R.; Docan, A.; Istrati, D.I.; Dima, F.M.; Stroe, M.D.; Vizireanu, C. The Antioxidant and Hepatoprotective Potential of Berberine and Silymarin on Acetaminophen Induced Toxicity in *Cyprinus carpio* L. *Animals* **2024**, *14*, 373. [CrossRef] [PubMed]
59. Desouky, H.E.; Jiang, G.-z.; Abasubong, K.P.; Dai, Y.-J.; Yuan, X.; Adjoumani, J.-J.Y.; Liu, W.-b. Plant-Based Additivities Improved the Growth Performance and Immune Response, and Mitigated the Inflammatory Signalling in Channel Catfish Fed a High-Fat Diet. *Aquac. Res.* **2023**, *2023*, 3525041. [CrossRef]
60. Deng, Y.; Tang, K.; Chen, R.; Nie, H.; Liang, S.; Zhang, J.; Zhang, Y.; Yang, Q. Berberine attenuates hepatic oxidative stress in rats with non-alcoholic fatty liver disease via the Nrf2/ARE signalling pathway. *Exp. Ther. Med.* **2019**, *17*, 2091–2098. [CrossRef] [PubMed]
61. Huang, Y.; Li, W.; Su, Z.-y.; Kong, A.-N.T. The complexity of the Nrf2 pathway: Beyond the antioxidant response. *J. Nutr. Biochem.* **2015**, *26*, 1401–1413. [CrossRef] [PubMed]
62. Ashrafizadeh, M.; Fekri, H.S.; Ahmadi, Z.; Farkhondeh, T.; Samarghandian, S. Therapeutic and biological activities of berberine: The involvement of Nrf2 signaling pathway. *J. Cell. Biochem.* **2020**, *121*, 1575–1585. [CrossRef] [PubMed]
63. Mahmoud, A.M.; Hozayen, W.G.; Ramadan, S.M. Berberine ameliorates methotrexate-induced liver injury by activating Nrf2/HO-1 pathway and PPARγ, and suppressing oxidative stress and apoptosis in rats. *Biomed. Pharmacother.* **2017**, *94*, 280–291. [CrossRef] [PubMed]
64. Han, C.Y.; Sun, T.T.; Xv, G.P.; Wang, S.S.; Gu, J.G.; Liu, C.Y. Berberine ameliorates CCl4-induced liver injury in rats through regulation of the Nrf2-Keap1-ARE and p53 signaling pathways. *Mol. Med. Rep.* **2019**, *20*, 3095–3102. [CrossRef] [PubMed]
65. Sagada, G.; Wang, L.; Xu, B.; Tegomo, F.A.; Chen, K.; Zheng, L.; Sun, Y.; Liu, Y.; Yang, Y.; Ullah, S.; et al. Synergistic Effect of Dietary Inactivated Lactobacillus plantarum and Berberine Supplementation on Growth Performance, Antioxidant Capacity, and Immune Function of Juvenile Black Sea Bream (*Acanthopagrus schlegelii*). *Aquac. Nutr.* **2022**, *2022*, 3053724. [CrossRef]
66. Khanmohammadi, S.; Kuchay, M.S. Toll-like receptors and metabolic (dysfunction)-associated fatty liver disease. *Pharmacol. Res.* **2022**, *185*, 106507. [CrossRef] [PubMed]
67. Ma, J.Q.; Li, Z.; Xie, W.R.; Liu, C.M.; Liu, S.S. Quercetin protects mouse liver against CCl_4-induced inflammation by the TLR2/4 and MAPK/NF-κB pathway. *Int. Immunopharmacol.* **2015**, *28*, 531–539. [CrossRef] [PubMed]
68. Liu, J.; Zhuang, Z.J.; Bian, D.X.; Ma, X.J.; Xun, Y.H.; Yang, W.J.; Luo, Y.; Liu, Y.L.; Jia, L.; Wang, Y.; et al. Toll-like receptor-4 signalling in the progression of non-alcoholic fatty liver disease induced by high-fat and high-fructose diet in mice. *Clin. Exp. Pharmacol. Physiol.* **2014**, *41*, 482–488. [CrossRef] [PubMed]
69. Miura, K.; Yang, L.; van Rooijen, N.; Brenner, D.A.; Ohnishi, H.; Seki, E. Toll-like receptor 2 and palmitic acid cooperatively contribute to the development of nonalcoholic steatohepatitis through inflammasome activation in mice. *Hepatology* **2013**, *57*, 577–589. [CrossRef] [PubMed]
70. Jang, H.-J.; Kim, H.-S.; Hwang, D.H.; Quon, M.J.; Kim, J.-A. Toll-like receptor 2 mediates high-fat diet-induced impairment of vasodilator actions of insulin. *Am. J. Physiol. Endocrinol. Metab.* **2013**, *304*, E1077–E1088. [CrossRef] [PubMed]
71. Wang, X.-P.; Lei, F.; Du, F.; Chai, Y.-S.; Jiang, J.-F.; Wang, Y.-G.; Yu, X.; Yan, X.-J.; Xing, D.-M.; Du, L.-J. Protection of gastrointestinal mucosa from acute heavy alcohol consumption: The effect of berberine and its correlation with TLR2, 4/IL1β-TNFα signaling. *PLoS ONE* **2015**, *10*, e0134044. [CrossRef] [PubMed]
72. Shan, J.-L.; Wei, R.-R.; Lu, W.; Ouyang, X.; Cheng, H.-Y.; Zhong, G.-Y.; Liu, J.-C.; Zhu, J.-X. Mechanism of anti-chronic alcoholic liver injury in rats of tibetan medicine Lagotis brachystachys extracts by TLR2/MyD88/NF-κB and NALP3 signaling pathway. *Chin. J. Exp. Tradit. Med. Formulae* **2020**, *026*, 80–85.
73. Ghezelbash, B.; Shahrokhi, N.; Khaksari, M.; Asadikaram, G.; Shahrokhi, M.; Shirazpour, S. Protective Roles of Shilajit in Modulating Resistin, Adiponectin, and Cytokines in Rats with Non-alcoholic Fatty Liver Disease. *Chin. J. Integr. Med.* **2022**, *28*, 531–537. [CrossRef] [PubMed]
74. Wang, L.; Jia, Z.; Wang, B.; Zhang, B. Berberine inhibits liver damage in rats with non-alcoholic fatty liver disease by regulating TLR4/MyD88/NF-κB pathway. *Turk. J. Gastroenterol.* **2020**, *31*, 902–909. [CrossRef]

75. Shailesh, S.; Sahoo, P.K. Lysozyme: An important defence molecule of fish innate immune system. *Aquac. Res.* **2008**, *39*, 223–239.
76. Bao, B.; Peatman, E.; Li, P.; He, C.; Liu, Z. Catfish hepcidin gene is expressed in a wide range of tissues and exhibits tissue-specific upregulation after bacterial infection. *Dev. Comp. Immunol.* **2005**, *29*, 939–950. [CrossRef]
77. Abasubong, K.P.; Li, X.F.; Adjoumani, J.J.Y.; Jiang, G.Z.; Desouky, H.E.; Liu, W.B. Effects of dietary xylooligosaccharide prebiotic supplementation on growth, antioxidant and intestinal immune-related genes expression in common carp Cyprinus carpio fed a high-fat diet. *J. Anim. Physiol. Anim. Nutr.* **2022**, *106*, 403–418. [CrossRef] [PubMed]
78. Abasubong, K.P.; Jiang, G.-Z.; Guo, H.-X.; Wang, X.; Huang, Y.-Y.; Dai, Y.-J.; Li, X.-F.; Dong, Y.-Z.; Gabriel, N.N.; Liu, W.-B. Oral bovine serum albumin administration alleviates inflammatory signals and improves antioxidant capacity and immune response under thioacetamide stress in blunt snout bream fed a high-calorie diet. *Fish Shellfish Immunol.* **2023**, *141*, 108996. [CrossRef] [PubMed]
79. Padda, R.S.; Gkouvatsos, K.; Guido, M.; Mui, J.; Vali, H.; Pantopoulos, K. A high-fat diet modulates iron metabolism but does not promote liver fibrosis in hemochromatotic Hjv$^{-/-}$ mice. *Am. J. Physiol.-Gastrointest. Liver Physiol.* **2015**, *308*, G251–G261. [CrossRef]
80. Li, Y.; Jiang, W.; Feng, Y.; Wu, L.; Jia, Y.; Zhao, R. Betaine Alleviates High-Fat Diet-Induced Disruption of Hepatic Lipid and Iron Homeostasis in Mice. *Int. J. Mol. Sci.* **2022**, *23*, 6263. [CrossRef]
81. Qian, Y.-C.; Wang, X.; Ren, J.; Wang, J.; Limbu, S.M.; Li, R.-X.; Zhou, W.-H.; Qiao, F.; Zhang, M.-L.; Du, Z.-Y. Different effects of two dietary levels of tea polyphenols on the lipid deposition, immunity and antioxidant capacity of juvenile GIFT tilapia (*Oreochromis niloticus*) fed a high-fat diet. *Aquaculture* **2021**, *542*, 736896. [CrossRef]
82. Doan, H.V.; Hoseinifar, S.H.; Jaturasitha, S.; Dawood, M.A.O.; Harikrishnan, R. The effects of berberine powder supplementation on growth performance, skin mucus immune response, serum immunity, and disease resistance of Nile tilapia (*Oreochromis niloticus*) fingerlings. *Aquaculture* **2020**, *520*, 734927. [CrossRef]

antioxidants

MDPI

Article

The Effectiveness of Four Nicotinamide Adenine Dinucleotide (NAD^+) Precursors in Alleviating the High-Glucose-Induced Damage to Hepatocytes in *Megalobrama amblycephala*: Evidence in NAD^+ Homeostasis, Sirt1/3 Activation, Redox Defense, Inflammatory Response, Apoptosis, and Glucose Metabolism

Yanzou Dong, Xi Wang, Luyao Wei, Zishang Liu, Xiaoyu Chu, Wei Xiong, Wenbin Liu and Xiangfei Li *

Key Laboratory of Aquatic Nutrition and Feed Science of Jiangsu Province, College of Animal Science and Technology, Nanjing Agricultural University, No. 1 Weigang Road, Nanjing 210095, China
* Correspondence: xfli@njau.edu.cn

Abstract: The administration of NAD^+ precursors is a potential approach to protect against liver damage and metabolic dysfunction. However, the effectiveness of different NAD^+ precursors in alleviating metabolic disorders is still poorly elucidated. The current study was performed to compare the effectiveness of four different NAD^+ precursors, including nicotinic acid (NA), niacinamide (NAM), nicotinamide riboside (NR), and nicotinamide mononucleotide (NMN) in alleviating high-glucose-induced injury to hepatocytes in a fish model, *Megalobrama amblycephala*. An in vitro high-glucose model was successfully established to mimic hyperglycemia-induced damage to the liver, which was evidenced by the reduced cell viability, the increased transaminase activity, and the depletion of cellular NAD^+ concentration. The NAD^+ precursors all improved cell viability, with the maximal effect observed in NR, which also had the most potent NAD^+ boosting capacity and a significant Sirt1/3 activation effect. Meanwhile, NR presented distinct and superior effects in terms of anti-oxidative stress, inflammation inhibition, and anti-apoptosis compared with NA, NAM, and NMN. Furthermore, NR could effectively benefit glucose metabolism by activating glucose transportation, glycolysis, glycogen synthesis and the pentose phosphate pathway, as well as inhibiting gluconeogenesis. Moreover, an oral gavage test confirmed that NR presented the most potent effect in increasing hepatic NAD^+ content and the NAD^+/NADH ratio among four NAD^+ precursors. Together, the present study results demonstrated that NR is most effective in attenuating the high-glucose-induced injury to hepatocytes in fish compared to other NAD^+ precursors.

Keywords: NAD^+ precursors; hepatocyte health; oxidative stress; apoptosis; glucose metabolism

Citation: Dong, Y.; Wang, X.; Wei, L.; Liu, Z.; Chu, X.; Xiong, W.; Liu, W.; Li, X. The Effectiveness of Four Nicotinamide Adenine Dinucleotide (NAD^+) Precursors in Alleviating the High-Glucose-Induced Damage to Hepatocytes in *Megalobrama amblycephala*: Evidence in NAD^+ Homeostasis, Sirt1/3 Activation, Redox Defense, Inflammatory Response, Apoptosis, and Glucose Metabolism. *Antioxidants* **2024**, *13*, 385. https://doi.org/10.3390/antiox13040385

Academic Editor: Evangelos Zoidis

Received: 26 February 2024
Revised: 19 March 2024
Accepted: 19 March 2024
Published: 22 March 2024

1. Introduction

For a number of years, metabolic diseases, including diabetes, obesity and non-alcoholic fatty liver disease (NAFLD) have been prevalent in both developed and developing countries, becoming a strong public health concern [1–3]. Abnormal hyperglycemia is the characteristic and distinct pathologic feature of diabetes, and it has also been frequently found in patients suffering obesity or/and NAFLD [3,4]. Longstanding high blood glucose has been found to be associated with a cascade of adverse health consequences. Among them, high-glucose-induced liver damage often worsens metabolic dysfunction and contributes to more severe and irreversible pathological consequences such as non-alcoholic steatohepatitis, cirrhosis, and even cancers [3]. Although the exact mechanism is not completely understood, oxidative stress has been reported to contribute to high-glucose-mediated injury to organs [5]. Chronic hyperglycemia could break the redox equilibrium.

Consequently, the accumulated reactive oxygen species (ROS) could accelerate liver damage by activating the inflammatory response and the apoptosis pathway [6,7]. This process further exacerbates insulin resistance and glucose metabolism dysfunction, thereby forming a vicious cycle [8,9].

Nicotinamide adenine dinucleotide (NAD^+) is one of the critical biomolecules for cellular redox balance. As an enzyme cofactor/co-substrate, NAD^+ participates in many critical physiological processes, including protein deacylation, DNA repair, inflammatory reactions, and defense against oxidative stress [10]. Previous studies in mammals have shown that the DNA damage induced by aberrant nutritional status exaggerates cellular NAD^+ consumption. Declines in NAD^+ levels consequently cause oxidative stress and contribute to the pathological processes of metabolic diseases [10,11]. The above facts suggest that rescuing NAD^+ deficiency is a potential approach to protect hepatocytes against damage induced by high glucose. Due to its high bioactivity, it is extremely difficult to directly provide exogenous NAD^+ to animals [11,12]. Vitamin B_3 (also known as nicotinic acid, NA) and niacinamide (NAM), as well as their derivatives like nicotinamide riboside (NR) and nicotinamide mononucleotide (NMN), are well-described NAD^+ precursor vitamins [13,14]. The administration of these NAD^+ donors could boost cellular NAD^+ content, thereby alleviating metabolic dysfunctions. It is noteworthy that emerging evidence has proposed different pharmacokinetics and functions of these NAD^+ precursors, which could be due to their diverse metabolic properties and biochemical characteristics [15]. Specifically, a previous study has shown that NA, NR, and NMN differ in the degree to which they improve NAD^+ concentration in several mammalian cell lines [11]. Moreover, a more recent study has demonstrated that the oral administration of NR displays a stronger ability to enhance hepatic NAD^+ concentration than NA or NAM in mice [16]. Despite this, to the best of our knowledge, no study is available comparing the effectiveness of different NAD^+ precursors in alleviating metabolic dysfunctions.

Considering that the related metabolic pathways are well conserved among different animal species, studies using lower animals are deemed to be able to provide new insights into the understanding of metabolism diseases in human counterparts [17]. Indeed, previous studies have revealed the similar pathophysiological pathways behind the metabolic disease in mammal and fish models [18–20]. Considering this fact, fish are deemed to be convenient, reliable, and low-cost experimental animals to substitute for mammals [21,22]. Blunt snout bream (*Megalobrama amblycephala*) is a kind of cyprinid fish with high commercial importance [18]. Due to its herbivorous feeding habit, this species is prone to the high-glucose-induced hepatocytes injury caused by the feeding of high-carbohydrate diets [23]. NAD^+ depletion has also notably been found in individuals this species suffering from hyperglycemia [24]. Thus, *M. amblycephala* is a suitable subject to study high-glucose-induced liver damage. Based on the above facts, the present study was conducted to compare the effectiveness of four different NAD^+ precursors in alleviating high-glucose-induced hepatocytes damage in *M. amblycephala*. To fulfill this goal, cell viability, transaminase activity, NAD^+ homeostasis, Sirt1/3 activation, redox equilibrium, inflammatory response, apoptosis, and glucose metabolism were all investigated. The results obtained may promote the development of effective nutritional interventions to alleviate metabolic dysfunctions in animals and humans.

2. Materials and Methods

2.1. Primary Hepatocytes Isolation and Culture

The isolation and culture procedures of primary *M. amblycephala* hepatocytes were carried out with reference to the published protocols [25] with slight modifications. Briefly, fish (body weight = 20.5 ± 0.5 g) were kept in freshwater, with a 1% streptomycin and penicillin solution added, and fasted overnight. Next, fish were anesthetized using MS-222 (100 mg/L, tricaine methanesulfonate, Sigma, St. Louis, MO, USA) and bled by cutting the gill arches. The liver was dissected under aseptic conditions, washed with a pH 7.0 phosphate buffer twice, and cut into 1 mm^3 pieces on an ice bed. Then, the pieces were digested with a 0.25% trypsin-EDTA solution at 28 °C for 40 min. The cell suspension was obtained after filtration through a 200 mesh cell strainer and centrifugation at 300× *g* for 5 min. The harvested hepatocytes were transferred to a cell culture plate and maintained in the DMEM/F12 medium supplemented with 10% fetal bovine serum and 2% streptomycin and penicillin (termed as complete medium) at 28 °C in a humidified atmosphere with 5% CO_2.

2.2. Cell Treatment

The adherent primary hepatocytes were serum-starved overnight after reaching 80% confluency. Then, the cell culture media were replaced with fresh media for further experiments. Cells cultured with a normal complete medium were used as the control group. The high-glucose medium was prepared by adding D-glucose to the complete medium, thereby obtaining the high-glucose group. Furthermore, four different NAD^+ precursors, namely nicotinic acid (NA), niacinamide (NAM), nicotinamide riboside (NR), and nicotinamide mononucleotide (NMN), were added to the high-glucose medium at doses of 10 and 50 μM, thereby obtaining the NAD^+-precursor treatment groups. The supplementation levels of the NAD^+ precursors were determined based on the results of our preliminary test, which obtained the highest cell viability at the dose of 50 μM.

2.3. Cell Viability

Primary hepatocytes were seeded into 24 well plates at a density of 1×10^6 cells per well. After treatments, cells were washed with a phosphate buffer and then induced with the DMEM/F12 medium containing 10% Cell Counting Kit-8 regent (CCK8, ApexBio Technology, Shanghai, China) for 2 h. Then, the absorbance was measured at a 450 nm wavelength. The value of the relative cell viability was calculated by normalizing to the control group.

2.4. Transaminase Activity Detection

The collected cell culture mediums were centrifugated at 800× *g* for 5 min with the harvested supernatants taken to measure the alanine aminotransferase (ALT) and aspartate transaminase (AST) activities using an enzymatic colorimetric method according to the published protocol [26]. Briefly, the sample was incubated with an a-Ketoglutarate buffer containing L-alanine (for ALT detection) or L-aspartate (for AST detection) at 37 °C for 60 min. Then, 2, 4-DNPH was added to the reaction systems, and the mixture was allowed to stand at room temperature for 20 min. After being mixed with 0.4 M NaOH solution and incubated for 10 min, the absorbance was measured at 505 nm.

2.5. Detection of NAD^+ and NADH Contents

The NAD^+ and NADH concentrations were determined based on the WST-8-dependent chromogenic reaction according to the published protocols [27]. The amounts of total NAD(H) or NADH in the samples were measured, and the NAD^+ amount was calculated by subtracting the NADH amount from that of total NAD(H). For total NAD(H) determination, the sample was mixed with a dehydrogenase working solution and incubated at 37 °C for 10 min. Next, a reaction buffer (20 mM Tris-HCl, 200 μM WST-8, and 8 μM PMS, pH 9.0) was added, and the reaction system was kept at 37 °C for another 30 min. Then, the absorbance was measured at 450 nm. For NADH measurement, the sample was first heating at 60 °C for 10 min. Then, the same procedure was performed again.

2.6. Antioxidant Status Examination

The hepatocytes were digested using a 0.25% trypsin-EDTA solution after different treatments. The harvested cells were homogenized in a phosphate buffer. Then, the cellular malonaldehyde (MDA) content was determined through the thiobarbituric acid (TBA) test based on published protocols [28]. Briefly, the sample was mixed with 20% trichloroacetic acid (TCA) and then dispersed in 0.05 M H_2SO_4 solution. Next, TBA (0.2% in sodium sulphate) and N-butanol were added to the mixture, which was subsequently heated at 100 °C for 1 h. After cooling on ice, the absorbance was measured at 530 nm.

The activities of antioxidant enzymes, including superoxide dismutase (SOD), catalase (CAT), and glutathione peroxidase (GPX), were all assayed using the methods described previously [29,30]. For the determination of SOD activity, the sample was added to the reaction system containing xanthine, WST-1, and Formosan dye. After being kept at 37 °C for 30 min, the absorbance of the reaction system was obtained at 450 nm. For the CAT activity assay, the sample was mixed with the reaction buffer (1% Triton X-100, 30% H_2O_2) and kept at 37 °C for 1 min. Then, the ammonium molybdate solution was added and the absorbance was measured at 405 nm. For GPX activity measurement, the sample was added to the reaction mixture containing 1 U glutathione reductase, 2 mM glutathione, 0.12 mM NADPH, 2 mM H_2O_2, and 6 mM cumene hydroperoxide and was incubated at 37 °C for 1 min. Then, dithiodinitrobenzoic acid was used to visualize the product in a color reaction, and the absorbance was measured at 412 nm.

2.7. Measurement of Inflammatory Cytokines

The concentrations of inflammatory cytokines including interleukin-6 (H007-1, IL6), interleukin-1β (H002-1, IL1β), and tumor necrosis factor-α (H052-1, TNFα) in the cell culture supernatants were all evaluated by cytokine-specific enzyme-linked immunosorbent assay (ELISA) kits according to the instructions provided by the manufacturer (Nanjing Jiancheng Institute of Bioengineering, Nanjing, China). Briefly, the sample was transferred to plates coated by specific antibodies labeled with horseradish peroxidase and was kept at 37 °C for 1 h. Next, the plates were washed 5 times with phosphate buffer containing 0.05% Tween-20. The chromogenic substrate solutions A and B were added to the plates to develop color. The absorbance was measured at 450 nm.

2.8. Caspase 3 Activity Assay

For the measurement of caspase 3 activity, the hepatocytes were washed with a phosphate buffer and were collected by a lysate buffer (10 mM PIPES, 2 mM EDTA, 1% NP40 and 4 mM DTT, pH 7.5) after treatments. Then, the cell lysates were centrifugated at 16,000× g for 15 min, and the obtained supernatants were incubated with 200 μmol of caspase 3 substrate Ac-DEVD-pNA (acetyl-Asp-Glu-Val-Asp p-nitroanilide) at 37 °C for 10 h. The absorbance was quantified at a wavelength of 405 nm.

2.9. Glucose Consumption Test

To test the glucose consumption of cells, primary hepatocytes were seeded into 24-well plates at a density of 1×10^6 cells per well. After the above-mentioned treatments, the cell culture supernatants were collected and used for the glucose concentration assay. At the same time, cells were digested using a 0.25% trypsin-EDTA solution and homogenized in a phosphate buffer for the quantification of cellular proteins. The media of each group collected before the experimental treatment were used to determine the initial glucose concentrations. Thereafter, the variations of glucose amounts in the cell culture medium during the treatments were calculated, and the values were normalized to the cellular protein contents. The glucose concentrations were then measured by the glucose oxidase method [31].

2.10. Glucose Production Assay

For cellular glucose production assay, cells after different treatments were rinsed with phosphate buffer twice and were inducted in glucose-free DMEM/F12 medium (Procell Life Science & Technology, Wuhan, China). After 6 h, the cell culture supernatants were collected and used for the glucose concentration test. The results were normalized by the protein concentration.

2.11. Glycogen Determination

For glycogen measurement, the hepatocytes were harvested through 0.25% trypsin-EDTA solution digestion and then subjected to base hydrolysis. Then, the hydrolysate was heated in a boiling water bath for 20 min. After being cooled down by flowing water, the reaction solution was incubated with a reaction mixture containing 50% sulfuric acid and 5 mM anthrone at 100 °C for 5 min. The absorbance was quantified at a wavelength of 620 nm after cooling.

2.12. Gene Expression Analysis

The primary hepatocytes were lysed using the AG RNAex Pro regent (Accurate Biotechnology (Hunan) Co., Ltd., Changsha, China) for 5 min at room temperature. Then, total RNA was extracted from the lysis using a SteadyPure RNA Extraction Kit with reference to the protocols provided by Accurate Biotechnology. Agarose gel electrophoresis was performed to confirm the quality of RNA samples with the purity determined by a Nanodrop UV spectrophotometer (Thermo Fisher Scientific, Waltham, MA, USA). Then, the first-strand complementary DNA (cDNA) was obtained by reverse transcription from 1 μg of total RNA using an Evo M-MLV RT Kit (Accurate Biotechnology (Hunan) Co., Ltd., Changsha, China). To quantify gene expression, the real-time PCR (RT-PCR) was conducted under the Bio-Rad CFX96 platform (Bio-Rad, Berkeley, CA, USA) using the SYBR Green Pro Taq HS premix acquired from Accurate Biotechnology. The $2^{-\Delta\Delta Ct}$ method was adopted to calculate the relative transcription with elongation factor 1α (*ef1a*) used as the housekeeping gene. The high expression stability of *ef1a* in *M. amblycephala* was demonstrated in a recent study [32]. Moreover, the sequences of primers used in the current study are listed in Table 1.

Table 1. Primers used in the present study.

Gene Abbreviations	Gene Full Names	Primer Sequences (5′-3′)	Accession Numbers
il6	*interleukin-6*	ACAAAGCGCTCTTCCTGTTTG GCCATTTCTCCTGGTCGTTCA	KJ755058.1
il1β	*interleukin-1β*	CGAAGGCATGTCGGAGCATT ACCACTTCCATACGACGCTC	XM_048181166.1
tnfα	*tumor necrosis factor-α*	GCATGCCAGTCAGGTAGTGT AGGGCCACAGAAAGAAGAGC	KU976426.1
bcl2	*b-cell lymphoma-2*	GATGAGCCCGTTAGTGGGAC TCTGCGAATCGCTCCCATC	XM_048179299.1
baxa	*bcl2 associated X, apoptosis regulator a*	TCCTATTTTGGCACCCCCAC CTCTCTGCTCCCCCTCATCT	XM_048196672.1
caspase9	-	TCCAGATGAGGTGGAACCCT CCAAAATGTCGCTGGGTGTG	KM604705.1
caspase3a	-	GGAGCCTGACAGCCATAACA TGAGCTCTAGTTGGTTGCCA	KY006115.1
caspase3b	-	TGGTATGTGCATGGGGAACA TATGTGCATGGGGAACAGGAC	XM_048187987.1
glut2	*glucose transporter 2*	ACGCACCCGATGTGAAAGT TTGGACAGCAGCATTGATT	KC513421.2
gs	*glycogen synthase*	CCTCCAGTAACAACTCACAACA CAGATAGATTGGTGGTTACGC	XM_048154697.1
gp	*glycogen phosphorylase*	CTGTCTACCAGCTGGGGTTG GGCCTTCTCCCAAGGGTTAC	XM_048205686.1
gk	*glucokinase*	ACTGGATCTTGGAGGGACGA AAGTCAGATATGCACCCGGC	KJ141202.1
pfka	*phosphofructokinase a*	AGGAAATTGCAGTGCAGTAAAG CTGCTTCTGCTTCTAAATCCGC	XM_048194728.1
pfkb	*phosphofructokinase b*	GAAACCGGCTCAGTCGAAGA ACGGTGTAAACCCTGTGACC	XM_048172371.1

Table 1. *Cont.*

Gene Abbreviations	Gene Full Names	Primer Sequences (5′-3′)	Accession Numbers
pk	*pyruvate kinase*	GCCGAGAAAGTCTTCATCGCACAG CGTCCAGAACCGCATTAGCCAC	XM_048152870.1
pepck	*phosphoenolpyruvatecarboxykinase*	CGGCTACAACTTCGGTCAGT ACGTGGAAGATCTTGGGCAG	XM_048198716.1
fbpase	*fructose-1,6-bisphosphatase*	TACCCAGATGTCACAGAAT CACTCATACAACAGCCTCA	KJ743995.1
g6pase	*glucose-6-phosphatas*	CAGGCATGATTGTTGCCGAG AATGGACCCAGGCTGGATTG	XM_048171060.1
g6pd	*glucose-6-phosphate dehydrogenase*	AGGTAAAGGTGCTGAAGT AAATGTAGCCTGAGTGGA	KJ743994.1
6pgd	*6-phosphogluconate dehydrogenase*	TCAAGGAAGCGTTTGACCGA CACTGTCATCTGTCAGGCGT	XM_048178257.1
ef1α	*elongation factor 1α*	CTTCTCAGGCTGACTGTGC CCGCTAGCATTACCCTCC	XM_048180512.1

2.13. Animal Experiment

In order to compare the effects of four NAD$^+$ precursors (NA, NAM, NR, and MNM) on the hepatic NAD$^+$ concentration at the in vivo level, an animal experiment was performed. *M. amblycephala* (body weight = 82.5 ± 1.5 g) were purchased from a national aquaculture hatchery located at Ezhou, Hubei province, China. All fish shared a consistent genetic background. Before the test, fish were transported to a recirculating aquaculture system for one week for acclimation. During this period, a commercial diet (containing 30% protein, 5% lipids, and 38% carbohydrates) was fed to the fish twice daily (8:30 and 17:30 h). After that, fish were fasted for 24 h and were slightly euthanized using a 50 mg/L tricaine methanesulfonate (MS-222, Sigma, USA) solution to reduce the stress response. Then, a total of 60 experimental fish were randomly divided into 5 groups with an oral administration of normal saline (5 mL/kg body weight, the vehicle group) and different NAD$^+$ precursors (NA, NAM, NR, and NMN, 50 μmol/kg body weight). The dose of NAD$^+$ precursors was determined based on the five-fold molar amount of the optimal daily dietary NA amount of *M. amblycephala* [33]. Three sampling time points were set at 1, 3, and 12 h after the gavage test referring to a previous study [16], and four fish forming each group were sampled at each sampling time point. Fish were euthanized using a 100 mg/L MS-222 solution before sampling and then were quickly dissected on an ice bed. The liver samples were collected, snap-frozen in liquid nitrogen, and transferred to −80 °C till use. Next, the hepatic NAD$^+$ and NADH concentrations were measured according to the above-mentioned method. Meanwhile, the water parameters were monitored during this test and maintained as follows: pH—7.2 to 7.4; dissolved oxygen—about 5 mg/L; and ammonia nitrogen—<0.2 mg/L. The animal study was approved by the Animal Care and Use Committee of Nanjing Agricultural University (SYXK (Su) 2011-0036, Nanjing, China). Fish were kept under a suitable water environment, fed a nutrient-balanced feed, and sacrificed after being euthanized following standard ethical procedures. All these efforts were conducted to reduce the stress response of fish and ensure the welfare of fish.

2.14. Western Blot

The liver sample was lysed in RIPA buffer, mixed with loading buffer, and boiled for 5 min. Then, the sample was subjected to SDS-PAGE electrophoresis. The gel and electrophoresis solution contained 0.05% SDS. The separated proteins were transferred to polyvinylidene fluoride membranes. Subsequently, the membrane was blocked with 5% (*w*/*v*) non-fat dry milk and was then incubated with the first antibody and the second antibody successively. Then, bands were visualized by an electro-chemiluminescence system. The primary antibodies of β-actin (20536-1-AP, Proteintech), Sirtuin 1 (SIRT1) (13161-1-AP, Proteintech), and SIRT3 (10099-1-AP, Proteintech) were used in the present study.

2.15. Statistical Analysis

In the present study, all tests were performed in quadruplicate, and the results are shown as means ± standard error (SE). Statistical analyses were carried out using the SPSS 20.0 software. For the in vitro experiment, comparisons between the control and the high-glucose group were conducted using the Student's *t*-test, while one-way ANOVA followed by the Tukey's post hoc test was performed to identify the differences among the high-glucose group and the groups supplemented with NAD^+ precursors. For the in vivo experiment, a two-way ANOVA test was used to analyze the differences among treatments means based on oral gavage, sampling times, and their interaction. Differences were considered statistically significant when p values were lower than 0.05.

3. Results

3.1. Establishment of the High-Glucose Model

To set up the high-glucose model, high-glucose cell culture media were prepared by additionally adding glucose to the complete medium at the concentrations of 20, 40, 60, and 80 mM. These high-glucose media were used to incubate the primary hepatocytes for 48 h. The 40, 60, and 80 mM supplemented glucose treatments markedly declined the cell viability to 67.8, 55.4, and 56.1% of the control group ($p < 0.05$), respectively. Meanwhile, the ALT and AST activities decreased significantly when glucose concentration increased up to 60 mM ($p < 0.05$) (Figure 1A). As shown in Figure 1B, the lowest cellular NAD^+ concentration and the NAD^+/NADH ratio, as well as the highest NADH contents, were all observed in the 60 and 80 mM glucose treatments ($p < 0.05$). Based on the above results, the glucose concentration in the high-glucose group was set at 60 mM in the following tests.

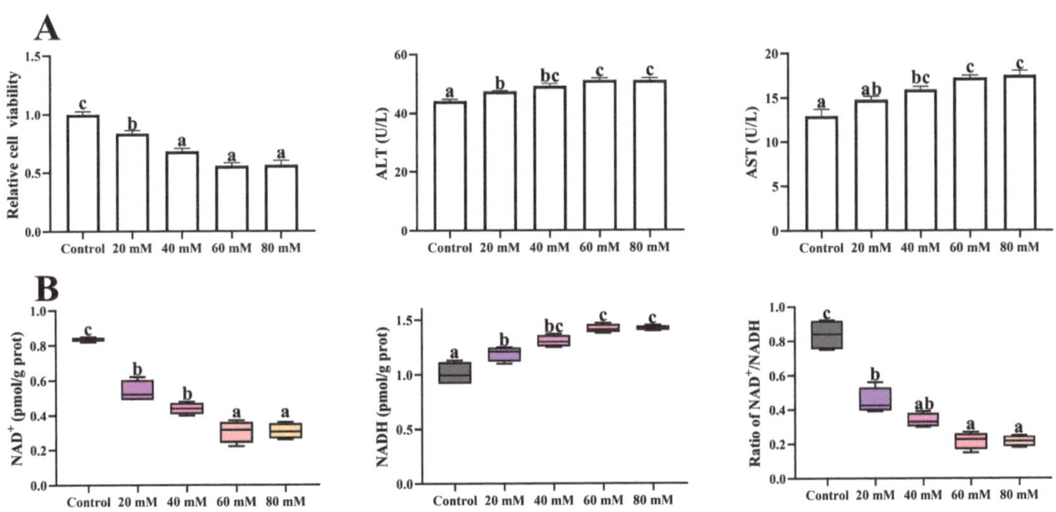

Figure 1. Hepatocellular damage indices (**A**) and NAD^+ homeostasis (**B**) of the primary hepatocytes exposed to high-glucose media (additional 20, 40, 60, and 80 mM glucose) for 48 h. Data are presented as mean ± SE. Data ($n = 4$) marked with different letters mean significant differences ($p < 0.05$, Tukey's test).

3.2. NAD^+ Precursors Improved the High-Glucose-Induced Hepatocyte Injury

As shown in Figure 2A, the cell viability of the high-glucose group was markedly decreased compared to the control ($p < 0.05$). The NA, NAM, and NR addition at the doses of 10 and 50 μM significantly increased the cell viability of hepatocytes exposed to high glucose ($p < 0.05$), with the 50 μM NR group exhibiting the highest value. The NMN administration presented similar effects, but the significance could only be observed at the dose of 50 μM ($p < 0.05$). Furthermore, the high-glucose group had markedly higher ALT and AST activities than the control group in cell culture supernatants ($p < 0.05$), while the addition of 50 μM NA and 10 and 50 μM NR significantly alleviated it ($p < 0.05$) (Figure 2B,C).

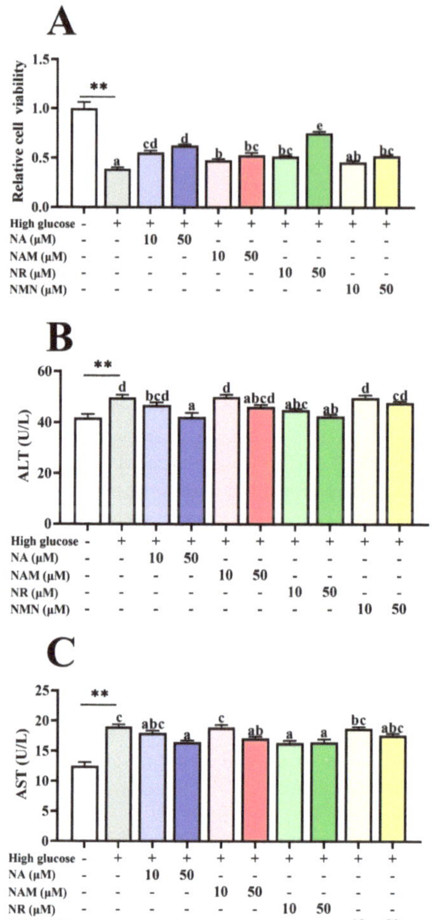

Figure 2. Effects of different NAD$^+$ precursors on the hepatocellular damage indices (**A–C**) of primary hepatocytes exposed to a high-glucose medium (additional 60 mM glucose) for 48 h. Data ($n = 4$) are presented as mean ± SE. The asterisk (*) means a significant difference between the control and the high-glucose group (** $p < 0.01$, *t*-test). Different letters mean significant differences among the high-glucose group and the groups supplemented with NAD$^+$ precursors ($p < 0.05$, Tukey's test).

3.3. NR Exhibited Superior NAD$^+$ Boosting and Sirt 1/3 Activation Effects

As shown in Figure 3, compared with the control, the high-glucose treatment markedly decreased the NAD$^+$ content and the NAD$^+$/NADH ratio but increased the NADH content. All groups supplemented with NAD$^+$ precursors had increased cellular NAD$^+$ levels and the NAD$^+$/NADH ratio with the highest value was noted in the 50 μM NR group ($p < 0.05$). Furthermore, all groups supplemented with NAD$^+$ precursors had a significantly lower NADH content compared to the high-glucose group, except the 10 μM NAM group ($p < 0.05$). As shown in Figure 4, the Sirt1 protein expression in the NR-treated groups was significantly higher than the high-glucose group and the NMN treatment ($p < 0.05$) but showed no statistical difference with the NA and NAM groups ($p > 0.05$). The highest abundance of Sirt3 was also noted in the NR group, but the significance was only found between the NR and NAM groups ($p < 0.05$).

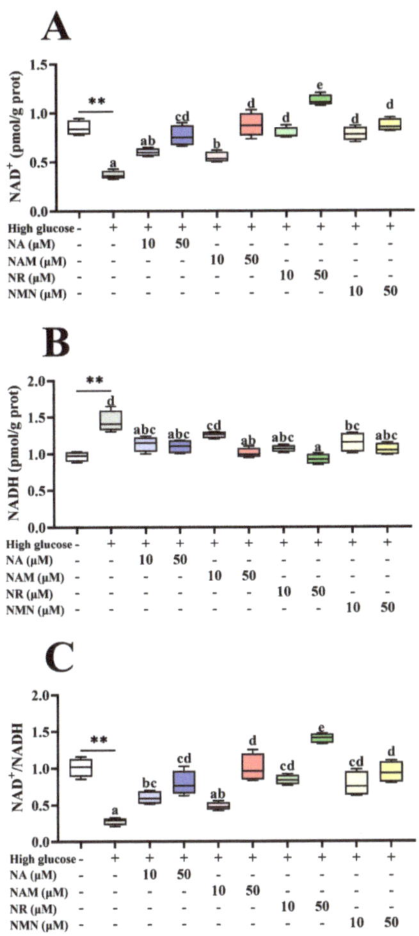

Figure 3. Effects of different NAD$^+$ precursors on the NAD$^+$ homeostasis (**A–C**) of primary hepatocytes exposed to a high-glucose medium (additional 60 mM glucose) for 48 h. Data (n = 4) are presented as mean ± SE. The asterisk (*) means a significant difference between the control and the high-glucose group (** $p < 0.01$, t-test). Different letters mean significant differences among the high-glucose group and the groups supplemented with NAD$^+$ precursors ($p < 0.05$, Tukey's test).

3.4. NAD$^+$ Precursors Attenuated High-Glucose-Induced Oxidative Stress

As shown in Figure 5, compared with the control, the high-glucose incubation markedly increased the MDA content but decreased the CAT and SOD activities ($p < 0.05$). The supplementation of 50 μM NR in the high-glucose medium markedly reduced the MDA content but promoted the GPX activity ($p < 0.05$). The supplementation of NA, NAM, and NR at 50 μM only significantly increased the SOD activity, while the CAT activity was markedly enhanced only by the 50 μM NA and NR treatments ($p < 0.05$).

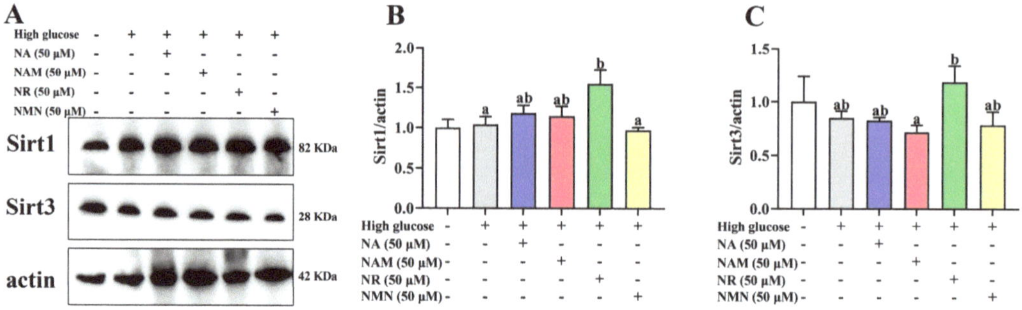

Figure 4. Effects of different NAD$^+$ precursors on the Sirt1 and Sirt3 protein abundance of primary hepatocytes exposed to a high-glucose medium (additional 60 mM glucose) for 48 h. (**A**) Image of bands, (**B**) relative level of Sirt1 abundance, (**C**) relative level of Sirt3 abundance. Data ($n = 4$) are presented as mean ± SE. Different letters mean significant differences among the high-glucose group and the group supplemented with NAD$^+$ precursors ($p < 0.05$, Tukey's test).

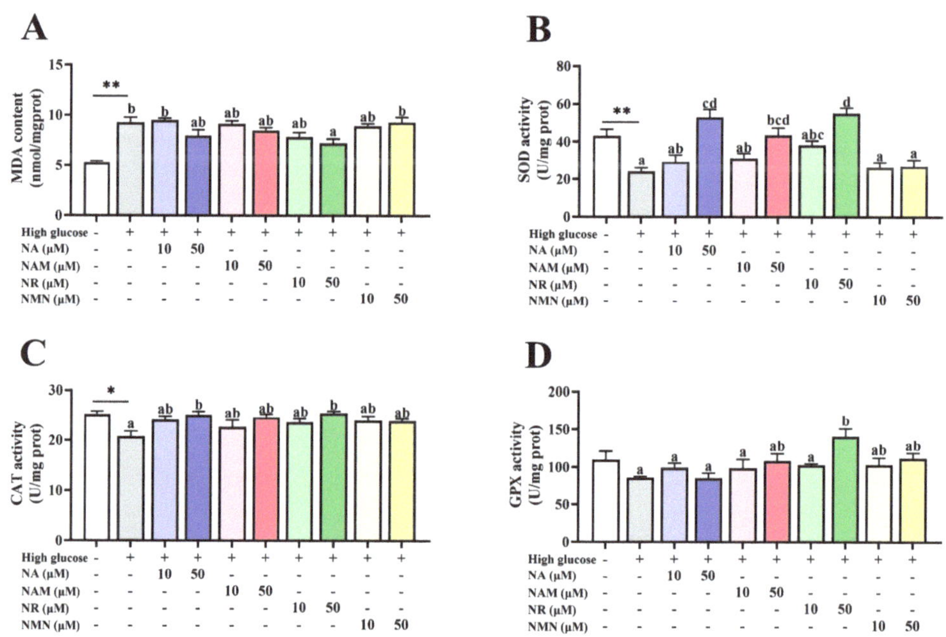

Figure 5. Effects of different NAD$^+$ precursors on the antioxidant status (**A–D**) of primary hepatocytes exposed to a high-glucose medium (additional 60 mM glucose) for 48 h. Data ($n = 4$) are presented as mean ± SE. The asterisk (*) means a significant difference between the control and the high-glucose group (* $p < 0.05$, ** $p < 0.01$, t-test). Different letters mean significant differences among the high-glucose group and the groups supplemented with NAD$^+$ precursors ($p < 0.05$, Tukey's test).

3.5. Anti-Inflammation Capacity of the Four NAD$^+$ Precursors

Compared with the control, the high-glucose incubation significantly up-regulated the gene expressions and concentrations of inflammatory cytokines, including IL1β, IL6, and TNFα in cell culture supernatants ($p < 0.05$). Only NR administration significantly decreased the IL1β and TNFα concentrations ($p < 0.05$) compared with the high-glucose group, while the IL6 concentration was obviously reduced by the treatments of 10 μM NR as well as 50 μM NA, NAM, and NR ($p < 0.05$) (Figure 6).

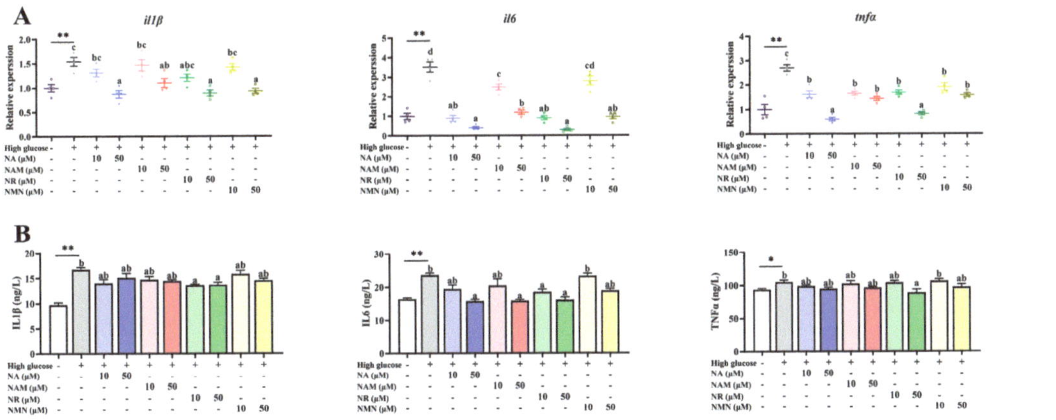

Figure 6. Effects of different NAD$^+$ precursors on the transcription (**A**) and concentration (**B**) of inflammatory cytokines in the cell culture supernatants of primary hepatocytes exposed to a high-glucose medium (additional 60 mM glucose) for 48 h. Data (n = 4) are presented as mean ± SE. The asterisk (*) means a significant difference between the control and the high-glucose group (* $p < 0.05$, ** $p < 0.01$, t-test). Different letters mean significant differences among the high-glucose group and the group supplemented with NAD$^+$ precursors ($p < 0.05$, Tukey's test).

3.6. Anti-Apoptosis Capacity of the Four NAD$^+$ Precursors

As presented in Figure 7, compared with the control, the transcriptions of *baxa*, *caspase9*, *caspase3a*, and *caspase3b*, as well as the caspase3 activity, were all significantly increased by the high-glucose incubation ($p < 0.05$). Compared with the high-glucose group, the supplementation of 10 μM NR as well as 50 μM NA and NR markedly up-regulated the transcription of *bcl2* ($p < 0.05$). Moreover, the transcription of *baxa* was significantly down-regulated by NR addition, while that of *caspase 9* was remarkably decreased by treatments of 10 μM NA as well as 50 μM NA, NR, and NMN ($p < 0.05$). All NAD$^+$ precursors reduced the transcriptions of *caspase3a* and *caspase3b*, with the lowest levels observed in the NR groups ($p < 0.05$). Furthermore, the caspase3 activity was significantly reduced by the treatments of 10 μM NR, as well as 50 μM NA, NAM, and NR ($p < 0.05$).

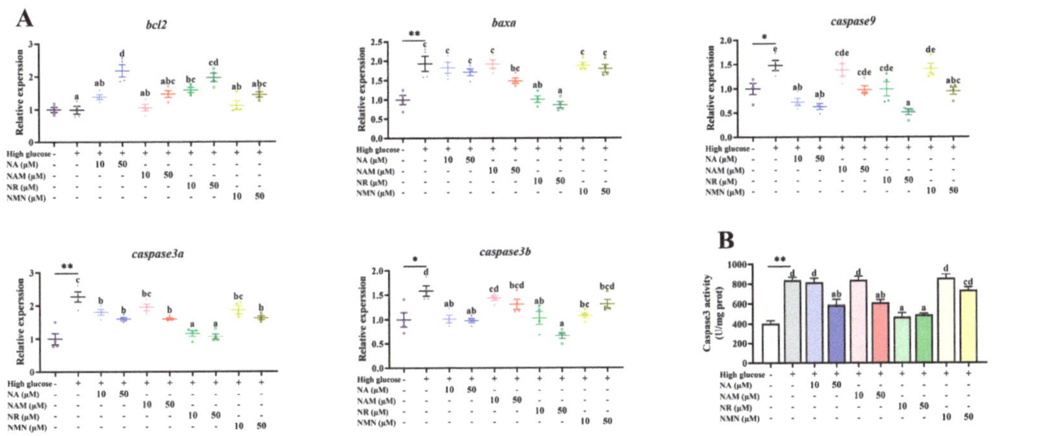

Figure 7. Effects of different NAD$^+$ precursors on the transcription of apoptosis-related genes (**A**) and the caspase 3 activity (**B**) of primary hepatocytes exposed to a high-glucose medium (additional 60 mM glucose) for 48 h. Data (n = 4) are presented as mean ± SE. The asterisk (*) means a significant

difference between the control and the high-glucose group (* $p < 0.05$, ** $p < 0.01$, t-test). Different letters mean significant differences among the high-glucose group and the groups supplemented with NAD⁺ precursors ($p < 0.05$, Tukey's test).

3.7. NAD⁺ Precursors Benefited Glucose Metabolism

As shown in Figure 8, the high-glucose treatment significantly increased the values of glucose consumption, production, and glycogen content of hepatocyte ($p < 0.05$) compared with the control. Compared to the high-glucose group, NA and NR administration significantly increased glucose consumption, while the supplementation of 10 μM NR and 50 μM of NA and NR markedly decreased glucose production, and 50 μM NR supplementation markedly increased glycogen content ($p < 0.05$).

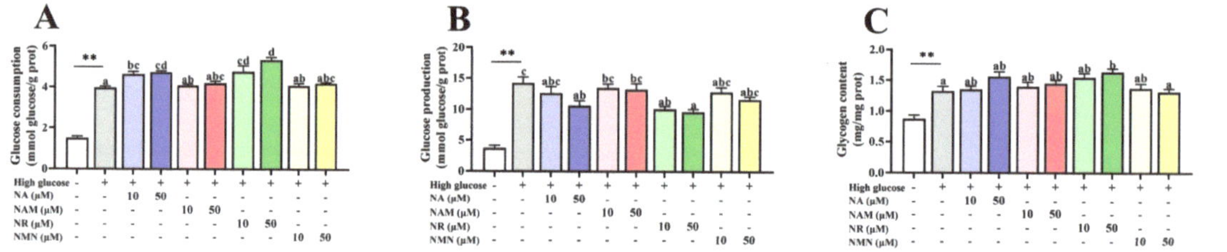

Figure 8. Effects of different NAD⁺ precursors on the glucose consumption (**A**) and production (**B**) and glycogen content (**C**) of primary hepatocytes exposed to a high-glucose medium (additional 60 mM glucose) for 48 h. Data ($n = 4$) are presented as mean ± SE. The asterisk (*) means a significant difference between the control and the high-glucose group (** $p < 0.01$, t-test). Different letters mean significant differences among the high-glucose group and the groups supplemented with NAD⁺ precursors ($p < 0.05$, Tukey's test).

As presented in Figures 9 and 10, compared with the control, the transcriptions of *glut2*, *gs*, *pkfa*, *pkfb*, *pk*, and *g6pd* were significantly up-regulated, while that of *gp* was markedly down-regulated ($p < 0.05$). Compared with the high-glucose group, 50 μM NR supplementation markedly increased the transcriptions of *glut2* and *gs* ($p < 0.05$). The transcriptions of *gk* and *g6pd* were up-regulated by the treatments of 50 μM NAM, as well as 10 and 50 μM NR. The addition of 10 μM NR significantly increased *pkfa* expression, while 10 and 50 μM NR both up-regulated *pk* expression. The administration of 50 μM NA and NR both decreased the transcription of *pepck*, while that of *6gpd* was increased by NAM, NR, and NMN additions ($p < 0.05$).

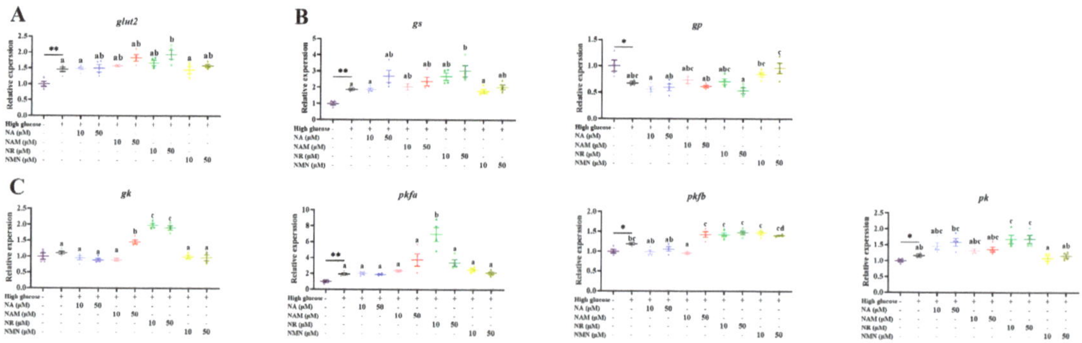

Figure 9. Effects of different NAD⁺ precursors on the transcription of glucose-transport- (**A**), glycogen-metabolism- (**B**), and glycolysis- (**C**) related genes of primary hepatocytes exposed to a high-glucose medium (additional 60 mM glucose) for 48 h. Data ($n = 4$) are presented as mean ± SE. The asterisk (*)

means a significant difference between the control and the high-glucose group (* $p < 0.05$, ** $p < 0.01$, *t*-test). Different letters mean significant differences among the high-glucose group and the groups supplemented with NAD$^+$ precursors ($p < 0.05$, Tukey's test).

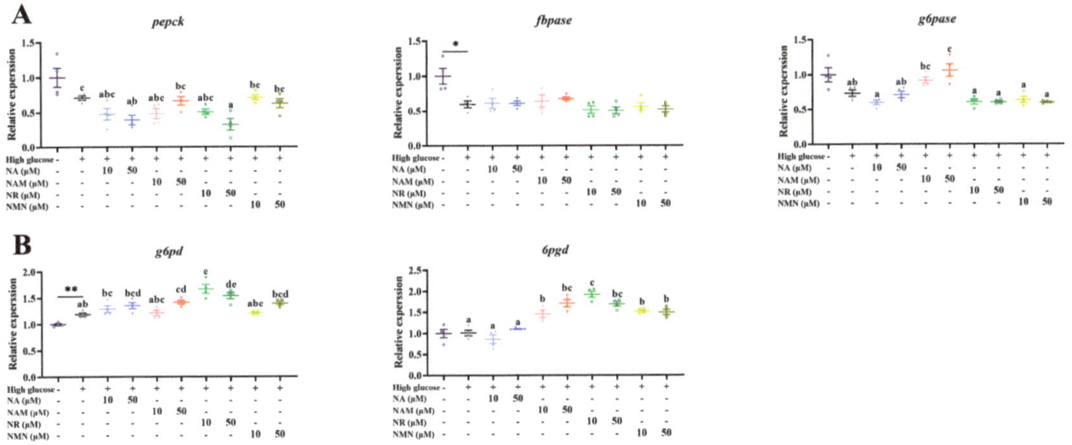

Figure 10. Effects of different NAD$^+$ precursors on the transcription of gluconeogenesis- (**A**) and the pentose phosphate pathway- (**B**) related genes of primary hepatocytes exposed to a high-glucose medium (additional 60 mM glucose) for 48 h. Data ($n = 4$) are presented as mean ± SE. The asterisk (*) means a significant difference between the control and the high-glucose group (* $p < 0.05$, ** $p < 0.01$, *t*-test). Different letters mean significant differences among the high-glucose group and the groups supplemented with NAD$^+$ precursors ($p < 0.05$, Tukey's test).

3.8. The Oral Gavage Test Showed That NR Had a Superior NAD$^+$ Promotion Effect In Vivo

As shown in Figure 11 and Table 2, in terms of sampling time, the NAD$^+$/NADH ratio at 3 h was significantly higher than those at 1 and 12 h ($p < 0.05$), while the NAD$^+$ concentration at 3 h was only markedly higher than that at 12 h ($p < 0.05$). In terms of oral gavage, the NR group exhibited the highest NAD$^+$ content among all the groups, revealing markedly higher values than the vehicle and NAM groups ($p < 0.05$). In addition, the NAD$^+$/NADH ratio of the NR group was significantly higher than that of the other groups ($p < 0.05$).

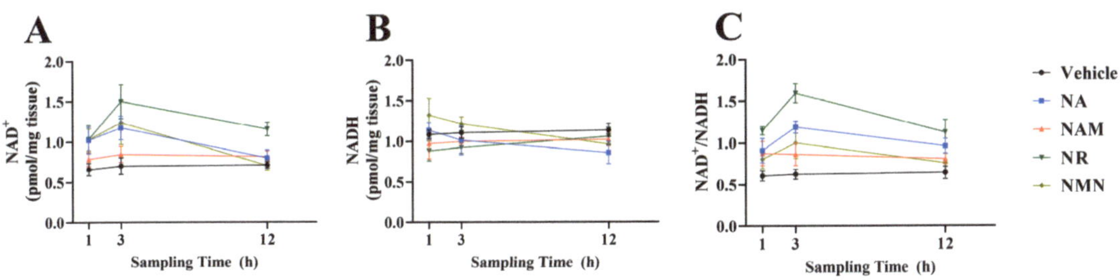

Figure 11. Hepatic NAD$^+$ (**A**) and NADH (**B**) concentrations and the NAD$^+$/NADH ratio (**C**) of fish after the oral administration of different NAD$^+$ precursors. Data ($n = 4$) were presented as mean ± SE.

Table 2. Two-way ANOVA analysis of the results obtained from the oral gavage test. Means of main effects followed by the same superscript letter (lower case letters (a, b, c) for oral gavage and upper case letters (A, B)) in the same column are not significantly different ($p > 0.05$).

	NAD+ (pmol/mg Tissue)	NADH (pmol/mg Tissue)	NAD+/NADH Ratio
Means of main effects			
Oral gavage			
Vehicle	0.693 ^{a}	1.110	0.626 ^{a}
NA	1.001 ^{ab}	1.005	1.022 ^{b}
NAM	0.822 ^{a}	0.999	0.852 ^{ab}
NR	1.230 ^{b}	0.957	1.291 ^{c}
NMN	0.994 ^{ab}	1.168	0.858 ^{ab}
Sampling time			
1 h	0.909 ^{AB}	1.082	0.871 ^{A}
3 h	1.095 ^{B}	1.054	1.057 ^{B}
12 h	0.840 ^{A}	1.008	0.861 ^{A}
p-values			
Oral gavage	<0.001	0.186	<0.001
Sampling time	0.014	0.620	0.012
Interaction	0.378	0.459	0.448

4. Discussion

In the present study, an in vitro high-glucose model was successfully constructed in fish to mimic the liver injury induced by high carbohydrate intake. The high-glucose incubation decreased cell viability and increased transaminase activity at the same time, evidencing the emergence of hepatocytes injury. Meanwhile, decreased NAD^+ content and $NAD^+/NADH$ ratio were also observed, as was in line with a previous study investigating the effects of high-carbohydrate feeding on the glycolipid metabolism of *M. amblycephala* [24]. Next, the effectiveness of four NAD^+ precursors was compared in alleviating the high-glucose-induced hepatocytes injury. Strikingly, the supplementation of all four NAD^+ precursors presented positive effects on the cell viability of hepatocytes exposed to high glucose, while NR showed the most profound effect. Furthermore, only NR significantly decreased the ALT and AST activities at the lower dose (10 μM). This indicates that NR may confer the most marked protection against the high-glucose-induced hepatocytes damage.

In the present study, the supplementation of NA exhibited a relatively low NAD^+ boosting capacity compared with NAM, NR, and NMN, which might be due to the fact that the formation of NAD^+ from NA needs more reaction steps. The NR-supplemented groups obtained higher NAD^+ concentrations than NAM-treated groups, which may imply that the activity of nicotinamide riboside kinases is higher than that of nicotinamide phosphoribosyltransferase in fish. Moreover, NMN supplementation exhibited a lower NAD^+ boosting capacity than NR. A possible explanation for this is that NMN cannot directly cross the cell membrane due to the lack of specific transporters [34]. Indeed, a previous study has proposed that NMN needs to be transformed into NR in the extracellular space before penetrating the cell membranes [35].

In the present study, NR administration increased the protein abundance of Sirt 1/3 compared with the high-glucose group, implying that NR might be most effective in regulating glucose metabolism and reducing oxidative stress among all four NAD^+ precursors. This is supported by the facts that (1) both Sirt1 and Sirt3 are NAD^+-dependent enzymes; (2) Sirt1 could regulate the activities of several transcription factors, which could consequently regulate glucose metabolism [36]; (3) while Sirt3 is related to the anti-oxidative defense of mitochondria [36]. The Sirt1/3 activation effects of NR have been attributed to be the enhancement of NAD^+ formation via the Salvage pathway [11,37,38].

It is well accepted that oxidative stress is closely involved in high-glucose-induced organ damage, while NAD^+ is an important participant in the regulation of redox balance [11,12]. In the present study, the MDA level of hepatocytes was exaggerated by high-glucose treatment coupled with decreased activities of SOD and CAT, indicating the occurrence of the dysfunction of the antioxidant system. Similar results have also been reported in the liver of *M. amblycephala* fed a high-carbohydrate diet [39], supporting the reliability of the in vitro high-glucose model in the present study. In addition, the supplementation of 50 μM NR markedly increased the activities of SOD, CAT, and GPX compared with the high-glucose group, while the treatment of 50 μM NA enhanced the activities of SOD and CAT, demonstrating that both NR and NA could activate the enzymatic antioxidant system. Notably, only the 50 μM NR group significantly reduced the MDA content, suggesting that NR displayed the most potent anti-oxidative property among four NAD^+ precursors. This beneficial effect may be ascribed to the decreased mitochondrial ROS formation via Sirt3 activation since mitochondria are the main sites of cellular ROS generation, while Sirt3-mediated anti-oxidative enzyme activation is one of the most important components of mitochondrial anti-oxidative defense [5,40]. In the present study, the NR treatment up-regulated the Sirt3 abundance, which further supported this.

IL1β, IL6, and TNFα are cytokines that are well identified as the inflammation markers involved in the metabolic disease of fish [23]. In this study, the increased transcriptions and concentrations of IL1β, IL6, and TNFα were found in the hepatocytes exposed to high glucose compared with the control, suggesting that high-glucose treatment resulted in inflammation. The inflammatory response is usually exaggerated as a consequence of oxidative stress, which would worsen organ damage through apoptosis activation [41,42]. Under oxidative stress, the key pro-inflammatory regulator, namely nuclear factor kappa B (NF-κB), could be activated by overproduced ROS, thereby stimulating the inflammatory cascade via promoting the gene transcription of inflammatory cytokines [43]. A recent in vivo study in *M. amblycephala* has illustrated that high-carbohydrate feeding could activate the NF-κB pathway [23]. Therefore, it was presumed that the high-glucose-induced inflammation in the present study was partly mediated by ROS-induced NF-κB activation. Moreover, only NR simultaneously reduced the transcriptions and concentrations of IL1β and IL6 at a lower level (10 μM). The markedly decreased TNFα concentration was only observed in the group supplemented with 50 μM NR. The above results indicate that NR exhibits the best anti-inflammatory potency among the four NAD^+ precursors in this study. This beneficial effect could be partly attributed to the antioxidant effect of NR due to the close relation between oxidative stress and inflammation. Furthermore, it should be noted that Sirt1 could directly suppress the NF-κB signaling through deacetylation [44]. Hence, the mechanism of the anti-inflammatory effects of NR may involve both Sirt1 and Sirt3 activation.

Oxidative stress and inflammatory actions are reported as strong inducers of the endogenous apoptotic pathway [45]. In the present study, the high-glucose treatment markedly up-regulated the transcriptions of *baxa*, *caspase9*, and *caspase3a/b* compared with the control. Cellular caspase3 activity was also enhanced by the high-glucose medium. In addition, only NR simultaneously up-regulated the transcription of the anti-apoptosis gene, *bcl2*, down-regulated the proapoptotic genes, *baxa* and *caspase 3a/b*, and reduced the caspase3 activity at the dose of 10 μM. Bcl2-associated X protein (Bax) family members are pivotal regulators of endogenous apoptosis. Among them, *bax* is a key proapoptotic gene, while *bcl2* is found to play an opposite role [46]. Furthermore, caspase 3 is the executioner of endogenous apoptosis, with its activity directly regulated by caspase 9 [47]. Therefore, the results obtained from the current study demonstrate that high-glucose treatment resulted in hepatocyte apoptosis, as was in line with a previous high-carbohydrate feeding study [23]. Meanwhile, NR has the most potent anti-apoptotic effect, which further emphasizes the most outstanding hepatocyte protection effects of NR among the four NAD^+ precursors. The mechanism behind the anti-apoptotic effect of NR is likely multifaceted. Firstly, the attenuated oxidative stress and inflammation may help to inhibit apoptosis. Secondly, NAD^+

acts as a substrate for poly (ADP-ribose) polymerases (PARPs), which play important roles in DNA repair [48]. NR administration increased cellular NAD^+ concentration, which could consequently increase the PARPs activity, thereby exerting the anti-apoptotic effect by alleviating genomic DNA damage [49]. Thirdly, Sirt3 could improve mitochondrial health, which could decrease the release of proapoptotic factors from mitochondria [5]. Therefore, the activation of Sirt3 may also promote the anti-apoptotic effect of NR.

Multiple NAD^+ precursors have been found to have favorable effects on glucose control [11,14]. However, limited evidence is available regarding the comparison of the effectiveness of different NAD^+ precursors. In the present study, the hepatocytes under high-glucose treatment had higher glucose consumption amounts and glycogen contents compared with the control, which might be an adaptive response to improve glucose utilization. This is supported by previous in vivo studies, where the high-carbohydrate feeding up-regulated the transcriptions of the genes involved in glucose transportation, glycolysis, and the pentose phosphate pathway [36]. Moreover, the glucose-production capacity was also up-regulated by high glucose. This is not surprising given the fact that fish often maintain active gluconeogenesis even in the presence of sufficient exogenous glucose [50,51]. This is actually an important manifestation of glucose metabolism dysregulation in fish. Furthermore, compared to the high-glucose group, increased glucose consumption and glycogen content, as well as decreased glucose production amount, were observed in all NAD^+ precursors groups, but to various degrees. Notably, the 50 μM NR treatment exhibited the highest glucose consumption level and cellular glycogen content, as well as the lowest glucose production. Furthermore, this treatment also resulted in the highest transcriptions of glucose transportation gene, *glut2*; glycogen synthesis gene, *gs*; glycolysis genes, *gk*, *pfka*, and *pk*; and the pentose phosphate pathway gene, *6pgd*. A down-regulation of gluconeogenesis genes *pepck* and *g6pase* was also observed in the NR-administrated group. These findings indicate that NR exhibits the most up-regulatory effect on glucose metabolism among the four NAD^+ precursors. Generally, NR may influence glucose metabolism in multiple manners as a NAD^+ precursor. On the one hand, NAD^+ directly participates in glycolysis, the citric acid cycle, and oxidative phosphorylation as an important electron carrier. Therefore, the enhanced NAD^+ level would activate these processes, thereby regulating glucose metabolism [10,14]. On the other hand, the elevated NAD^+ level could activate Sirt1, which could modulate glucose metabolism pathways at the gene-expression level by regulating the related transcription factors involved in glucose metabolism [11,14].

The above results demonstrate that NR is the most effective NAD^+ precursor to alleviate the high-glucose-induced hepatocytes injury. To further validate the in vitro results, an in vivo test was undertaken. As expected, the oral gavage of the four NAD^+ precursors all increased hepatic NAD^+ contents and the NAD^+/NADH ratio with the highest values obtained in the NR treatment. Similarly, a previous study has indicated that NR administration exhibits a greater capability of NAD^+ enhancement than NA and NAM in a mammal model [16]. Furthermore, the NAD^+ boosting effect of NR is also reported in *Caenorhabditis elegans* and *Drosophila melanogaster* [52]. These results indicate that the cellular NAD^+-boosting effect of NR might be highly conserved among lower and higher animals.

5. Conclusions

In summary, the present study compared the effectiveness of four NAD^+ precursors (NA, NAM, NR, and NMN) in alleviating high-glucose-induced hepatocyte damage in a fish model. The four NAD^+ precursors all revealed beneficial effects, while NR exhibited the most potent effect in terms of cell viability, NAD^+ boosting, oxidative defense, anti-inflammation, and anti-apoptosis. Moreover, NR could effectively regulate glucose metabolism by activating glucose transportation, glycolysis, glycogen synthesis, and the pentose phosphate pathway, as well as inhibiting gluconeogenesis at the transcriptional level. The above beneficial effects of NR can be partly attributed to the activation of Sirt1/3.

Author Contributions: Conceptualization, Y.D. and X.L.; Methodology, X.W., L.W., Z.L., X.C. and W.X.; Validation, X.W.; Formal analysis, Y.D.; Investigation, Y.D.; Data curation, Y.D.; Writing—original draft, Y.D.; Writing—review & editing, X.L.; Supervision, W.L. and X.L.; Funding acquisition, W.L. and X.L. All authors have read and agreed to the published version of the manuscript.

Funding: This research was funded by the Natural Science Foundation of Jiangsu Province (BK20221514) and the earmarked fund for Agriculture Research System of China (CARS-45-12).

Institutional Review Board Statement: The current study was carried out under the supervision of the Animal Care and Use Committee in Nanjing Agricultural University (SYXK (Su) 2011-0036, Nanjing, China), approval date: 6 April 2023.

Informed Consent Statement: Not applicable.

Data Availability Statement: The data presented in this study are available on request from the corresponding author.

Conflicts of Interest: The authors declare no conflict of interest.

References

1. Szkudelska, K.; Szkudelski, T. Resveratrol, obesity and diabetes. *Eur. J. Pharmacol.* **2010**, *635*, 1–8. [CrossRef] [PubMed]
2. Petro, A.E.; Cotter, J.; Cooper, D.A.; Peters, J.C.; Surwit, S.J.; Surwit, R.S. Fat, carbohydrate, and calories in the development of diabetes and obesity in the C57BL/6J mouse. *Metabolism* **2004**, *53*, 454–457. [CrossRef] [PubMed]
3. Mohamed, J.; Nazratun Nafizah, A.H.; Zariyantey, A.H.; Budin, S.B. Mechanisms of Diabetes-Induced Liver Damage: The role of oxidative stress and inflammation. *Sultan Qaboos Univ. Med. J.* **2016**, *16*, e132–e141. [CrossRef]
4. Oezcan, U.; Yilmaz, E.; Oezcan, L.; Furuhashi, M.; Vaillancourt, E.; Smith, R.O.; Goerguen, C.Z.; Hotamisligil, G.S. Chemical chaperones reduce ER stress and restore glucose homeostasis in a mouse model of type 2 diabetes. *Science* **2006**, *313*, 1137–1140. [CrossRef] [PubMed]
5. Zhang, L.; Ge, Y.P.; Huang, Y.Y.; Chen, W.L.; Liu, W.B.; Zhang, D.D.; Li, X.F. Benfotiamine attenuates the high-carbohydrate diet-induced mitochondrial redox imbalance in fish *Megalobrama amblycephala* by activating SIRT3. *Aquaculture* **2023**, *572*, 739553. [CrossRef]
6. Chen, Z.; Tian, R.; She, Z.; Cai, J.; Li, H. Role of oxidative stress in the pathogenesis of nonalcoholic fatty liver disease. *Free Radic. Biol. Med.* **2020**, *152*, 116–141. [CrossRef]
7. Klaunig, J.E.; Wang, Z.; Pu, X.; Zhou, S. Oxidative stress and oxidative damage in chemical carcinogenesis. *Toxicol. Appl. Pharmacol.* **2011**, *254*, 86–99. [CrossRef]
8. Wang, N.; Liu, Y.; Ma, Y.; Wen, D. Hydroxytyrosol ameliorates insulin resistance by modulating endoplasmic reticulum stress and prevents hepatic steatosis in diet-induced obesity mice. *J. Nutr. Biochem.* **2018**, *57*, 180–188. [CrossRef]
9. Pereira, S.; Moore, J.; Li, J.-X.; Yu, W.Q.; Ghanim, H.; Vlavcheski, F.; Joseph, Y.D.; Dandona, P.; Volchuk, A.; Cummins, C.L.; et al. 4-Phenylbutyric acid improves free fatty acid-induced hepatic insulin resistance in vivo. *Endocr. Connect.* **2021**, *10*, 861–872. [CrossRef] [PubMed]
10. Katsyuba, E.; Romani, M.; Hofer, D.; Auwerx, J. NAD^+ homeostasis in health and disease. *Nat. Metab.* **2020**, *2*, 9–31. [CrossRef] [PubMed]
11. Canto, C.; Houtkooper, R.H.; Pirinen, E.; Youn, D.Y.; Oosterveer, M.H.; Cen, Y.; Fernandez-Marcos, P.J.; Yamamoto, H.; Andreux, P.A.; Cettour-Rose, P.; et al. The NAD^+ Precursor Nicotinamide Riboside Enhances Oxidative Metabolism and Protects against High-Fat Diet-Induced Obesity. *Cell Metab.* **2012**, *15*, 838–847. [CrossRef]
12. Verdin, E. NAD^+ in aging, metabolism, and neurodegeneration. *Science* **2015**, *350*, 1208–1213. [CrossRef] [PubMed]
13. Amjad, S.; Nisar, S.; Bhat, A.A.; Shah, A.R.; Frenneaux, M.P.; Fakhro, K.; Haris, M.; Reddy, R.; Patay, Z.; Baur, J.; et al. Role of NAD^+ in regulating cellular and metabolic signaling pathways. *Mol. Metab.* **2021**, *49*, 101195. [CrossRef] [PubMed]
14. Yang, S.J.; Choi, J.M.; Kim, L.; Park, S.E.; Rhee, E.J.; Lee, W.Y.; Oh, K.W.; Park, S.W.; Park, C.-Y. Nicotinamide improves glucose metabolism and affects the hepatic NAD-sirtuin pathway in a rodent model of obesity and type 2 diabetes. *J. Nutr. Biochem.* **2014**, *25*, 66–72. [CrossRef] [PubMed]
15. Canto, C. NAD^+ Precursors: A Questionable Redundancy. *Metabolites* **2022**, *12*, 630. [CrossRef] [PubMed]
16. Trammell, S.A.J.; Schmidt, M.S.; Weidemann, B.J.; Redpath, P.; Jaksch, F.; Dellinger, R.W.; Li, Z.; Abel, E.D.; Migaud, M.E.; Brenner, C. Nicotinamide riboside is uniquely and orally bioavailable in mice and humans. *Nat. Commun.* **2016**, *7*, 12948. [CrossRef]
17. Schlegel, A.; Stainier, D.Y.R. Lessons from "lower" organisms: What worms, flies, and zebrafish can teach us about human energy metabolism. *PLoS Genet.* **2007**, *3*, 2037–2048. [CrossRef] [PubMed]
18. Prisingkorn, W.; Prathomya, P.; Jakovlic, I.; Liu, H.; Zhao, Y.H.; Wang, W.M. Transcriptomics, metabolomics and histology indicate that high-carbohydrate diet negatively affects the liver health of blunt snout bream (*Megalobrama amblycephala*). *BMC Genom.* **2017**, *18*, 856. [CrossRef]
19. Kamel, M.; Ninov, N. Catching new targets in metabolic disease with a zebrafish. *Curr. Opin. Pharm.* **2017**, *37*, 41–50. [CrossRef]
20. Zang, L.; Maddison, L.A.; Chen, W. Zebrafish as a Model for Obesity and Diabetes. *Front. Cell Dev. Biol.* **2018**, *6*, 91. [CrossRef]

21. Asaoka, Y.; Terai, S.; Sakaida, I.; Nishina, H. The expanding role of fish models in understanding non-alcoholic fatty liver disease. *Dis. Model. Mech.* **2013**, *6*, 905–914. [CrossRef]
22. Oka, T.; Nishimura, Y.; Zang, L.; Hirano, M.; Shimada, Y.; Wang, Z.; Umemoto, N.; Kuroyanagi, J.; Nishimura, N.; Tanaka, T. Diet-induced obesity in zebrafish shares common pathophysiological pathways with mammalian obesity. *BMC Physiol.* **2010**, *10*, 21. [CrossRef]
23. Xu, C.; Liu, W.B.; Shi, H.J.; Mi, H.F.; Li, X.F. Benfotiamine ameliorates high-carbohydrate diet-induced hepatic oxidative stress, inflammation and apoptosis in *Megalobrama amblycephala*. *Aquacult. Res.* **2021**, *52*, 3174–3185. [CrossRef]
24. Shi, H.J.; Xu, C.; Liu, M.Y.; Wang, B.K.; Liu, W.B.; Chen, D.H.; Zhang, L.; Xu, C.Y.; Li, X.F. Resveratrol Improves the Energy Sensing and Glycolipid Metabolism of Blunt Snout Bream *Megalobrama amblycephala* Fed High-Carbohydrate Diets by Activating the AMPK-SIRT1-PGC-1α Network. *Front. Physiol.* **2018**, *9*, 1258. [CrossRef]
25. Zhou, W.; Rahimnejad, S.; Tocher, D.R.; Lu, K.; Zhang, C.; Sun, Y. Metformin attenuates lipid accumulation in hepatocytes of blunt snout bream (*Megalobrama amblycephala*) via activation of AMP-activated protein kinase. *Aquaculture* **2019**, *499*, 90–100. [CrossRef]
26. Reitman, S.; Frankel, S. A Colorimetric Method for the Determination of Serum Glutamic Oxalacetic and Glutamic Pyruvic Transaminases. *Am. J. Clin. Pathol.* **1957**, *28*, 56–63. [CrossRef]
27. Chamchoy, K.; Pakotiprapha, D.; Pumirat, P.; Leartsakulpanich, U.; Boonyuen, U. Application of WST-8 based colorimetric NAD(P)H detection for quantitative dehydrogenase assays. *BMC Biochem.* **2019**, *20*, 4. [CrossRef] [PubMed]
28. Uchiyama, M.; Mihara, M. Determination of malonaldehyde precursor in tissues by thiobarbituric acid test. *Anal. Biochem.* **1978**, *86*, 271–278. [CrossRef]
29. Cakmak, I.; Marschner, H. Magnesium Deficiency and High Light Intensity Enhance Activities of Superoxide Dismutase, Ascorbate Peroxidase, and Glutathione Reductase in Bean Leaves 1. *Plant Physiol.* **1992**, *98*, 1222–1227. [CrossRef] [PubMed]
30. Ōyanagui, Y. Reevaluation of assay methods and establishment of kit for superoxide dismutase activity. *Anal. Biochem.* **1984**, *142*, 290–296. [CrossRef] [PubMed]
31. Wilson, R.; Turner, A.P.F. Glucose oxidase: An ideal enzyme. *Biosens. Bioelectron.* **1992**, *7*, 165–185. [CrossRef]
32. Xu, C.; Li, Y.Y.; Brown, P.B.; Liu, W.B.; Gao, L.L.; Ding, Z.R.; Li, X.F.; Xie, D.Z. Interactions between dietary carbohydrate and thiamine: Implications on the growth performance and intestinal mitochondrial biogenesis and function of *Megalobrama amblycephala*. *Br. J. Nutr.* **2022**, *127*, 321–334. [CrossRef] [PubMed]
33. Li, X.F.; Wang, T.J.; Qian, Y.; Jiang, G.Z.; Zhang, D.D.; Liu, W.B. Dietary niacin requirement of juvenile blunt snout bream *Megalobrama amblycephala* based on a dose-response study. *Aquacult. Nutr.* **2017**, *23*, 1410–1417. [CrossRef]
34. Covarrubias, A.J.; Perrone, R.; Grozio, A.; Verdin, E. NAD^+ metabolism and its roles in cellular processes during ageing. *Nat. Rev. Mol. Cell Biol.* **2021**, *22*, 119–141. [CrossRef] [PubMed]
35. Nikiforov, A.; Dölle, C.; Niere, M.; Ziegler, M. Pathways and Subcellular Compartmentation of NAD Biosynthesis in Human Cells: From entry of extracellular precursors to mitochondrial NAD generation. *J. Biol. Chem.* **2011**, *286*, 21767–21778. [CrossRef] [PubMed]
36. Shi, H.J.; Li, X.F.; Xu, C.; Zhang, D.; Zhang, L.; Xia, S.L.; Liu, W. Nicotinamide improves the growth performance, intermediary metabolism and glucose homeostasis of blunt snout bream *Megalobrama amblycephala* fed high-carbohydrate diets. *Aquacult. Nutr.* **2020**, *26*, 1311–1328. [CrossRef]
37. Wang, S.; Wan, T.; Ye, M.; Qiu, Y.; Pei, L.; Jiang, R.; Pang, N.; Huang, Y.; Liang, B.; Ling, W.; et al. Nicotinamide riboside attenuates alcohol induced liver injuries via activation of SirT1/PGC-1α/mitochondrial biosynthesis pathway. *Redox Biol.* **2018**, *17*, 89–98. [CrossRef]
38. Cerutti, R.; Pirinen, E.; Lamperti, C.; Marchet, S.; Sauve, A.A.; Li, W.; Leoni, V.; Schon, E.A.; Dantzer, F.; Auwerx, J.; et al. NAD^+-Dependent Activation of Sirt1 Corrects the Phenotype in a Mouse Model of Mitochondrial Disease. *Cell Metab.* **2014**, *19*, 1042–1049. [CrossRef]
39. Ge, Y.P.; Zhang, L.; Chen, W.L.; Sun, M.; Liu, W.B.; Li, X. Resveratrol Modulates the Redox Response and Bile Acid Metabolism to Maintain the Cholesterol Homeostasis in Fish *Megalobrama amblycephala* Offered a High-Carbohydrate Diet. *Antioxidants* **2023**, *12*, 121. [CrossRef]
40. Zhang, J.; Wang, X.; Vikash, V.; Ye, Q.; Wu, D.; Liu, Y.; Dong, W. ROS and ROS-Mediated Cellular Signaling. *Oxidative Med. Cell. Longev.* **2016**, *2016*, 4350965. [CrossRef]
41. Liguori, I.; Russo, G.; Curcio, F.; Bulli, G.; Aran, L.; Della-Morte, D.; Gargiulo, G.; Testa, G.; Cacciatore, F.; Bonaduce, D.; et al. Oxidative stress, aging, and diseases. *Clin. Interv. Aging* **2018**, *13*, 757–772. [CrossRef]
42. Schieber, M.; Chandel, N.S. ROS Function in Redox Signaling and Oxidative Stress. *Curr. Biol.* **2014**, *24*, R453–R462. [CrossRef]
43. Tian, L.; Zhou, X.Q.; Jiang, W.-D.; Liu, Y.; Wu, P.; Jiang, J.; Kuang, S.Y.; Tang, L.; Tang, W.N.; Zhang, Y.A.; et al. Sodium butyrate improved intestinal immune function associated with NF-κB and p38MAPK signalling pathways in young grass carp (*Ctenopharyngodon idella*). *Fish Shellfish Immunol.* **2017**, *66*, 548–563. [CrossRef]
44. Kauppinen, A.; Suuronen, T.; Ojala, J.; Kaarniranta, K.; Salminen, A. Antagonistic crosstalk between NF-κB and SIRT1 in the regulation of inflammation and metabolic disorders. *Cell. Signal.* **2013**, *25*, 1939–1948. [CrossRef]
45. Vringer, E.; Tait, S.W.G. Mitochondria and cell death-associated inflammation. *Cell Death Differ.* **2023**, *30*, 304–312. [CrossRef]
46. Reed, J.C. Mechanisms of apoptosis. *Am. J. Pathol.* **2000**, *157*, 1415–1430. [CrossRef]
47. Cohen, G.M. Caspases: The executioners of apoptosis. *Biochem. J.* **1997**, *326*, 1–16. [CrossRef] [PubMed]

48. Gibson, B.A.; Kraus, W.L. New insights into the molecular and cellular functions of poly(ADP-ribose) and PARPs. *Nat. Rev. Mol. Cell Biol.* **2012**, *13*, 411–424. [CrossRef]
49. Hassa, P.O.; Hottiger, M.O. The diverse biological roles of mammalian PARPS, a small but powerful family of poly-ADP-ribose polymerases. *Front. Biosci.* **2008**, *13*, 3046–3082. [CrossRef] [PubMed]
50. Polakof, S.; Panserat, S.; Soengas, J.L.; Moon, T.W. Glucose metabolism in fish: A review. *J. Comp. Physiol. B* **2012**, *182*, 1015–1045. [CrossRef] [PubMed]
51. Wilson, R.P. Utilization of dietary carbohydrate by fish. *Aquaculture* **1994**, *124*, 67–80. [CrossRef]
52. Fang, E.F.; Hou, Y.; Lautrup, S.; Jensen, M.B.; Yang, B.; SenGupta, T.; Caponio, D.; Khezri, R.; Demarest, T.G.; Aman, Y.; et al. NAD^+ augmentation restores mitophagy and limits accelerated aging in Werner syndrome. *Nat. Commun.* **2019**, *10*, 5284. [CrossRef] [PubMed]

antioxidants

Article

Phosphatidylserine Counteracts the High Stocking Density-Induced Stress Response, Redox Imbalance and Immunosuppression in Fish *Megalobrama ambylsephala*

Yangyang Jiang [1,†], Zishang Liu [2,†], Ling Zhang [2], Wenbin Liu [2], Haiyang Li [1] and Xiangfei Li [2,*]

1 Anhui Province Key Laboratory of Aquaculture and Stock Enhancement, Fisheries Research Institute, Anhui Academy of Agricultural Sciences, Hefei 230031, China
2 Key Laboratory of Aquatic Nutrition and Feed Science of Jiangsu Province, College of Animal Science and Technology, Nanjing Agricultural University, No. 1 Weigang Road, Nanjing 210095, China
* Correspondence: xfli@njau.edu.cn; Tel.: +86-025-8439-5382
† These authors contributed equally to this work.

Abstract: This study was conducted to investigate the effects of dietary phosphatidylserine (PS) supplementation on the growth performance, stress response, non-specific immunity and antioxidant capacity of juvenile blunt snout bream (*Megalobrama ambylcephala*) cultured under a high stocking density. A 2 × 2 two-factorial design was adopted, including two stocking densities (10 and 20 fish/m^3) and two dietary PS levels (0 and 50 mg/kg). After the 12-week feeding trial, the high stocking density significantly decreased the final weight; weight gain rate; specific growth rate; feed intake; nitrogen retention efficiency; plasma complement 3 (C3) level; albumin/globulin (ALB/GLB, A/G) ratio; activity of myeloperoxidase, lysozyme (LZM) and glutathione peroxidase (GPX); *gpx* transcription; and abundance of sirtuin3 (Sirt3) and nuclear factor erythroid-2-related factor 2 (Nrf2). However, it significantly increased the plasma levels of cortisol, glucose (GLU), lactic acid (LD), total protein and GLB; hepatic malondialdehyde (MDA) content; and *sirt1* transcription. PS supplementation significantly increased the plasma ALB and C4 levels; the A/G ratio; the activity of LZM, CAT and GPX; the transcription of *sirt1*, *nrf2*, manganese-containing superoxide dismutase and catalase; and the Nrf2 abundance. However, it significantly decreased the plasma levels of cortisol, GLU and GLB, as well as the hepatic MDA content. In addition, there was a significant interaction between the stocking density and PS supplementation regarding the effects on the plasma LD, ALB, GLB and C3 levels; A/G ratio; hepatic CAT activity; and protein abundance of Sod2. In conclusion, PS supplementation can counteract the high stocking density-induced stress response, redox imbalance and immunosuppression in blunt snout bream.

Keywords: stocking density; phosphatidylserine; stress response; antioxidant; fish culture

Citation: Jiang, Y.; Liu, Z.; Zhang, L.; Liu, W.; Li, H.; Li, X. Phosphatidylserine Counteracts the High Stocking Density-Induced Stress Response, Redox Imbalance and Immunosuppression in Fish *Megalobrama ambylsephala*. *Antioxidants* **2024**, *13*, 644. https://doi.org/10.3390/antiox13060644

Academic Editors: Alessandra Napolitano and Erchao Li

Received: 19 April 2024
Revised: 14 May 2024
Accepted: 23 May 2024
Published: 25 May 2024

1. Introduction

Recently, high-density intensive culture has become a common practice in modern aquaculture due to the increased demand for aquatic products and the pursuit of high profits [1]. However, high stocking densities can trigger a stress response in aquatic animals and compromise their health, thereby raising serious concerns about animal welfare. For example, water and air pollution problems are generally exacerbated under the high-density culture mode [2]. In addition, high-density culture can result in reduced growth performance, an altered body composition, a reduced antioxidant capacity and intestinal inflammation in fish [1,3–6]. This inevitably hinders the sustainable development of the aquaculture industry. Therefore, developing effective nutritional interventions to reduce the stress response and promote the health status of aquatic animals is crucial.

Phosphatidylserine (PS), also known as serine phospholipid or complex neuronic acid, is an active substance that constitutes the inner layer of biological cell membranes [7].

Previous studies have shown that PS is mainly found in brain cells and can (1) repair damaged nerve cell membranes and enhance the activity of nerve growth factors [8]; (2) change the fluidity of nerve cell membranes and increase the synthesis of acetylcholine [9]; and (3) reduce the levels of stress hormones like cortisol, thereby alleviating brain fatigue and mental stress [10]. To date, PS has been widely used as a nutritional supplement in various types of food and pharmaceuticals for sleep aids and the relief of low mood in humans [7,11]. However, its use in the aquaculture industry is still rarely reported. To the best of our knowledge, only one study has reported that PS can activate hemocyanin phenoloxidase activity to modulate the immune response in Atlantic horseshoe crab (*Limulus polyphemus*) [12]. However, whether it can alleviate the stress response in aquatic animals, thereby promoting their health, is still unknown.

Blunt snout bream (*Megalobrama ambylsephala*) is an herbivorous freshwater fish that is widely farmed in China due to its high survival rate, fast growth rate, disease resistance and flavorful meat, with annual production of 767,343 tons in 2022 [13,14]. Because of the increased demand for its products, high-density culture has become a common practice in the practical farming of this species [3]. This inevitably leads to a stress response in this species and a decreased antioxidant capacity and causes an inflammatory response [1,5]. Therefore, developing nutritional interventions to promote its health status is crucial. Considering this, the present experiment was designed to investigate the effects of dietary PS supplementation on the growth performance, stress response, non-specific immune function and antioxidant capabilities of blunt snout bream cultured under a high stocking density. The results can advance the development of effective nutritional interventions to ensure the welfare of aquatic species subjected to the high-density aquaculture mode.

2. Materials and Methods

2.1. Experimental Design, Fish and Feeding Trial

A 2 × 2 factorial design was employed in this study with two stocking densities (10 and 20 fish/m^3) and two dietary levels (0 and 50 mg/kg) of PS designated. Accordingly, a total of 4 experimental groups were set up, comprising a normal-density (10 fish/m^3) group without PS supplementation (ND), an ND group supplemented with 50 mg/kg PS (NDPS), a high-density (20 fish/m^3) group without PS supplementation (HD) and a HD group supplemented with 50 mg/kg PS (HDPS). The stocking densities were designated following a previous study that defined 240 g/m^3 fish as the high-density group of blunt snout bream at the start of the feeding trial [3]. The high-density group in this study was subjected to 380 g/m^3, taking into consideration the initial weight of the fish. Therefore, they were cultured under the high stocking density during the whole feeding trial. The dietary PS dose referred to the intake amount (600 mg per day) of PS in humans recommended by the National Health Commission of the People's Republic of China [15] and was adjusted taking into consideration the body weight and ration size of the fish.

Blunt snout bream was procured from an Ezhou (Hubei, China) fish hatchery and was short-term bred in several flowing cages (4 × 3 × 3 m, length–width–height) located in an artificial earthen pond. During this period, the fish were fed with a commercial feed (no. 122, Shuaifeng Feed Co., Ltd., Nanjing, China) for domestication for 2 weeks. Then, 720 healthy fish (initial weight, 19.73 ± 0.17 g) were randomly assigned to 12 cages (2 × 1 × 1 m, length–width–height) located in an artificial earthen pond with 20 fish per cage in the ND group and 40 fish per cage in the HD group. Then, the fish were fed the experimental diets (Table 1) three times (7:30, 12:00, 16:30 h) a day to visual satiation during a 12-week culture trial. Each group was tested in triplicate. Before conducting this study, we checked the feed intake data of several previous studies [16,17] using blunt snout bream as the target species. Then, an average ration size throughout the feeding trial was obtained. Accordingly, the dietary PS dosage was designated. This allowed the fish to receive the targeted amounts of PS during the whole feeding trial without frequently monitoring their body weight, which may trigger the stress response in fish and is not conducive to their health. During the feeding trial, the fish were cultured under the following conditions: the pH was maintained between 7.1 and 7.3, the water temperature varied between 26 and 28 °C, the dissolved oxygen was kept between 5.0 and 6.0 mg/L, and the total ammonia nitrogen was kept under 0.04 mg/L.

Table 1. Formulation and proximate composition of the experimental diets.

Ingredients (%)	Diets (Phosphatidylserine Level, mg/kg)	
	0	50
Fish meal	5.00	5.00
Soybean meal	18.00	18.00
Rapeseed meal	20.00	20.00
Cottonseed meal	15.00	15.00
Fish oil	2.16	2.16
Soybean oil	2.16	2.16
Wheat flour	24.00	24.00
Wheat bran	4.50	4.50
Cellulose	5.99	5.99
$Ca(H_2PO_4)_2$	1.80	1.80
Salt	0.40	0.40
Phosphatidylserine [1]	0.00	50.00
Premix [2]	1.00	1.00
Proximate composition (%)		
Moisture	8.98	8.87
Crude protein	30.49	30.51
Crude lipid	5.94	5.99
Ash	6.62	6.58
Gross energy (MJ/kg)	18.32	18.34

[1] Phosphatidylserine (Cat#S832149-5g, Macklin, Shanghai, China) with purification of 50% was used in a double dose (100 mg/kg) to achieve the target concentration (50 mg/kg) in this study. [2] The premix supplied the following minerals (g/kg) and vitamins (IU or mg/kg): $CuSO_4 \cdot 5H_2O$, 2.0 g; $FeSO_4 \cdot 7H_2O$, 25 g; $ZnSO_4 \cdot 7H_2O$, 22 g; $MnSO_4 \cdot 4H_2O$, 7 g; Na_2SeO_3, 0.04 g; KI, 0.026 g; $CoCl_2 \cdot 6H_2O$, 0.1 g; vitamin A, 900,000 IU; vitamin D, 200,000 IU; vitamin E, 4500 mg; vitamin K_3, 220 mg; vitamin B_1, 320 mg; vitamin B_2, 1090 mg; vitamin B_5, 2000 mg; vitamin B_6, 500 mg; vitamin B_{12}, 1.6 mg; vitamin C, 5000 mg; pantothenate, 1000 mg; folic acid, 165 mg; choline, 60,000 mg.

2.2. Sample Collection

The fish in each cage were numbered with their weights and lengths measured after a 24 h fast to determine the growth performance parameters. Next, 4 randomly chosen fish from each cage were anesthetized with 100 mg/L of tricaine methanesulfonate (Sigma, Saint Louis, MO, USA). Shortly after, blood samples were taken from the tail veins using disposable medical syringes. Following centrifugation (3000× g at 4 °C for 10 min), plasma samples were collected and kept at −20 °C for subsequent analysis. To determine the biometric parameters, samples of the liver, viscera, and intraperitoneal fat were obtained and weighed and then stored in liquid nitrogen for subsequent analysis.

2.3. Analytical Procedures

2.3.1. Growth Performance and Feed Utilization Formula

$$\text{Survival rate (SR, \%)} = 100 \times \frac{\text{final number of fish in each cage}}{\text{initial number of fish in each cage}}$$

$$\text{Weight gain rate (WGR, \%)} = 100 \times \frac{[\text{final body weight (g)} - \text{initial body weight (g)}]}{\text{initial body weight (g)}}$$

$$\text{Specific growth rate (SGR, \%/d)} = 100 \times \frac{(\ln W_t - \ln W_0)}{\text{days}}$$

$$\text{Feed intake(FI, g per fish)} = \frac{\text{total feed intake of each cage (g)}}{\text{total number of fish in each cage}}$$

$$\text{Feed conversion rate (FCR)} = \frac{\text{feed intake (g)}}{\text{weight gain (g)}}$$

$$\text{Protein efficiency ratio (PER)} = \frac{\text{weight gain (g)}}{\text{protein intake (g)}}$$

$$\text{Nitrogen/energy retention efficiency (NRE/ERE, \%)} = 100 \times \frac{[(W_t \times C_t) - (W_0 \times C_0)]}{C_{diet} \times \text{feed intake (g)}}$$

$$\text{Hepatosomatic index (HSI, \%)} = 100 \times \frac{\text{liver weight (g)}}{\text{body weight (g)}}$$

$$\text{Condition factor (CF)} = 100 \times \frac{\text{body weight(g)}}{\text{body length (cm)}^3}$$

In the formulas, W_0 and W_t are the initial and final weights, C_0 and C_t are the initial and final nutrient content in the body, and C_{diet} is the nutrient content in the diet.

2.3.2. Proximate Composition Analysis

The proximate composition of the experimental diets was measured according to the AOAC [18]. To determine the moisture content, samples were dried at 105 °C until they reached a consistent weight. The crude protein content (nitrogen content × 6.25) was estimated using an automated Kjeldahl nitrogen instrument (FOSS KT260, Herisau, Switzerland) to measure the nitrogen concentration. By using an ether extraction method, a Soxhlet system (Soxtec System HT6, Tecator, Höganäs, Sweden) was utilized to measure the lipid content. The samples were burned for four hours at 550 °C to determine the amount of ash. Finally, the gross energy content was determined using a bomb calorimeter (PARR 1281, Parr Instrument Company, Moline, IL, USA).

2.3.3. Plasma Indicator Analysis

The plasma level of cortisol was measured following the protocols reported by Winberg and Lepage [19]. The glucose (GLU) level was measured using the glucose oxidase method [20]. The lactic acid (LD) level was determined by the method reported by Shuang, et al. [21]. The plasma total protein (TP) and albumin (ALB) content were both determined by the method reported by Li et al. [22]. By deducting the albumin values from the total protein, the globulin (GLB) content was determined. Moreover, by dividing the ALB values by the GLB values, the A/G ratio was computed. The plasma complement 3 (C3) and 4 (C4) levels were measured using an enzyme-linked immunosorbent assay method [23]. Aspartate aminotransferase (AST) and alanine aminotransferase (ALT) activity were both measured by the method reported by Yuan et al. [24]. Lysozyme (LZM) and myeloperoxidase (MPO) activity were both measured according to Zhang et al. [25]. The temperature of the LZM enzymatic reaction was 25 °C, and the substrate concentration was the concentration when the *Lysomicrococcus Substration* reached 0.65–0.75 under absorbance at 450 nm at 25 °C. The substrate of the MPO enzymatic reaction was 3% H_2O_2 and the enzymatic reaction temperature was 37 °C.

2.3.4. Hepatic Antioxidant Analysis

Liver samples were prepared according to LYGREN and WAAGBØ [26]. The content of malondialdehyde (MDA) was measured using the procedures outlined by Satoh [27]. The activity of catalase (CAT), glutathione peroxidase (GPX), and superoxide dismutase (SOD) was measured by the method reported by Zhang et al. [28]. The soluble protein content of liver homogenates was measured using the technique reported by Bradford [29].

2.3.5. Real-Time Quantitative PCR

Using an RNA purification kit (Invitrogen, Carlsbad, CA, USA), the total RNA from the liver was extracted. Then, the extracted RNA's purity and concentration were evaluated using the absorbance at 260 and 280 nm, respectively. The RNA's reverse transcription was performed using an RT-PCR kit (SYBR® Prime Script™, Accurate Biology, Changsha, China). Next, using ddH_2O, the resulting cDNA was diluted to 10%. On a QuantStudio7 Flex Real-time PCR instrument (Thermo Fisher, Waltham, MA, USA), the polymerase chain reaction was carried out using the SYBR® Green II fluorescence kit (Accurate Biology, Changsha, China). Next, using ddH_2O, the resulting cDNA was diluted to 10%. Ten microliters of 2 × SYBR® Green real-time PCR master mix (Accurate Biology, Changsha, China), 7.2 microliters of water treated with DEPC, two microliters of DNA, and 0.4 microliters of each forward and reverse primer, forming a total volume of 20 microliters, were included in the reaction system. One cycle at 95 °C for 30 s and forty cycles from 95 °C maintained for 5 s to 60 °C maintained for 30 s were used to perform the reaction. Melting curve analysis was carried out with warming to 95 °C for 15 s and from 95 °C to 60 °C for 1 min. The primers were designed and synthesized based on the available sequences (Table 2). Using the $2^{-\Delta\Delta CT}$ approach [30], the transcription of the target genes was standardized by elongation factor 1 alpha, which served as a reference gene.

Table 2. Nucleotide sequences of primers used to assay gene expression by RT-PCR.

Gene Name	Forward and Reverse Primers (5′-3′)	Accession Number or Reference
sirt1	CAAACGACTCGGAGCCTCAC GGTCTCGTCTTCCGAACTGG	MT518159.1
nrf2	CTTTGATGGATGCCTTCGGC TCTGGGTAACGGGTGAATGC	[31]
keap1	TGAGGAGATCGGCTGCACTG TGGCAATGGGACAAGCTGAA	[31]
mnsod	TGTTGGAGGCCATTAAGCGT AAAGGGTCTTGGTTAGCGCA	KF195932.1
cuznsod	CACGCTCAACTTTGGCACAT TGTCAACAGGGAGACCATGC	KF479046.1
cat	CCGGGGGATATCAGTTGGGT TCCAAACCACTGAACTCGGG	KF378714.1
gpx	GAACGCCCACCCTCTGTTTG CGATGTCATTCCGGTTCACG	[32]
ef1a	CTTCTCAGGCTGACTGTGC CCGCTAGCATTACCCTCC	X77689.1

sirt1, sirtuin 1; *nrf2*, nuclear factor erythroid-2-related factor 2; *keap1*, recombinant kelch-like ech-associated protein 1; *mnsod*, manganese superoxide dismutase; *cuznsod*, copper-zinc superoxide dismutase; *cat*, catalase; *gpx*, glutathione peroxidase; *ef1a*, elongation factor 1 alpha.

2.3.6. Western Blotting Assay

Phenylmethanesulfonyl fluoride (PMSF, Cat#20104ES03, Yeasen Biotechnology, Shanghai, China) was mixed with the RIPA lysis buffer (Cat#20101ES60, Yeasen Biotechnology, Shanghai, China) to produce a protein lysate containing 1 mM of PMSF. Then, the liver tissue was homogenized at 4 °C using this lysate. After this, the protein concentrations in the supernatants were measured using a BCA kit (Cat#E112-01, Vazyme, Nanjing, China). Using the lysates mentioned above, the protein concentrations in the samples were normalized. The proteins in the lysate were separated using 4–20% precast protein plus gel (Cat#36250ES10, Yeasen Biotechnology Co., Ltd., Shanghai, China) electrophoresis prior to being deposited onto polyvinylidene fluoride membranes. The membranes were blocked for 15 min in a fast-blocking solution (Cat#36122ES60, Yeasen Biotechnology, Shanghai, China) and then were treated with primary antibodies overnight at 4 °C. Anti-beta-actin (42 KDa, 1:1000, 66009-1-Ig, Proteintech, Wuhan, China), anti-sirtuin 3 (Sirt3) (28 KDa, 1:2000, Proteintech, 10099-1-AP, Wuhan, China), anti-superoxide dismutase 2 (Sod2) (25 KDa, 1:1000, Proteintech, 24127-1-AP, Wuhan, China), and anti-nuclear factor erythroid-2-related factor 2 (Nrf2) (68 KDa, 1:2000, Proteintech, 16396-1-AP, Wuhan, China) were all used. After 3 TBST washes, the membranes were treated for 2 h with secondary antibodies (1:5000, BA1054, Boster, Wuhan, China). The immunoreactive bands were found using a high-sensitivity chemiluminescence kit (E412-01/02, Vazyme, Nanjing, China). The bands were visualized and quantified using the Image J (Image J 1.53t, Bethesda, MD, USA) software with beta-actin used to standardize the protein expression.

2.4. Statistical Analysis

Two-way ANOVA (SPSS 22.0, SPSS Inc., Chicago, IL, USA) was used to evaluate all data in order to identify any significant differences between the treatments in terms of the stocking density, the dietary PS supplement, and their interactions. If the interactive effects were significant ($p < 0.05$), one-way ANOVA was conducted for further analysis, accounting for the normality and chi-square of the data distribution. If significance ($p < 0.05$) was detected, the means were subsequently ordered using Tukey's HSD multiple range test. All data are expressed as the mean ± S.E. (standard error of the mean).

3. Results

3.1. Growth Performance and Feed Utilization

As shown in Table 3, neither the culture density nor the PS supplementation affected the survival rate (SR), feed conversion ratio (FCR), hepatopancreas somatic index (HSI), condition factor (CF), protein efficiency ratio (PER), and energy retention efficiency (ERE) ($p > 0.05$). However, a high stocking density significantly reduced ($p < 0.01$) the final weight (FW), weight gain rate (WGR), specific growth rate (SGR), feed intake (FI), and nitrogen retention efficiency (NRE).

Table 3. Growth performance and feed utilization of blunt snout bream subjected to different stocking densities and phosphatidylserine supplementation.

	IW (g)	FW (g)	SR (%)	WGR (%)	SGR (%/d)	FI (g)	FCR	PER	NRE (%)	ERE (%)	HSI (%)	CF
ND	19.73 ± 0.64	82.39 ± 3.04	96.67 ± 3.33	317.49 ± 7.41	1.70 ± 0.02	114.81 ± 7.52	1.84 ± 0.16	1.81 ± 0.15	38.33 ± 4.09	20.26 ± 2.67	1.23 ± 0.04	1.96 ± 0.03
NDPS	19.67 ± 0.41	94.44 ± 2.86	90.00 ± 5.77	380.01 ± 4.68	1.87 ± 0.01	128.06 ± 5.10	1.71 ± 0.13	1.92 ± 0.01	44.21 ± 2.53	22.07 ± 2.33	1.22 ± 0.02	2.09 ± 0.05
HD	19.60 ± 0.00	69.19 ± 1.65	91.67 ± 3.33	253.01 ± 8.42	1.50 ± 0.03	92.38 ± 1.25	1.87 ± 0.04	1.76 ± 0.04	32.73 ± 1.34	19.10 ± 1.82	1.26 ± 0.05	1.96 ± 0.06
HDPS	19.93 ± 0.26	70.27 ± 4.79	91.67 ± 6.01	253.23 ± 28.54	1.49 ± 0.09	97.91 ± 7.67	1.96 ± 0.11	1.68 ± 0.09	32.60 ± 3.24	16.48 ± 1.50	1.26 ± 0.04	1.97 ± 0.03
Two-way ANOVA												
SD	ns	***	ns	***	***	**	ns	ns	*	ns	ns	ns
PS	ns	ns	ns	ns	ns	ns	ns	ns	ns	ns	ns	ns
Interaction	ns	ns	ns	ns	ns	ns	ns	ns	ns	ns	ns	ns

SD, stocking density; PS, phosphatidylserine; ND, normal density; NDPS, ND supplemented with 50 mg/kg PS; HD, high density; HDPS, HD supplemented with 50 mg/kg PS; IW, initial weight; FW, final weight. The two-way ANOVA result is indicated by asterisks. * $p < 0.05$, ** $p < 0.01$, *** $p < 0.001$, ns: not significant.

3.2. Stress Response Indicators

As shown in Figure 1, the high stocking density significantly elevated ($p < 0.001$) the plasma cortisol, GLU, and LD concentrations. However, PS supplementation significantly reduced ($p < 0.001$) the cortisol and GLU levels. In addition, a significant interactive effect ($p < 0.05$) between the stocking density and PS supplementation was noted for the plasma LD level.

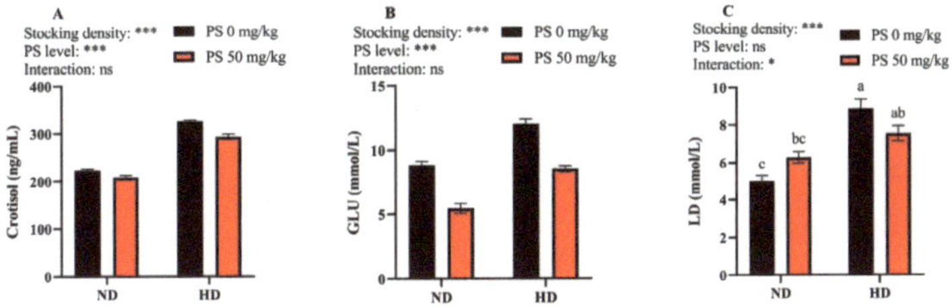

Figure 1. Plasma concentrations of cortisol (**A**), glucose (GLU, **B**), and lactic acid (LD, **C**) in juvenile blunt snout bream subjected to different stocking densities and phosphatidylserine supplementation. SD, stocking density; PS, phosphatidylserine; ND, normal density; HD, high density. Each datum represents the mean of four replicates. The two-way ANOVA result is indicated by asterisks. * $p < 0.05$, *** $p < 0.001$, ns: not significant. Significant differences ($p < 0.05$) between groups are indicated with different letters.

3.3. Hepatic Injury and Non-Specific Immunity Indicators

Figure 2 shows that neither the culture density nor the PS supplementation affected the plasma ALT and AST activity ($p > 0.05$). The plasma TP and GLB concentrations increased significantly ($p < 0.01$) under the high-density culture, while the opposite result was observed in the plasma A/G ratio as well as the C3, MPO, and LZM activity ($p < 0.05$). In addition, PS supplementation significantly increased ($p < 0.05$) the plasma ALB and C4 levels, the A/G ratio, and the LZM activity, while it significantly ($p < 0.05$) decreased the GLB levels. Furthermore, the plasma ALB, GLB, and C3 levels, as well as the A/G ratio, were significantly ($p < 0.05$) affected by the interaction between the stocking density and PS supplementation.

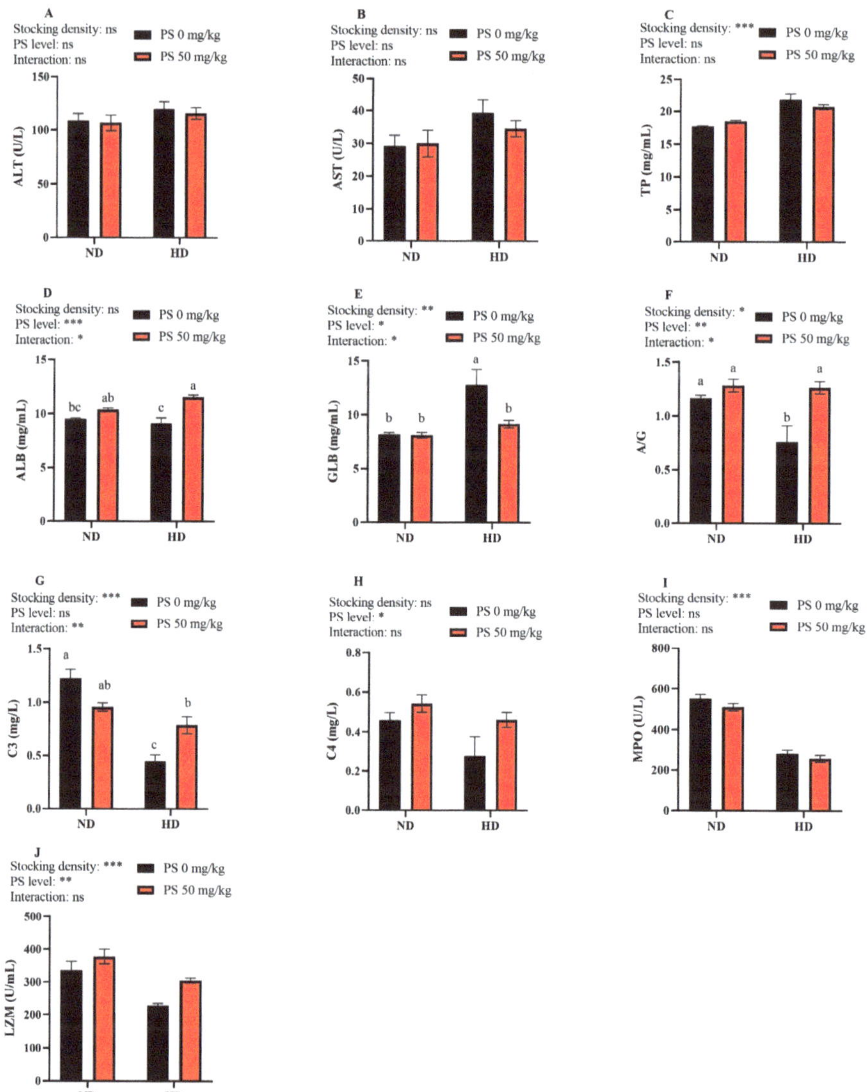

Figure 2. Plasma levels of total protein (TP, **C**), albumin (ALB, **D**), globulin (GLB, **E**), ALB/GLB (A/G, **F**), complement 3 (C3, **G**), and complement 4 (C4, **H**), as well as the activity of alanine

aminotransferase (ALT, **A**), aspartate aminotransferase (AST, **B**), myeloperoxidase (MPO, **I**), and lysozyme (LZM, **J**), in juvenile blunt snout bream subjected to different stocking densities and phosphatidylserine supplementation. SD, stocking density; PS, phosphatidylserine; ND, normal density; HD, high density. Each datum represents the mean of four replicates. The two-way ANOVA result is indicated by asterisks. * $p < 0.05$, ** $p < 0.01$, *** $p < 0.001$, ns: not significant. Significant differences ($p < 0.05$) between groups are indicated with different letters.

3.4. Hepatic Antioxidant Indices

Figure 3 shows that neither the culture density nor the PS supplementation affected the hepatic SOD activity ($p > 0.05$). The high-density culture significantly ($p < 0.01$) reduced the hepatic GPX activity but elevated the MDA content. The dietary supplementation of PS significantly ($p < 0.05$) increased the liver CAT and GPX activity, but reduced the MDA content. Moreover, a significant interaction ($p < 0.05$) between the culture density and PS supplementation was noted for the hepatic CAT activity.

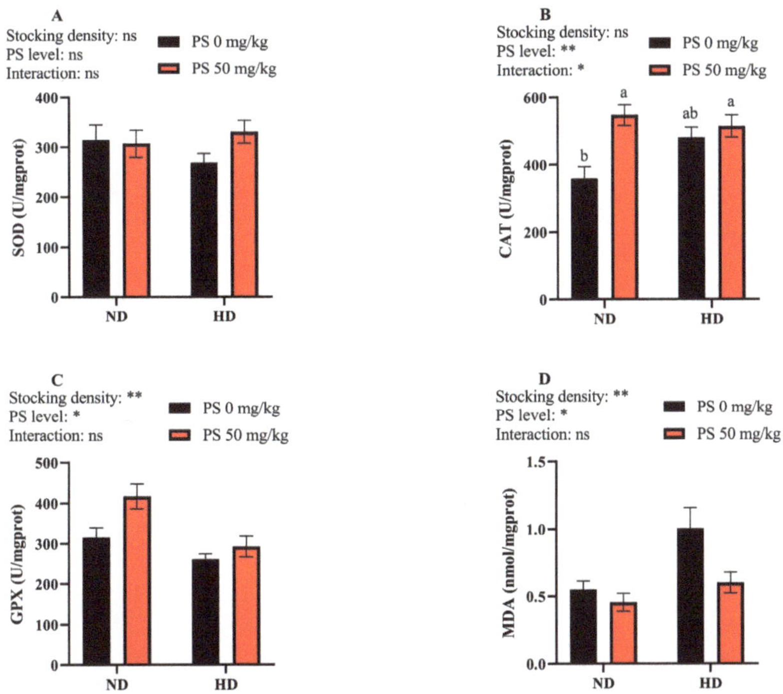

Figure 3. Hepatic activity of superoxide dismutase (SOD, **A**), catalase (CAT, **B**), and glutathione peroxidase (GPX, **C**), as well as the malondialdehyde (MDA, **D**) content, in blunt snout bream subjected to different stocking densities and phosphatidylserine supplementation. SD, stocking density; PS, phosphatidylserine; ND, normal density; HD, high density. Each datum represents the mean of four replicates. The two-way ANOVA result is indicated by asterisks. * $p < 0.05$, ** $p < 0.01$, ns: not significant. Significant differences ($p < 0.05$) between groups are indicated with different letters.

3.5. Expression Levels of Hepatic Antioxidant-Related Genes

As shown in Figure 4, neither the culture density nor the PS supplementation affected the hepatic transcription of *keap1* and *cuznsod* ($p > 0.05$). The high stocking density significantly ($p < 0.05$) inhibited *gpx* expression but promoted *sirt1* expression. The dietary supplementation of PS significantly ($p < 0.05$) promoted the transcription of *sirt1*, *nrf2*, *mnsod*, and *cat*.

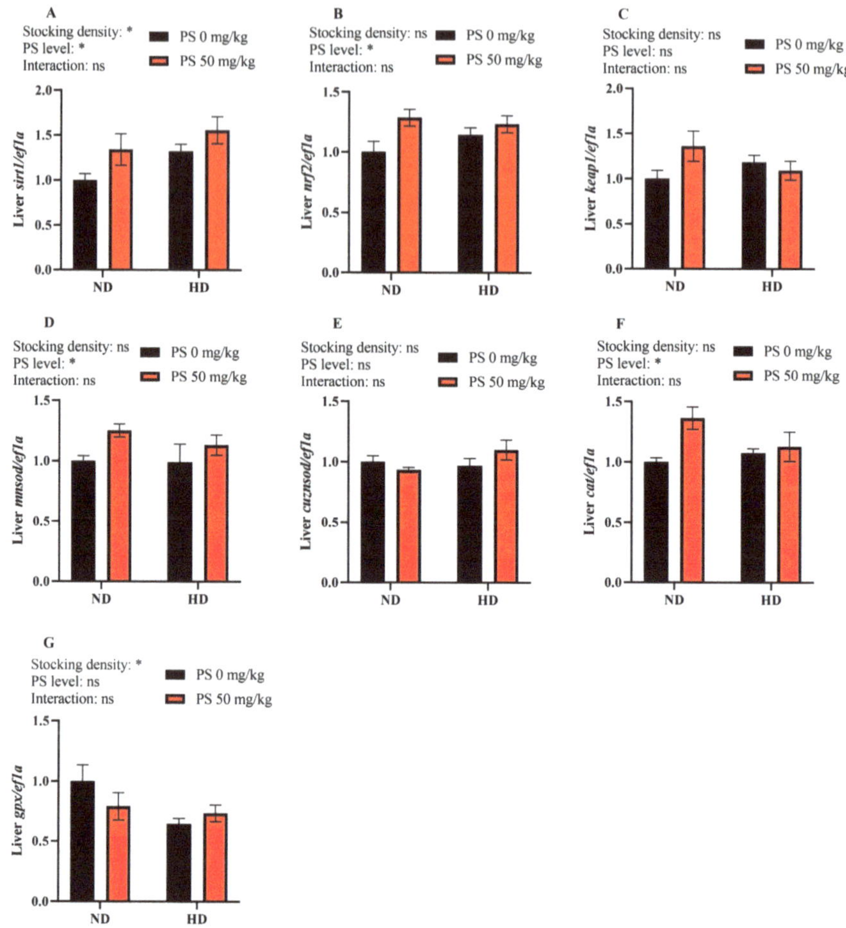

Figure 4. Hepatic transcription of sirtuin 1 (*sirt1*, **A**), nuclear factor erythroid-2-related factor 2 (*nrf2*, **B**), recombinant kelch-like ech-associated protein 1 (*keap1*, **C**), manganese superoxide dismutase (*mnsod*, **D**), copper-zinc superoxide dismutase (*cuznsod*, **E**), catalase (*cat*, **F**), and glutathione peroxidase (*gpx*, **G**) in blunt snout bream subjected to different stocking densities and phosphatidylserine supplementation. SD, stocking density; PS, phosphatidylserine; ND, normal density; HD, high density. Each datum represents the mean of four replicates. The two-way ANOVA result is indicated by asterisks. * $p < 0.05$, ns: not significant.

3.6. Expression Levels of Hepatic Antioxidant-Related Proteins

As shown in Figure 5, neither the culture density nor the PS supplementation affected the protein abundance of Sod2 ($p > 0.05$). However, the high stocking density significantly ($p < 0.01$) inhibited the protein abundance of Sirt3 and Nrf2, and PS significantly ($p < 0.001$) elevated the abundance of Nrf2. In addition, the stocking density and PS supplementation exerted a significant ($p < 0.05$) interactive effect on the Sod2 abundance, but no significance was observed among the four groups ($p > 0.05$).

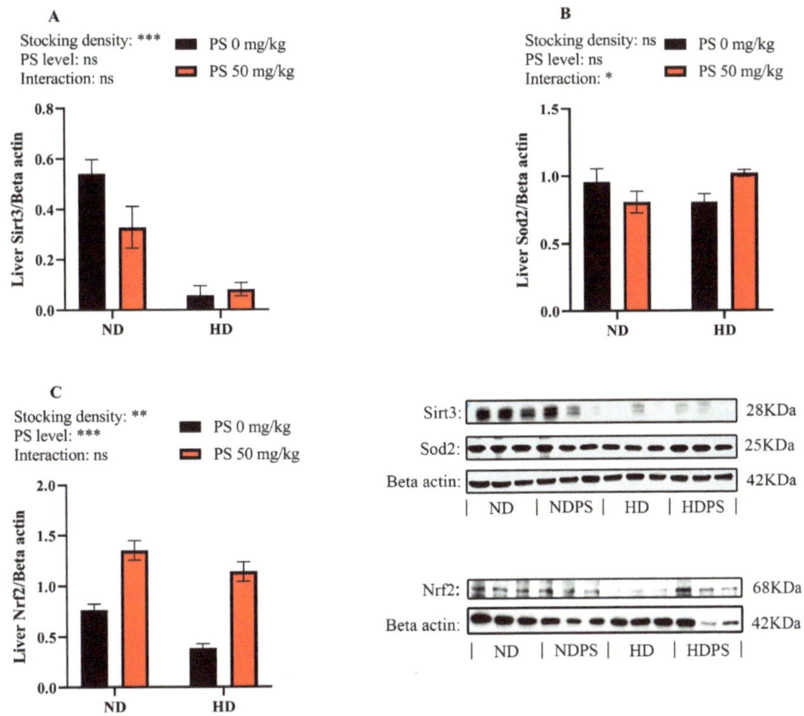

Figure 5. Hepatic protein expressions of sirtuin 3 (Sirt3, **A**), superoxide dismutase 2 (Sod2, **B**), and nuclear factor erythroid-2-related factor 2 (Nrf2, **C**) in blunt snout bream subjected to different stocking densities and phosphatidylserine supplementation. SD, stocking density; PS, phosphatidylserine; ND, normal density; NDPS, ND supplemented with 50 mg/kg PS; HD, high density; HDPS, HD supplemented with 50 mg/kg PS. Each datum represents the mean of three replicates. The two-way ANOVA result is indicated by asterisks. * $p < 0.05$, ** $p < 0.01$, *** $p < 0.001$, ns: not significant.

4. Discussion

In the present study, a high stocking density significantly reduced the FW, WGR, SGR, FI, and NRE of blunt snout bream, indicating that the growth performance and appetite of the fish were suppressed at a high stocking density. This result was parallel to those of previous studies on the same species [5,33]. The following reasons can be given for these results: (1) high-density aquaculture causes the competition of fish for food and space, which ultimately leads to a reduction in growth performance and food intake [3]; (2) a high stocking density can reduce the digestive enzyme activity of fish, thereby resulting in poor feed utilization and growth retardation [1]; and (3) high-density aquaculture can inhibit the expression of growth-related genes like growth hormones, growth hormone receptors, and insulin-like growth factor 1 in fish, thus reducing their growth rates [33]. However, neither the stocking density nor the PS supplementation affected the SR, FCR, HSI, CF, PER, and ERE of the fish, indicating that both the stocking density and PS administration had no significant influence on their survival and feed utilization. The following are the possible reasons for this: (1) when confronted with stimuli induced by the culture density, the stress response of fish enables them to maintain their homeostasis without disruption [34], and (2) the stocking density and PS supplementation in this study were not high enough to affect the survival and feed utilization of blunt snout bream.

In the current study, the high stocking density markedly elevated the plasma cortisol, GLU and LD concentrations, suggesting that high-density culture causes a stress response in blunt snout bream. The following theories support this conclusion: (1) cortisol is generally released when fish are subjected to stressful conditions [34]; (2) elevated cortisol levels can enhance the gluconeogenic pathways of fish, thereby producing high amounts of GLU to provide sufficient energy to cope with stress [35]; and (3) anaerobic metabolism is enhanced in stressed fish, thereby leading to LD accumulation [36]. Moreover, stress reduces the appetite and feed intake of fish [37,38], which may account for the decrease in FI and feed utilization at high densities. However, the administration of PS markedly lowered the plasma cortisol and GLU levels, suggesting that PS can alleviate the stress response induced by crowding stress in blunt snout bream. This is not surprising, as a previous study has reported that the prolonged administration of PS counteracts the stress-induced activation of the hypothalamic–pituitary–adrenal axis in humans [39]. In addition, a significant interaction between the stocking density and PS supplementation was noted in the plasma LD levels, with PS elevating the LD level at a normal density but decreasing it at a high density. It is hypothesized that PS, as a nutraceutical, is metabolized through the liver, and the long-term administration of PS to healthy fish would increase the metabolic burden on the liver, which is also the site of LD metabolism [40]. However, at high-density conditions, PS can alleviate the stress response in fish, which in turn reduces LD levels [39].

In the present study, the high stocking density elevated the plasma activity of ALT and AST, although no significant difference was noted. This implies that the high-density aquaculture mode might lead to minor liver injury in blunt snout bream. In addition, PS supplementation in the ND group elevated the plasma AST activity, but it decreased in the HD group, confirming the conjecture about the metabolic burden on the liver caused by PS supplementation at a normal density. The high stocking density significantly decreased the plasma C3 levels and MPO and LZM activity, suggesting that the high-density culture reduced the non-specific immunity of blunt snout bream. It is widely acknowledged that (1) C3 is the most abundant component of the complement system, and its level can reflect the body's immune capability [41]; (2) MPO is an enzyme involved in phagocytosis for bactericidal purposes [42]; and (3) LZM has antibacterial and antiviral effects by attacking the cell walls of bacteria for lysis, as well as inactivating viruses by binding directly to viral proteins [43]. However, the levels of TP and GLB showed opposite results, which might have been caused by the liver injury [44]. The decrease in the A/G ratio further supported this speculation. In addition, the dietary administration of PS markedly elevated the plasma ALB and C4 levels, the A/G ratio, and the LZM activity, suggesting that PS can enhance the non-specific immunity of fish. According to Coates, Kelly and Nairn [12], PS can enhance the immunity of Atlantic horseshoe crab (*Limulus polyphemus*) by activating hemocyanin. Furthermore, a significant interaction between the stocking density and PS supplementation was noted in the ALB, GLB, and C3 levels, as well as the A/G ratio. At a normal density, PS exerted no significant effect; however, PS significantly elevated these parameters at a high density. It was implied that the immune-enhancing role of PS might only be displayed under stressful conditions.

In this study, the high stocking density significantly reduced the GPX activity and elevated the MDA content, suggesting that the high-density aquaculture mode could reduce the antioxidant capacity of bunt snout bream. According to Yu, Yang, Liang, Ren, Ge, and Ji [4], the culture density can affect the mRNA expression of *gpx1* in this fish species, thereby influencing the redox defense. In addition, the administration of PS markedly enhanced the antioxidant capacity, as was revealed by the increased CAT and GPX activity and the decreased MDA content This result was expected, since PS has been reported to repair damaged cell membranes and improve the antioxidant capacity of the body [7,8,45]. In addition, there was a significant interaction between the stocking density and PS supplementation regarding CAT activity. PS significantly elevated the CAT activity at a normal density but exerted a limited effect at a high density. This difference is difficult to explain due to the fact that the relevant literature is lacking.

To further unveil the molecular mechanisms underlying the antioxidant defense of fish, the transcription of the Keap1–Nrf2 pathway-related genes and proteins, as well as the abundance of the mitochondrial antioxidant Sirt3–Sod2 pathway-related proteins, was measured. The results showed that the high stocking density significantly elevated the *sirt1* transcription but decreased the *gpx* transcription and the protein levels of Sirt3 and Nrf2, suggesting that the fish were under oxidative stress when subjected to the high-density aquaculture mode, as *sirt1* is generally activated to reduce damage when animals are subjected to oxidative stress [46,47]. In addition, PS supplementation markedly increased the transcription of *sirt1*, *nrf2*, *mnsod*, and *cat* as well as the Nrf2 abundance, suggesting an enhancement in the antioxidant capacity. Previous studies have shown that under a normal state, Nrf2 binds to Keap1 and is inactivated; however, when the organism is subjected to oxidative stress, Nrf2 is released and activated by Sirt1 and subsequently binds to the antioxidant response elements, thereby exerting antioxidant effects by targeting several downstream effectors, including Mnsod [48,49]. Furthermore, an interactive effect between the stocking density and PS supplementation was noted in the protein abundance of Sod2, as was manifested by a decrease in Sod2 expression by PS at a normal density and an increase at a high density. This suggests that PS might only be effective in enhancing the mitochondrial antioxidant capabilities under stressful conditions.

5. Conclusions

In summary, the present study indicated that a high stocking density induced a stress response in blunt snout bream and reduced its growth performance, non-specific immunity, and antioxidant capacity. Dietary supplementation with 50 mg/kg PS can counteract these side effects caused by the high-density aquaculture mode. Based on this, PS has high potential to be used as a functional feed additive to promote the health status of fish cultured under high-density conditions, thereby improving the economic efficiency of intensive fish farming.

Author Contributions: Y.J. and Z.L. contributed equally to the article regarding the methodology, software, validation, formal analysis, investigation, writing—original draft and visualization; L.Z.: software and investigation; W.L.: resources, project administration and funding acquisition; H.L.: project administration; X.L.: writing—review and editing, supervision, project administration and funding acquisition. All authors have read and agreed to the published version of the manuscript.

Funding: This research was finally supported by earmarked funds for the China Agriculture Research System (CARS-45) and the Carbon Peak and Carbon Neutrality Technology Innovation Foundation of Jiangsu Province (BE2022421).

Institutional Review Board Statement: The permission to handle animals in this work was given by Nanjing Agricultural University's Animal Care and Use Committee (license number: SYXK (Su) 2011-0036).

Informed Consent Statement: Not applicable.

Data Availability Statement: The data generated during the current study are available from the first author.

Conflicts of Interest: The authors declare no conflicts of interest.

References

1. Wang, Y.W.; Zhu, J.; Ge, X.p.; Sun, S.M.; Su, Y.L.; Li, B.; Hou, Y.R.; Ren, M.C. Effects of stocking density on the growth performance, digestive enzyme activities, antioxidant resistance, and intestinal microflora of blunt snout bream (*Megalobrama amblycephala*) juveniles. *Aquac. Res.* **2018**, *50*, 236–246. [CrossRef]
2. Hu, Y.; Yang, T.; Liu, Y.; Li, F.; Xu, C.; Fang, F.; Feng, J. High Fish Stocking Density Weakens the Effects of Rice-Fish Co-culture on Water Eutrophication and Greenhouse Gas Emissions. *Water Air Soil Pollut.* **2022**, *233*, 222. [CrossRef]
3. Yadata, G.W.; Ji, K.; Liang, H.; Ren, M.; Ge, X.; Yang, Q. Effects of dietary protein levels with various stocking density on growth performance, whole body composition, plasma parameters, nitrogen emission and gene expression related to TOR signaling of juvenile blunt snout bream (*Megalobrama ambylcephala*). *Aquaculture* **2020**, *519*, 734730. [CrossRef]
4. Yu, H.; Yang, Q.; Liang, H.; Ren, M.; Ge, X.; Ji, K. Effects of stocking density and dietary phosphorus levels on the growth performance, antioxidant capacity, and nitrogen and phosphorus emissions of juvenile blunt snout bream (*Megalobrama amblycephala*). *Aquac. Nutr.* **2020**, *27*, 581–591. [CrossRef]

5. Qi, C.; Xie, C.; Tang, R.; Qin, X.; Wang, D.; Li, D. Effect of Stocking Density on Growth, Physiological Responses, and Body Composition of Juvenile Blunt Snout Bream, *Megalobrama amblycephala*. *J. World Aquac. Soc.* **2016**, *47*, 358–368. [CrossRef]
6. Lund, I.; Steenfeldt, S.J.; Herrmann, B.; Pedersen, P.B. Feed intake as explanation for density related growth differences of common sole Solea solea. *Aquac. Res.* **2013**, *44*, 367–377. [CrossRef]
7. Schmitt, H. Phosphatidylserine A natural brain nutrient. *Agro Food Ind. Hi-Tech* **2009**, *20*, 16–18.
8. Leventis, P.A.; Grinstein, S. The Distribution and Function of Phosphatidylserine in Cellular Membranes. *Annu. Rev. Biophys.* **2010**, *39*, 407–427. [CrossRef] [PubMed]
9. Kidd, P.M. Attention deficit/hyperactivity disorder (ADHD) in children: Rationale for its integrative management. *Altern. Med. Rev.* **2000**, *5*, 402–428.
10. Hellhammer, J.; Fries, E.; Buss, C.; Engert, V.; Tuch, A.; Rutenberg, D.; Hellhammer, D. Effects of soy lecithin phosphatidic acid and phosphatidylserine complex (PAS) on the endocrine and psychological responses to mental stress. *Stress* **2004**, *7*, 119–126. [CrossRef]
11. Toita, R.; Fujita, S.; Kang, J.-H. Macrophage Uptake Behavior and Anti-inflammatory Response of Bovine Brain- or Soybean-derived Phosphatidylserine Liposomes. *J. Oleo Sci.* **2018**, *67*, 1131–1135. [CrossRef] [PubMed]
12. Coates, C.J.; Kelly, S.M.; Nairn, J. Possible role of phosphatidylserine–hemocyanin interaction in the innate immune response of *Limulus polyphemus*. *Dev. Comp. Immunol.* **2011**, *35*, 155–163. [CrossRef] [PubMed]
13. Ji, K.; Liang, H.; Chisomo-Kasiya, H.; Mokrani, A.; Ge, X.; Ren, M.; Liu, B. Effects of dietary tryptophan levels on growth performance, whole body composition and gene expression levels related to glycometabolism for juvenile blunt snout bream, *Megalobrama amblycephala*. *Aquac. Nutr.* **2018**, *24*, 1474–1483. [CrossRef]
14. *China Fishery Statistical Yearbook*; Chinese Agricultural Press: Beijing, China, 2023.
15. National Health Commission of the People's Republic of China. 1 November 2010. Available online: www.nhc.gov.cn/sps/s7891/201010/3b5fec5548404e0b965ca9605854d0ba.shtml (accessed on 15 April 2022).
16. He, C.F.; Li, X.F.; Jiang, G.Z.; Zhang, L.; Sun, M.; Ge, Y.P.; Chen, W.L.; Liu, W.B. Feed types affect the growth, nutrient utilization, digestive capabilities, and endocrine functions of *Megalobrama amblycephala*: A comparative study between pelleted and extruded feed. *Fish Physiol. Biochem.* **2022**, *48*, 1025–1038. [CrossRef] [PubMed]
17. He, C.F.; Liu, W.B.; Shi, H.J.; Zhang, L.; Zhang, L.; Li, X.F. Utilization of pelleted and extruded feed by blunt snout bream *Megalobrama amblycephala*: Insights from growth performance, health status and feed cost. *J. Anim. Physiol. Anim. Nutr.* **2021**, *105*, 1203–1213. [CrossRef] [PubMed]
18. AOAC. *Official Methods of Analysis*; AOAC International: Rockville, MD, USA, 2006; Available online: https://www.aoac.org (accessed on 20 December 2022).
19. Winberg, S.; Lepage, O. Elevation of brain 5-HT activity, POMC expression, and plasma cortisol in socially subordinate rainbow trout. *Am. J. Physiol.-Regul. Integr. Comp. Physiol.* **1998**, *274*, R645–R654. [CrossRef] [PubMed]
20. Asadi, F.; Hallajian, A.; Asadian, P.; Shahriari, A.; Pourkabir, M. Serum lipid, free fatty acid, and proteins in juvenile sturgeons: *Acipenser persicus* and *Acipenser stellatus*. *Comp. Clin. Pathol.* **2009**, *18*, 287–289. [CrossRef]
21. Shuang, L.; Chen, S.-L.; Ren, C.; Su, X.-L.; Xu, X.-N.; Zheng, G.-D.; Zou, S.-M. Effects of hypoxia and reoxygenation on oxidative stress, histological structure, and apoptosis in a new hypoxia-tolerant variety of blunt snout bream (*Megalobrama amblycephala*). *Comp. Biochem. Physiol. Part A Mol. Integr. Physiol.* **2023**, *278*, 111358. [CrossRef] [PubMed]
22. Li, X.-F.; Xu, C.; Tian, H.-Y.; Jiang, G.-Z.; Zhang, D.-D.; Liu, W.-B. Feeding rates affect stress and non-specific immune responses of juvenile blunt snout bream *Megalobrama amblycephala* subjected to hypoxia. *Fish Shellfish Immunol.* **2016**, *49*, 298–305. [CrossRef]
23. Zwirner, J.; Dobos, G.; Götze, O. A novel ELISA for the assessment of classical pathway of complement activation in vivo by measurement of C4-C3 complexes. *J. Immunol. Methods* **1995**, *186*, 55–63. [CrossRef]
24. Yuan, X.-Y.; Liu, W.-B.; Wang, C.-C.; Huang, Y.-Y.; Dai, Y.-J.; Cheng, H.-H.; Jiang, G.-Z. Evaluation of antioxidant capacity and immunomodulatory effects of cottonseed meal protein hydrolysate and its derivative peptides for hepatocytes of blunt snout bream (*Megalobrama amblycephala*). *Fish Shellfish Immunol.* **2020**, *98*, 10–18. [CrossRef] [PubMed]
25. Zhang, C.-N.; Li, X.-F.; Jiang, G.-Z.; Zhang, D.-D.; Tian, H.-Y.; Li, J.-Y.; Liu, W.-B. Effects of dietary fructooligosaccharide levels and feeding modes on growth, immune responses, antioxidant capability and disease resistance of blunt snout bream (*Megalobrama amblycephala*). *Fish Shellfish Immunol.* **2014**, *41*, 560–569. [CrossRef] [PubMed]
26. LYGREN, B.; WAAGBØ, R. Nutritional impacts on the chemiluminescent response of Atlantic salmon (*Salmo salar* L.) head kidney phagocytes, in vitro. *Fish Shellfish Immunol.* **1999**, *9*, 445–456. [CrossRef]
27. Satoh, K. Serum lipid peroxide in cerebrovascular disorders determined by a new colorimetric method. *Clin. Chim. Acta Int. J. Clin. Chem.* **1978**, *90*, 37–43. [CrossRef]
28. Zhang, L.; Zheng, X.-C.; Huang, Y.-Y.; Ge, Y.-P.; Sun, M.; Chen, W.-L.; Liu, W.-B.; Li, X.-F. Carbonyl cyanide 3-chlorophenylhydrazone induced the imbalance of mitochondrial homeostasis in the liver of *Megalobrama amblycephala*: A dynamic study. *Comp. Biochem. Physiol. Part C Toxicol. Pharmacol.* **2021**, *244*, 109003. [CrossRef]
29. Bradford, M.M. A rapid and sensitive method for the quantitation of microgram quantities of protein utilizing the principle of protein-dye binding. *Anal. Biochem.* **1976**, *72*, 248–254. [CrossRef] [PubMed]
30. Livak, K.J.; Schmittgen, T.D. Analysis of relative gene expression data using real-time quantitative PCR and the $2^{-\Delta\Delta CT}$ method. *Methods* **2001**, *25*, 402–408. [CrossRef] [PubMed]

31. Zhao, Z.; Liu, B.; Ge, X.; Li, Z.; Yang, X.; Zhou, Z.; Zhao, F. Emodin attenuates CY-induced oxidative injury in PBLs of the blunt snout bream (*Megalobrama amblycephala*) though the Nrf2-Keap1 signaling pathway. *Aquaculture* **2021**, *545*, 737201. [CrossRef]
32. Mokrani, A.; Ren, M.; Liang, H.; Yang, Q.; Ji, K.; Kasiya, H.C.; Ge, X. Effect of the total replacement of fishmeal with plant proteins and supplemental essential amino acids in the extruded diet on antioxidants genes, enzyme activities, and immune response in juvenile blunt snout bream. *Aquac. Int.* **2020**, *28*, 555–568. [CrossRef]
33. Liu, Z.-S.; Zhang, L.; Chen, W.-L.; He, C.-F.; Qian, X.-Y.; Liu, W.-B.; Li, X.-F. Insights into the interaction between stocking density and feeding rate in fish *Megalobrama ambylcephala* based on growth performance, innate immunity, antioxidant activity, and the GH-IGF1 axis. *Aquaculture* **2024**, *580*, 740355. [CrossRef]
34. Barton, B.A. Stress in fishes: A diversity of responses with particular reference to changes in circulating corticosteroids. *Integr. Comp. Biol.* **2002**, *42*, 517–525. [CrossRef]
35. Martínez-Porchas, M.; Martínez-Córdova, L.R.; Ramos-Enriquez, R. Cortisol and glucose: Reliable indicators of fish stress? *Pan-Am. J. Aquat. Sci.* **2009**, *4*, 158–178.
36. Grutter, A.; Pankhurst, N. The effects of capture, handling, confinement and ectoparasite load on plasma levels of cortisol, glucose and lactate in the coral reef fish *Hemigymnus melapterus*. *J. Fish Biol.* **2000**, *57*, 391–401. [CrossRef]
37. Conde-Sieira, M.; Chivite, M.; Míguez, J.M.; Soengas, J.L. Stress effects on the mechanisms regulating appetite in teleost fish. *Front. Endocrinol.* **2018**, *9*, 416277. [CrossRef]
38. Pawlak, P.; Burren, A.; Seitz, A.; Pietsch, C. Effects of different acute stressors on the regulation of appetite genes in the carp (*Cyprinus carpio* L.) brain. *R. Soc. Open Sci.* **2023**, *10*, 230040. [CrossRef]
39. Ma, X.; Li, X.; Wang, W.; Zhang, M.; Yang, B.; Miao, Z. Phosphatidylserine, inflammation, and central nervous system diseases. *Front. Aging Neurosci.* **2022**, *14*, 975176. [CrossRef]
40. Calzada, E.; Onguka, O.; Claypool, S.M. Phosphatidylethanolamine metabolism in health and disease. *Int. Rev. Cell Mol. Biol.* **2016**, *321*, 29–88. [CrossRef] [PubMed]
41. Sun, Y.Z.; Yang, H.L.; Ma, R.L.; Song, K.; Li, J.S. Effect of Lactococcus lactis and Enterococcus faecium on growth performance, digestive enzymes and immune response of grouper *Epinephelus coioides*. *Aquac. Nutr.* **2012**, *18*, 281–289. [CrossRef]
42. Wang, B.; Wang, Y.; Jia, T.; Feng, J.; Qu, C.; Wu, X.; Yang, X.; Zhang, Q. Changes in physiological responses and immunity of blunt snout bream *Megalobrama amblycephala* from transport stress. *Fish Physiol. Biochem.* **2022**, *48*, 1183–1192. [CrossRef]
43. Xia, S.L.; Li, X.F.; Abasubong, K.P.; Xu, C.; Shi, H.J.; Liu, W.B.; Zhang, D.D. Effects of dietary glucose and starch levels on the growth, apparent digestibility, and skin-associated mucosal non-specific immune parameters in juvenile blunt snout bream (*Megalobrama amblycephala*). *Fish Shellfish Immunol.* **2018**, *79*, 193–201. [CrossRef]
44. Li, X.-F.; Liu, W.-B.; Lu, K.-L.; Xu, W.-N.; Wang, Y. Dietary carbohydrate/lipid ratios affect stress, oxidative status and non-specific immune responses of fingerling blunt snout bream, *Megalobrama amblycephala*. *Fish Shellfish Immunol.* **2012**, *33*, 316–323. [CrossRef]
45. Chaung, H.-C.; Chang, C.-D.; Chen, P.-H.; Chang, C.-J.; Liu, S.-H.; Chen, C.-C. Docosahexaenoic acid and phosphatidylserine improves the antioxidant activities in vitro and in vivo and cognitive functions of the developing brain. *Food Chem.* **2013**, *138*, 342–347. [CrossRef]
46. Fusi, J.; Bianchi, S.; Daniele, S.; Pellegrini, S.; Martini, C.; Galetta, F.; Giovannini, L.; Franzoni, F. An in vitro comparative study of the antioxidant activity and SIRT1 modulation of natural compounds. *Biomed. Pharmacother.* **2018**, *101*, 805–819. [CrossRef]
47. Chung, J.-Y.; Chen, H.; Zirkin, B. Sirt1 and Nrf2: Regulation of Leydig cell oxidant/antioxidant intracellular environment and steroid formation†. *Biol. Reprod.* **2021**, *105*, 1307–1316. [CrossRef]
48. Li, W.; Kong, A.N. Molecular mechanisms of Nrf2-mediated antioxidant response. *Mol. Carcinog.* **2009**, *48*, 91–104. [CrossRef]
49. Chew, L.Y.; Zhang, H.; He, J.; Yu, F. The Nrf2-Keap1 pathway is activated by steroid hormone signaling to govern neuronal remodeling. *Cell Rep.* **2021**, *36*, 109466. [CrossRef]

antioxidants

Article

Effects of Alkalinity Exposure on Antioxidant Status, Metabolic Function, and Immune Response in the Hepatopancreas of *Macrobrachium nipponense*

Shubo Jin [1,2], Mingjia Xu [2], Xuanbin Gao [2], Sufei Jiang [1], Yiwei Xiong [1], Wenyi Zhang [1], Hui Qiao [1], Yan Wu [1] and Hongtuo Fu [1,2,*]

[1] Key Laboratory of Freshwater Fisheries and Germplasm Resources Utilization, Ministry of Agriculture and Rural Affairs, Freshwater Fisheries Research Center, Chinese Academy of Fishery Sciences, Wuxi 214081, China; jinsb@ffrc.cn (S.J.); jiangsf@ffrc.cn (S.J.); xiongyw@ffrc.cn (Y.X.); zhangwy@ffrc.cn (W.Z.); qiaoh@ffrc.cn (H.Q.); wuyan@ffrc.cn (Y.W.)

[2] Wuxi Fisheries College, Nanjing Agricultural University, Wuxi 214081, China; xumingjia8562@163.com (M.X.); gaoxuanbin@163.com (X.G.)

* Correspondence: fuht@ffrc.cn

Abstract: The oriental river prawn *Macrobrachium nipponense* is an important freshwater economic species in China, producing huge economic benefits. However, *M. nipponense* shows lower alkali tolerance than fish species, thus genetic selection is urgently needed in order to improve alkali tolerance in this species. In the present study, the effects of alkalinity exposure on the hepatopancreas of *M. nipponense* were measured under the alkali concentrations of 0 (control), 4, 8, and 12 mmol/L with the exposure time of 96 h through histological observations, measurement of antioxidant enzymes, metabolic profiling analysis, and transcriptome profiling analysis. The present study identified that the low concentration of alkali treatment (<4 mmol/L) did not result in morphological changes in the hepatopancreas and activity changes in antioxidant enzymes, while high-alkali treatment (>8 mmol/L) damaged the normal structures of the lumen and vacuoles and significantly stimulated the levels of superoxide dismutase, catalase, and total antioxidant capacity, indicating these antioxidant enzymes play essential roles in the protection of the body from the damage caused by the alkali treatment. Metabolic profiling analysis revealed that the main enriched metabolic pathways of differentially expressed metabolites in the present study were consistent with the metabolic pathways caused by environmental stress in plants and other aquatic animals. Transcriptome profiling analysis revealed that the alkali concentration of <8 mmol/L did not lead to significant changes in gene expression. The main enriched metabolic pathways were selected from the comparison between 0 mmol/L vs. 12 mmol/L, and some significantly up-regulated genes were selected from these metabolic pathways, predicting these selected metabolic pathways and genes are involved in the adaptation to alkali treatment in *M. nipponense*. The expressions of Ras-like GTP-binding protein, Doublesex and mab-3 related transcription factor 1a, and Hypothetical protein JAY84 are sensitive to changes in alkali concentrations, suggesting these three genes participated in the process of alkali adaptation in *M. nipponense*. The present study identified the effects of alkalinity exposure on the hepatopancreas of *M. nipponense*, including the changes in antioxidant status and the expressions of metabolites and genes, contributing to further studies of alkali tolerance in this species.

Keywords: *Macrobrachium nipponense*; hepatopancreas; alkali treatment; metabolic profiling analysis; transcriptome profiling analysis

Citation: Jin, S.; Xu, M.; Gao, X.; Jiang, S.; Xiong, Y.; Zhang, W.; Qiao, H.; Wu, Y.; Fu, H. Effects of Alkalinity Exposure on Antioxidant Status, Metabolic Function, and Immune Response in the Hepatopancreas of *Macrobrachium nipponense*. *Antioxidants* **2024**, *13*, 129. https://doi.org/10.3390/antiox13010129

Academic Editor: Erchao Li

Received: 13 December 2023
Revised: 15 January 2024
Accepted: 18 January 2024
Published: 21 January 2024

1. Introduction

The oriental river prawn, *Macrobrachium nipponense*, is widely distributed in China and other Asian countries [1]. It is an important commercial freshwater prawn species in China with annual production of over two hundred thousand tons, accounting for 5.72% of the total production of freshwater prawns. The main regions for *M. nipponense* culture include Jiangsu Province, Anhui Province, Zhejiang Province, and Jiangxi Province, producing huge economic benefits [2]. The main culture region of *M. nipponense* is in the southeast part of China, while the production in the north part of China is limited. A reasonable reason for this is that the water in the north part of China is mainly saline–alkali water and *M. nipponense* cannot adapt to this water environment.

Alkali tolerance has been identified in many fish and crustacean species (Table 1). The fish species include *Ctenopharyngodon idellus*, *Hypophthalmichthys molitrix*, *Aristichthys nobolis*, *Tribolodon brandti*, and *Gymnocypris przewalskii* [3–5]. The crustacean species include *Penaeus chinensis*, *Penaeus vannamei*, and *Palaemon przewalskii* [6–8]. Previous study has shown LC_{50} values of alkalinity of 27.66 mmol/L at 12 h, 26.94 mmol/L at 24 h, 22.51 mmol/L at 48 h, 15.00 mmol/L at 72 h, and 14.42 mmol/L at 96 h with a safety value of 4.71 mmol/L under conditions of water temperature of (23.1 ± 1.48) °C, pH = (8.9 ± 0.30), salinity of (0.62 ± 0.27), and dissolved oxygen level of (7.2 ± 0.30) mg/L, using Taihu No2 as the research species (a new variety of *M. nipponense* through genetic selection) [9]. Alkalinity tolerance in crustacean species was generally lower than that of fish species. There are extensive saline–alkali water resources in China. However, the alkali tolerance of *M. nipponense* is insufficient to adapt to water environments with high alkali concentrations. Thus, it is important for the sustainable development of the *M. nipponense* industry if the alkali tolerance can be improved in this species. Therefore, studies on the mechanism of alkali tolerance in *M. nipponense* are urgently needed, including the identification of alkali-tolerance-related genes and SNPs.

Table 1. Alkali tolerance in fishes and crustaceans.

Species	LC_{50} Value at 24 h (mmol/L)	Safe Alkali Value (mmol/L)
Ctenopharyngodon idellus	82.2	
Hypophthalmichthys molitrix	95	
Aristichthys nobolis	65.7	
Tribolodon brandti	89.31	18.79
Gymnocypris przewalskii		64
Penaeus chinensis	3.28	
Penaeus vannamei	12.40	
Palaemon przewalskii		3.5
Macrobrachium nipponense	14.42	4.71

Transcriptome-profiling analyses have been conducted in many aquatic animals in order to select alkali-tolerance-related genes, including *Leuciscus waleckii* [10], *Lateolabrax maculatus* [11], *Luciobarbus capito* [12], and *Leuciscus waleckii* [13]. These studies suggested that pathways related to stress response and extreme environment adaptation are the main enriched metabolic pathways of differentially expressed genes, including phenylalanine, tyrosine and tryptophan biosynthesis, cell cycle, and DNA replication.

In the present study, we aimed to analyze the effects of alkalinity exposure on the morphological changes in the hepatopancreas and the levels of antioxidants in the hepatopancreas after exposure of the prawns to water environments with different alkali concentrations (0, 4, 8, and 12 mmol/L). Furthermore, the integrated analysis of the transcriptome and metabolome was also performed in order to select genes and metabolites in response to the treatment of alkalinity.

2. Materials and Methods

2.1. Sample Collection

All of the wild prawns (*M. nipponense*) from the Yangtze River used in the present study were provided by the Dapu *M. nipponense* Breeding Base in Wuxi, China (120°13′44″ E, 31°28′22″ N). A total of 1200 prawns were collected with a body weight of 3.79–4.21 g for males and 2.31–3.14 for females and randomly divided into four groups. The prawns were kept in aerated fresh water with dissolved oxygen content $\geq$ 6 mg/L for 3 days prior to the alkali treatment. Previous study has identified LC_{50} values of alkalinity as 14.42 mmol/L at 96 h in *M. nipponense* [9]. Thus, four alkali concentrations were prepared through adding $NaHCO_3$ to the aerated fresh water in the present study, including 0 (control, water without $NaHCO_3$), 4, 8, and 12 mmol/L under conditions of water temperature of (28.3 $\pm$ 1.26) °C, pH = (7.81–8.32), and dissolved oxygen level of >6.0 mg/L. The alkali concentrations were measured according to the criterion of SC/T9406-2012 [14]. Each alkali concentration was prepared in three tanks, and 100 prawns were maintained in each tank. All prawns were maintained in the different alkali concentrations for 96 h, and then hepatopancreases were collected for histological observations, measurement of antioxidant enzymes, metabolic profiling analysis, transcriptome-profiling analysis, and qPCR analysis. Five hepatopancreases were collected from each alkali concentration and pooled together to form a biological replicate. Eight biological replicates were performed for metabolic profiling analysis, while three biological replicates were performed for the measurement of the activities of antioxidant enzymes, transcriptome-profiling analysis, and qPCR analysis.

2.2. Hematoxylin and Eosin (HE) Staining of Hepatopancreas

First, 4% paraformaldehyde was used to fix the tissues used for the histological observations. The hepatopancreases were collected from three individuals of each alkali concentration in order to analyze the morphological changes in the hepatopancreas caused by the alkali treatment. All three hepatopancreases from each concentration were sliced (three biological replicates), and two slices were prepared from each tissue (two technique replicates). The detailed procedures of HE staining have been well described in previous studies [15,16]. Briefly, hepatopancreases were dehydrated in varying ethanol concentrations. The dehydrated hepatopancreases were then transparent and embedded by using different percentages of xylene/wax mixture. The embedded hepatopancreases were finally sliced to 5 μm thickness using a slicer (Leica, Wetzlar, Germany). HE was used to stain the slices for 3–8 min. An Olympus SZX16 microscope (Olympus Corporation, Tokyo, Japan) was used to view the morphological changes.

2.3. Measurement of the Activities of Antioxidant Enzymes

The activities of antioxidant enzymes were measured in the hepatopancreas by using commercial kits purchased from the Nanjing Jiancheng Bioengineering Institute, including malondialdehyde (MAD), superoxide dismutase (SOD), catalase (CAT), glutathione (GSH), glutathione peroxidase (GSH-PX), and total antioxidant capacity (T-AOC). All the antioxidant indexes were measured by using a microplate reader (Bio-rad iMark, San Francisco, CA, USA), following the manufacturer's instructions.

2.4. Metabolic Profiling Analysis

Metabolic profiling analysis was performed to select the differentially expressed metabolites (DEMs) in *M. nipponense* caused by the alkali treatment, which were determined by liquid chromatography–mass spectrometry (LC/MS) analysis [17]. The detailed procedures for the metabolic profiling analysis have been well described in a previous published paper [18]. The metabolic profiling was analyzed by an ACQUITY UHPLC system (Waters Corporation, Milford, CT, USA) and an AB SCIEX Triple TOF 5600 System (AB SCIEX, Framingham, MA, USA) in both ESI positive and ESI negative ion modes. The criterion of a seven-fold cross-validation was employed to ensure the robustness and predictive ability of the model, and permutation tests were employed to perform further validation.

2.5. Transcriptome-Profiling Analysis

Transcriptome-profiling analysis was performed to select the differentially expressed genes (DEGs) in *M. nipponense* caused by the alkali treatment, which were sequenced by an Illumina Hiseq-2500 sequencing platform. The detailed procedures for the RNA-Seq and analysis have been well described in previous published papers [19,20]. Briefly, the total RNA was extracted from each biological replicate, conducted by using RNAiso Plus Reagent (TaKaRa, San Jose, CA, USA), according to the manufacturer's instructions. A spectrophotometer (Eppendorf, Hamburg, Germany) was employed to measure the concentration of total RNA. A 2100 Bioanalyzer (Agilent Technologies, Inc., Santa Clara, CA, USA) was employed to measure the integrity of total RNA, and RNA integrity number (RIN) value should be >7.0. A total of 4 μg of total RNA was used to construct the library. The sequencing was conducted by using the Illumina Hiseq-2500 sequencing platform under the parameter of PE150.

Fastp software (version 0.11.5) was employed to remove the low-quality raw reads with the default parameters [21]. The HISAT2 software (version 2.2.1.0) was then employed to map the obtained clean reads to the *M. nipponense* reference genome (GenBank access numbers: GCA_015110555.1 and GCA_015104395.1) [22]. Genes were annotated in the Gene Ontology (GO) (http://www.geneontology.org/, accessed on 15 August 2023) [23], Cluster of Orthologous Groups (COG) (http://www.ncbi.nlm.nih.gov/COG/, accessed on 15 August 2023) [24], and Kyoto Encyclopedia of Genes and Genomes (KEGG) databases (http://www.genome.jp/kegg/, accessed on 15 August 2023) [25], using an E-value of 10^{-5} [19]. Gene expression was calculated using the FPKM method, where FPKM = cDNA fragments/mapped fragments (millions)/transcript length (kb), using HTSeq-count [26]. DESeq2 was used to perform the differential expression analysis [27]. The Benjamini–Hochberg correction method was used to calculate the false discovery rate (FDR) [28] with q-value < 0.05. Fold change > 2 was considered to show up-regulated differentially expressed genes (DEGs), and fold change < 0.5 was considered to show down-regulated DEGs.

2.6. qPCR Analysis

qPCR was used to measure the expression of DEGs selected from the present study in order to verify the accuracy of RNA-Seq. Previously published studies have described the detailed procedures [29,30]. Briefly, total RNA was extracted from the hepatopancreas of each biological replicate, using the UNIQ-10 Column TRIzol Total RNA Isolation Kit (Sangon, Shanghai, China). A total of 1 μg total RNA from each tissue was used to synthesize the cDNA template, according to the manufacturer's instructions for the PrimeScript™ RT reagent kit (Takara Bio Inc., Shiga, Japan). The UltraSYBR Mixture (CWBIO, Beijing, China) was used to measure the expression level of each tissue, according to the manufacturer's instructions. The Bio-Rad iCycler iQ5 Real-Time PCR System (Bio-Rad) was used to conduct the qPCR analysis, which can carry out SYBR Green RT-qPCR assay. Table 2 lists all of the primers used in the present study for qPCR analysis. The eukaryotic translation initiation factor 5A (EIF) has been proven to be a suitable and stable reference gene under various conditions in *M. nipponense* and was used in this study [31]. The $2^{-\Delta\Delta CT}$ method was used to determine the relative expression levels [32].

2.7. Statistical Analysis

SPSS Statistics 23.0 was employed to carry out the statistical analysis in the present study, estimated by one-way ANOVA, followed by Duncan's multiple range test [29,30]. A probability level of 0.05 was used to indicate significance ($p < 0.05$). The homogeneity of variances was measured in prior to ANOVA (Sig. > 0.05). Meanwhile, a linear regression analysis was performed on each set of data. The linear regression analysis revealed that the residual deviation is close to 1, while the mean residual of each group of data is close to 0, indicating that the residuals of the data are normally distributed and can be analyzed. The confidence intervals were calculated at the 95% level. Quantitative data were expressed as the mean ± SD.

Table 2. Primers used for the qPCR validation in the present study.

Gene	Forward Primer	Reverse Primer	Efficiency (%)	Product Size (bp)
HP-JAY84	CGCTCTAGATCCGTGAGCAG	ACGAGGCCAGAAACTCTTGG	95.4	232
RaG	GTCCTCAAGATCGTGGCTGT	TGCATCTGTCGAAACCCTCC	97.1	196
Dmrt1- a	GTTGGCTTCGTCCCAGAAGA	TGATCACACTCCACGCTGAC	94.8	185
TARF6	TGCTCCATAGTCCGGCATTT	GCAGATCTGGCTTGCTTACC	101.5	240
CHMP7	ATTCCGCAGTGGTGTAGAGG	GAGCTCACTCCCGAAGCAAG	99.2	144
ADRH-WM6	GGTCCAGGAGAGATTCGACG	CCACACAAAATGCAGCCACA	97.9	117
NXF1	TAGGACACCACTTGCTGCTG	TGCATTCGCTTGTCTGAGGT	102.5	153
Nup107	AGGAGGAAACTGGCTCTCCA	TCCAGCGATCAAGTTCACCC	97.5	150
Hsp90	ATGCCCGAGGAACCAATGAC	AGAGCTCCTTGCCAGATTCG	98.2	208
eIF2	TGGAAAGACCGAACCAGTCG	AAAAGCTCCCCTACGTGTCG	103.6	162
Ap4m1	TGGAATGGGCACAGTATCACC	CCTCCAGAGTCTACAAGCCG	96.8	201
NPC2	CTCTGCTACAGCCTTCCAGG	TTCGACTCCCTTGCAAGCAT	97.1	232
CDC23	AGGCTCAGAAATTGCGACCA	GGCACGTTCACCTTCACCTA	97.4	189
E3-FANCL	GTGGCAATGAAGAATGGGGC	GTCTGCTATTCGGAAGCCCA	98.6	150
39S-RPL32	TGAGCATGAAGTCTTTCCCGT	CAGTCTGCGAGAAAACCACTG	101.5	121
39S-RPL33	GGCCAACGTTTTTCGATCTGG	AAAGACGCACACCTGACGGA	97.8	199
60S-RPL19	CAATGCTCGTGCCAAGATGT	TCTGCCTTCCTTTGGGCAAC	95.3	105
RP-L21	GAGACGCCACAACTCAAGGA	TTCCGTCTACCCCTTCCACT	104.8	122
InR	TCCTCGGTGCCTCAAGAAAC	CCACTGCAGACCTCGAATGT	101.9	173
GATOR-WDR59	CACATCCATCCACCCCTGTG	ACAGCCTGTTGGGCATTCAT	97.6	263
Cbc-10	CCTCTGGAGTGCAATGGGAG	TTGCTGCTGAACCCAGTCTC	99.7	131
Cbc-7	AGGAGGAAAGTCGAAAGCCG	AGATGACGAGAAGCACTGGC	99.6	141
HP-798	GACGTTCTTCGCACACTTCA	TCATGCGTTCCGTTTCCAGA	102.4	113
ATP-CF6	CAAGGTTGCTCGCCAGTATG	TTTTGCAAACAGTTCAGGTGGT	95.6	120
EIF	CATGGATGTACCTGTGGTGAAAC	CTGTCAGCAGAAGGTCCTCATTA	98.5	157

3. Results

3.1. Survival Rate under Different Alkaline Concentrations

The survival rate gradually decreased with the increase in alkali concentration after the 96 h of alkaline treatment. The survival rate of 0 mmol/L was 91.33%, compared to that of 48.66% under the alkaline concentration of 12 mmol/L (Figure 1).

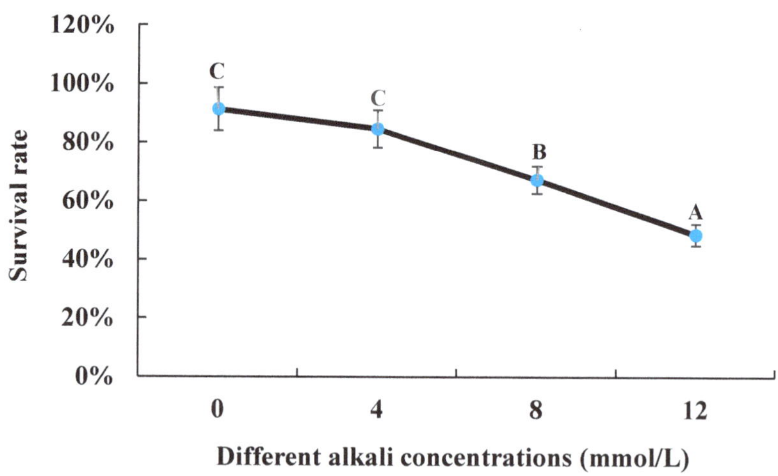

Figure 1. The survival rate of *M. nipponense* under the treatment of different alkali concentrations. Letters indicated the differences of survival rate between different alkali concentrations.

3.2. Histological Observations

The morphological changes in the hepatopancreas caused by the alkali treatment were revealed by the histological observations (Figure 2). Histological observations revealed that the hepatopancreas included secretory cells, basement membrane, lumen, storage cells, and vacuoles. The tissue morphology of the hepatopancreas was normal without significant damage at concentrations of 0 mmol/L and 4 mmol/L. However, the alkalinity at the concentration of 8 mmol/L resulted in the increase in the lumen and vacuoles, and secretory cells and storage cells were decreased. When the alkaline concentration reached 12 mmol/L, the lumen and vacuoles of the hepatopancreas were significantly increased, and the basement membrane was severely damaged, affecting the morphology of secretory cells and storage cells in the hepatopancreas.

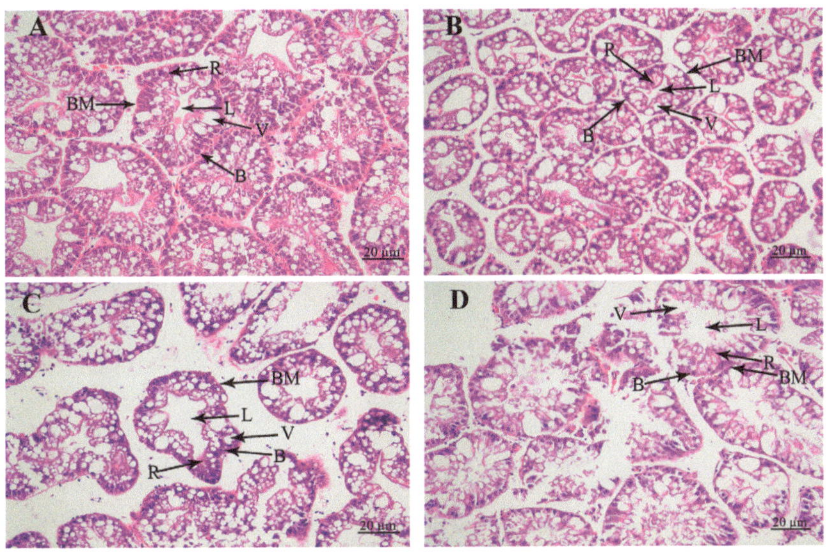

Figure 2. The changes in hepatopancreas under the treatment of different alkali concentrations by histological observations. B: secretory cells of type B; BM: basement membrane; L: lumen; R: storage cells of type R; V: vacuoles. Scale bars = 20 μm. (**A**) The histological observation of hepatopancreas under the alkali concentration of 0 mmol/L; (**B**) the histological observation of hepatopancreas under the alkali concentration of 4 mmol/L; (**C**) the histological observation of hepatopancreas under the alkali concentration of 8 mmol/L; (**D**) the histological observation of hepatopancreas under the alkali concentration of 12 mmol/L.

3.3. Measurement of the Activities of Antioxidant Enzymes

The activities of antioxidant enzymes were also measured in the hepatopancreas after the treatment with different alkali concentrations (Figure 3). The activities of SOD were gradually increased with the increase in alkali concentrations. The activities at the concentrations of 8 mmol/L and 12 mmol/L were significantly higher than those of 0 mmol/L and 4 mmol/L ($p < 0.05$), while the activities between 0 mmol/L and 4 mmol/L and between 8 mmol/L and 12 mmol/L showed no significant difference ($p > 0.05$). The highest activities of CAT and T-AOC were observed at the concentration of 8 mmol/L, which showed a significant difference from those of 12 mmol/L and 4 mmol/L, respectively ($p < 0.05$). However, the activities of MDA, GSH, and GSH-PX showed no difference after the treatment of different concentrations of alkali. Interestingly, all of these six enzymes showed no difference between 0 mmol/L and 4 mmol/L ($p > 0.05$).

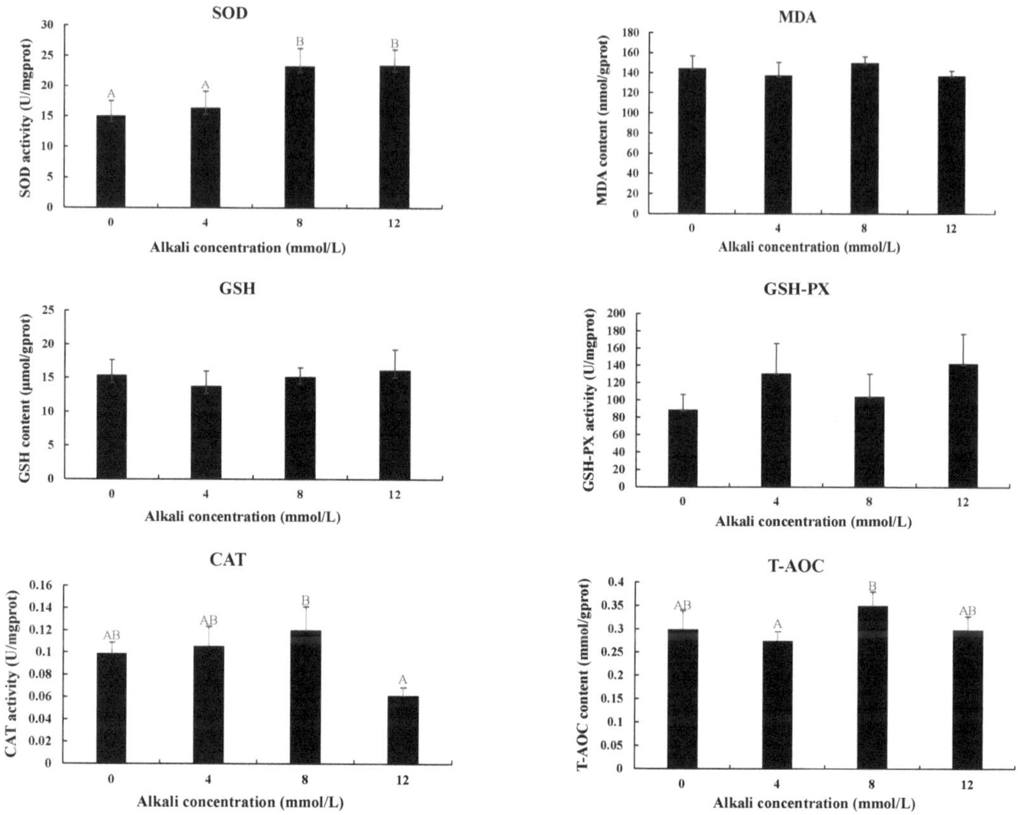

Figure 3. The measurements of the activities of antioxidant enzymes in the hepatopancreas under the treatment of different alkali concentrations. CAT: catalase; GSH: glutathione; GSH-PX: glutathione peroxidase; MDA: malondialdehyde; SOD: superoxide dismutase; T-AOC: total antioxidant capacity. Data are shown as mean ± standard deviation (SD) of tissues from three biological replicates. Capital letters indicated the significant difference of the activities of antioxidant enzymes between different alkali concentrations.

3.4. Metabolome-Profiling Analysis

Latent structure discriminant analysis was used to analyze the overall quality of the metabolic profiling analysis in the present study (Figure 4), suggesting a robust and reliable model to identify the different metabolic patterns in the hepatopancreas of *M. nipponense* after the treatment of different alkali concentrations.

The differentially expressed metabolites (DEMs) were selected based on the criterion of >2.0 for up-regulated metabolites and <0.5 for down-regulated metabolites. A total of 114 metabolites were differentially expressed between the alkali concentration of 0 mmol/L and 4 mmol/L, of which 85 metabolites were up-regulated and 29 metabolites were down-regulated. Sixty-eight metabolites showed differential expression at the alkali concentrations of 0 mmol/L and 8 mmol/L, including forty-five up-regulated metabolites and twenty-three down-regulated metabolites. A total of 139 DEMs were identified between the alkali concentrations of 0 mmol/L and 12 mmol/L, of which 115 metabolites were up-regulated and 24 metabolites were down-regulated. KEGG analysis revealed that metabolic pathways, biosynthesis of secondary metabolites, biosynthesis of plant secondary metabolites, biosynthesis of amino acids, and microbial metabolism in diverse environments represented the main enriched metabolic pathways of DEMs of all three comparisons in the present study (Table 3).

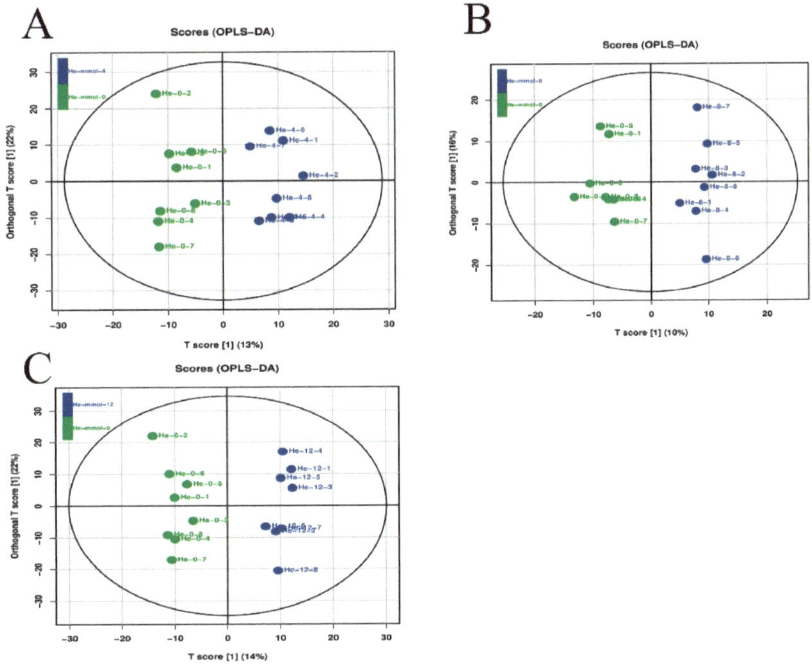

Figure 4. Orthogonal projections of latent structure discriminate analysis (OPLS-DA) of hepatopancreas after the treatment of different alkali concentrations. The LC-MS spectra were used to measure the OPLS-DA score. (**A**) The OPLS-DA analysis of 0 mmol/L vs. 4 mmol/L; (**B**) the OPLS-DA analysis of 0 mmol/L vs. 8 mmol/L; (**C**) the OPLS-DA analysis of 0 mmol/L vs. 12 mmol/L.

Table 3. The main metabolic pathways of DEMs.

Metabolic Pathways (0 vs. 4)	DEMs	Metabolic Pathways (0 vs. 8)	DEMs	Metabolic Pathways (0 vs. 12)	DEMs
Metabolic pathways	44	Metabolic pathways	38	Metabolic pathways	59
Biosynthesis of secondary metabolites	21	Biosynthesis of secondary metabolites	13	Biosynthesis of secondary metabolites	31
Biosynthesis of plant secondary metabolites	14	Microbial metabolism in diverse environments	12	Biosynthesis of amino acids	18
Biosynthesis of amino acids	12	Biosynthesis of plant secondary metabolites	11	Microbial metabolism in diverse environments	16
Microbial metabolism in diverse environments	11	Biosynthesis of cofactors	7	Biosynthesis of plant secondary metabolites	18
Protein digestion and absorption	11	Nucleotide metabolism	7	Central carbon metabolism in cancer	13
Aminoacyl-tRNA biosynthesis	9	Biosynthesis of amino acids	6	Biosynthesis of cofactors	12
Central carbon metabolism in cancer	9	Carbon metabolism	6	Protein digestion and absorption	12
Mineral absorption	8	Cysteine and methionine metabolism	6	ABC transporters	10
ABC transporters	8	Biosynthesis of plant hormones	6	Carbon metabolism	10
2-Oxocarboxylic acid metabolism	7	Biosynthesis of alkaloids derived from histidine and purine	6	Aminoacyl-tRNA biosynthesis	10
Glucosinolate biosynthesis	7	Purine metabolism	6	2-Oxocarboxylic acid metabolism	10
Biosynthesis of various plant secondary metabolites	7	D-Amino acid metabolism	5	Glycine, serine, and threonine metabolism	9
D-Amino acid metabolism	6	Taste transduction	5	D-Amino acid metabolism	9
Biosynthesis of plant hormones	6	Glyoxylate and dicarboxylate metabolism	5	Mineral absorption	8

3.5. Transcriptome-Profiling Analysis

A total of 44,084 genes matched the known genes in the *M. nipponense* genome, which is mostly consistent with the number of genes (44,086) in the *M. nipponense* genome. However, 4938 novel isoforms were also predicted in this transcriptome analysis, of which the gene functions need further investigation.

The DEGs were selected based on the criterion of >2.0 for up-regulated genes and <0.5 for down-regulated genes in the present study. A total of 184, 149, and 3949 DEGs were identified in the hepatopancreas between 0 mmol/L vs. 4 mmol/L, 0 mmol/L vs. 8 mmol/L, and 0 mmol/L vs. 12 mmol/L, respectively. Sixty-seven down-regulated DEGs and one hundred and seventeen up-regulated DEGs were identified between 0 mmol/L vs. 4 mmol/L. The comparison between 0 mmol/L vs. 8 mmol/L identified 57 up-regulated

DEGs and 92 down-regulated DEGs. A total of 1630 up-regulated DEGs and 2319 down-regulated DEGs were identified between 0 mmol/L vs. 12 mmol/L.

A total of 157, 130, and 3637 DEGs were annotated in the GO database between 0 mmol/L vs. 4 mmol/L, 0 mmol/L vs. 8 mmol/L, and 0 mmol/L vs. 12 mmol/L, respectively. Cells, cell parts, binding, cellular processes, catalytic activity, and metabolic processes were the main enriched functional groups in all of these three comparisons, indicating the genes enriched in these functional groups may play essential roles in the adaptation to alkaline stress in this species (Table 4).

Table 4. The main functional groups of DEGs by GO analysis.

0 mmol/L vs. 4 mmol/L	0 mmol/L vs. 8 mmol/L	0 mmol/L vs. 12 mmol/L
Binding	Cell	Cell
Catalytic activity	Cell part	Cell part
Cellular process	Binding	Cellular process
Cell	Cellular process	Binding
Cell part	Catalytic activity	Metabolic process
Metabolic process	Metabolic process	Organelle
Organelle	Membrane	Catalytic activity
Membrane	Organelle	Biological regulation
Membrane part	Membrane part	Organelle part
Extracellular region	Biological regulation	Developmental process
Multicellular organismal process	Response to stimulus	Multicellular organismal process
Developmental process	Organelle part	Membrane
Response to stimulus	Developmental process	Cellular component organization or biogenesis
Biological regulation	Multicellular organismal process	Response to stimulus
Localization	Localization	Protein-containing complex

A total of 32 and 41 DEGs were annotated in the KEGG database between 0 mmol/L vs. 4 mmol/L and 0 mmol/L vs. 8 mmol/L, respectively. Peroxisome was the most enriched metabolic pathway between 0 mmol/L vs. 4 mmol/L, of which five DEGs were enriched. Retinol metabolism, pentose and glucuronate interconversions, and metabolism of xenobiotics by cytochrome P450 with four DEGs were identified as the most enriched metabolic pathways between 0 mmol/L vs. 8 mmol/L. The number of DEGs between 0 mmol/L vs. 12 mmol/L reached 1045, which were annotated in the KEGG database. Endocytosis, RNA transport, protein processing in endoplasmic reticulum, lysosome, ubiquitin mediated proteolysis, ribosome, mTOR signaling pathway, and oxidative phosphorylation represent the most enriched metabolic pathways between 0 mmol/L vs. 12 mmol/L, of which the number of DEGs was $\geq$40. The main metabolic pathways in each comparison are listed in Table 5.

Table 5. The main metabolic pathways of DEGs by KEGG analysis.

Metabolic Pathways (0 vs. 4)	DEGs	Metabolic Pathways (0 vs. 8)	DEGs	Metabolic Pathways (0 vs. 12)	DEGs
Peroxisome	5	Retinol metabolism	4	**Endocytosis**	59
Arginine and proline metabolism	4	Pentose and glucuronate interconversions	4	**RNA transport**	53
Fatty acid degradation	3	**Metabolism of xenobiotics by cytochrome P450**	4	**Protein processing in endoplasmic reticulum**	48
Drug metabolism—other enzymes	3	Glycerophospholipid metabolism	3	**Lysosome**	44
Ascorbate and aldarate metabolism	3	Drug metabolism—cytochrome P450	3	**Ubiquitin mediated proteolysis**	43
Metabolism of xenobiotics by cytochrome P450	2	Glutathione metabolism	3	**Ribosome**	43
Tryptophan metabolism	2	Fructose and mannose metabolism	3	**mTOR signaling pathway**	42
Pentose and glucuronate interconversions	2	**Phagosome**	3	**Oxidative phosphorylation**	40
Drug metabolism—cytochrome P450	2	**Pyruvate metabolism**	3	**Ribosome biogenesis in eukaryotes**	39
Pyruvate metabolism	2	**Citrate cycle (TCA cycle)**	2	**Amino sugar and nucleotide sugar metabolism**	37

Note: The metabolic pathways related to oxidative stress and cellular organization are bolded.

A total of 25 genes were considered as the strong candidate genes that were predicted to be involved in the mechanism of alkaline tolerance in *M. nipponense*. Three genes were differentially expressed among all of these three comparisons, indicating these three genes are sensitive to changes in alkaline concentrations. These three genes included Ras-like GTP-binding protein (*RaG*), Doublesex and mab-3 related transcription factor 1a (*Dmrt1-a*), and Hypothetical protein JAY84 (*HP-JAY84*). The other 22 genes were significantly differentially expressed between 0 mmol/L vs. 12 mmol/L, which were enriched in the main enriched metabolic pathways (Table 6).

Table 6. The main DEGs from the transcriptome-profiling analysis.

Gene	Accession Number	Species	Fold Change		
			0 vs. 4	4 vs. 8	8 vs. 12
Hypothetical protein (*HP-JAY84*)	JAY84_18770	*Candidatus Thiodiazotropha*	0.004	292.04	0.009
Ras-like GTP-binding protein (*RaG*)	XP_053656102.1	*Cherax quadricarinatus*	2.35	6.96	12.91
Doublesex and mab-3 related transcription factor 1a (*Dmrt1-a*)	QDE10512.1	*Macrobrachium rosenbergii*	91.77	0.10	14.32
Gene	Accession number	Species	Metabolic pathway	Fold change	
				0 vs. 12	
TNF receptor associated factor 6 (*TARF6*)	ASM46956.1	*Macrobrachium nipponense*	Endocytosis	3.58	
Charged multivesicular body protein 7 (*CHMP7*)	XP_027209977.1	*Penaeus vannamei*	Endocytosis	6.92	
ATP-dependent RNA helicase WM6 (*ADRH-WM6*)	RXG50776.1	*Armadillidium vulgare*	RNA transport	6.15	
Ribonuclease P protein subunit p29	XP_027220920.1	*Penaeus vannamei*	RNA transport	12.21	
Nuclear RNA export factor 1 (*NXF1*)	XP_045623594.1	*Procambarus clarkii*	RNA transport	4.32	
Nuclear pore protein Nup107 (*Nup107*)	XP_027211930.1	*Penaeus vannamei*	RNA transport	4.76	
Hsp90 protein	ROT76137.1	*Penaeus vannamei*	Protein processing in endoplasmic reticulum	3.97	
Eukaryotic translation initiation factor 2 (*eIF2*)	XP_053634587.1	*Cherax quadricarinatus*	Protein processing in endoplasmic reticulum	3.68	
AP-4 complex subunit mu-1-like isoform X2 (*Ap4m1*)	XP_027222938.1	*Penaeus vannamei*	Lysosome	7.11	
NPC intracellular cholesterol transporter 2 (*NPC2*)	XP_047502679.1	*Penaeus chinensis*	Lysosome	6.96	
Cell division cycle protein 23 (*CDC23*)	XP_027236079.1	*Penaeus vannamei*	Ubiquitin-mediated proteolysis	4.53	
E3 ubiquitin-protein ligase FANCL (*E3-FANCL*)	XP_027214794.1	*Penaeus vannamei*	Ubiquitin-mediated proteolysis	48.84	
39S ribosomal protein L32 (*39S-RPL32*)	XP_027217747.1	*Penaeus vannamei*	Ribosome	14.32	
39S ribosomal protein L33 (*39S-RPL33*)	XP_047997684.1	*Leguminivora glycinivorella*	Ribosome	8.11	
60S ribosomal protein L19 (*60S-RPL19*)	XP_027212740.1	*Penaeus vannamei*	Ribosome	10.41	
Ribosomal prokaryotic L21 protein (*RP-L21*)	XP_042878336.1	*Penaeus japonicus*	Ribosome	8.51	
Insulin-like receptor (*InR*)	XP_027218065.1	*Penaeus vannamei*	mTOR signaling pathway	15.35	
GATOR complex protein WDR59 (*GATOR-WDR59*)	XP_027223649.1	*Penaeus vannamei*	mTOR signaling pathway	4.56	
Cytochrome b-c1 complex subunit 10 (*Cbc-10*)	KZC10939.1	*Dufourea novaeangliae*	Oxidative phosphorylation	4.53	
Cytochrome b-c1 complex subunit 7 (*Cbc-7*)	XP_027231282.1	*Penaeus vannamei*	Oxidative phosphorylation	7.84	
Hypothetical protein L798_02749 (*HP-798*)	KDR07695.1	*Zootermopsis nevadensis*	Oxidative phosphorylation	5.21	
ATP synthase-coupling factor 6 (*ATP-CF6*)	XP_042857694.1	*Penaeus japonicus*	Oxidative phosphorylation	4.66	

3.6. qPCR Analysis

qPCR analyses were used to verify the expressions of DEGs selected from this study (Figure 5). qPCR analyses showed the same expression trends as RNA-Seq. *RaG*, *Dmrt1-a*, and *HP-JAY84* showed differential expressions in all three comparisons (0 mmol/L vs. 4 mmol/L, 4 mmol/L vs. 8 mmol/L, and 8 mmol/L vs. 12 mmol/L) ($p < 0.05$), of which *RaG* was gradually increased with the increase in alkali concentration. Interestingly, qPCR analysis also identified that the *InR* expressions were differentially expressed between all three comparisons ($p < 0.05$). The expressions of fifteen DEGs reached the peak at the alkali concentration of 12 mmol/L ($p < 0.05$), while the expressions at 0 mmol/L, 4 mmol/L and 8 mmol/L remained stable ($p > 0.05$). Two DEGs (*eIF2* and *60S-RPL19*) showed higher expressions at 8 mmol/L and 12 mmol/L than at 0 mmol/L and 4 mmol/L ($p < 0.05$), while the expressions showed no difference between 0 mmol/L and 4 mmol/L and between 8 mmol/L and 12 mmol/L ($p > 0.05$). The expressions of *Hsp90* and *GATOR-WDR59* gradually increased from 0 mmol/L to 12 mmol/L, while the expression showed no significant difference between 0 mmol/L and 4 mmol/L for *Hsp90* and between 4 mmol/L and 8 mmol/L for *GATOR-WDR59* ($p > 0.05$).

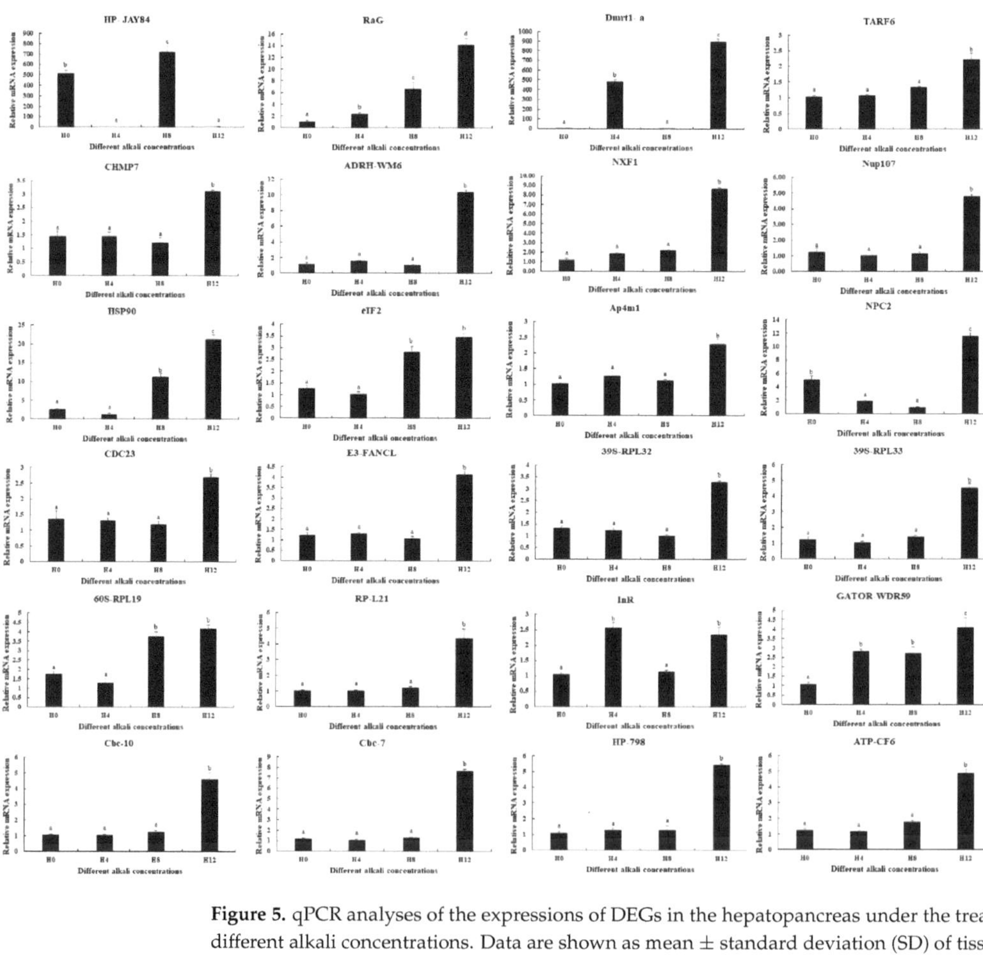

Figure 5. qPCR analyses of the expressions of DEGs in the hepatopancreas under the treatment of different alkali concentrations. Data are shown as mean ± standard deviation (SD) of tissues from three biological replicates. Letters indicate a significant difference in the expressions of DEGs between different alkali concentrations.

4. Discussion

Previous study has identified that the alkaline LC_{50} at 12 h, 24 h, 48 h, 72 h, and 96 h in juvenile prawns of "Taihu No2" (a new variety of *M. nipponense*, selected through the hybridization of *M. nipponense* and *M. hainanense*) were 27.66 mmol/L, 26.94 mmol/L, 22.51 mmol/L, 15.00 mmol/L, and 14.42 mmol/L, respectively [9]. Compared with other prawn or shrimp species, juvenile *M. nipponense* showed stronger alkali resistance and can be cultured in appropriate saline and alkali water. However, the tolerance of carbonate alkalinity of this species is dramatically lower than those of freshwater fish species. Thus, the long-term goal is to find out the mechanism of alkali tolerance in *M. nipponense* in order to culture a new strain of this species with stronger alkali tolerance. In the present study, we investigated the effects of different alkali concentrations on the hepatopancreas of *M. nipponense* through histological observations, measuring the activities of antioxidant enzymes, and performing metabolic profiling analysis and transcriptome-profiling analyses in the hepatopancreas.

The survival rate of *M. nipponense* gradually decreased from 0 mmol/L (91.33%) to 48.33% under the concentration of 12 mmol/L after 96 h of alkali treatment. Previous study has shown that the LC_{50} value of alkali treatment at 96 h was 14.42 mmol/L, using juvenile "Taihu No2" as the research species [9]. In the present study, over half of the prawns were dead under the alkali concentration of 12 mmol/L after 96 h of treatment. The above results indicated that "Taihu No2" showed stronger abilities to resist the stress of alkali treatment than Yangtze River wild populations, or stronger abilities to resist the stress of alkali treatment were observed in the juvenile prawns compared to adult prawns.

Some previous publications have identified the effects of alkali treatment on the morphological changes in gills in aquatic animals [33–36], while related reports on the morphological changes in the hepatopancreas are rare. Alkali treatment leads to the detachment of the basement membrane of liver tubules from epithelial cells in *Eriocheir sinensis* [37]. In the present study, alkali treatment resulted in the significant damage to the lumen, vacuoles, secretory cells, and storage cells, thus affecting the normal physiological functions of the hepatopancreas.

The measurement of antioxidant enzymes has been widely used to analyze the effects of stress on the behaviors of prawns [38,39]. The effects of alkali stress on antioxidant enzymes have been widely analyzed in many plants [40–42], while the study of the effects on aquatic animals is rare. A pH of 7.8 stimulated the transcript levels of CAT and GPx and the activity of GPx, while strong alkalization (pH 8.8) has negative effects on the activities of antioxidant enzymes, suggesting alkaline exposure has more harmful effects on antioxidant activity in the liver of hybrid tilapia than acidic exposure [43]. The activities of SOD reached the peak at 3 days in the liver of *Gymnocypris przewalskii* after alkaline treatment at concentrations of 32 mmol/L and 64 mmol/L [5]. The activities of SOD and CAT gradually increased and then decreased to a normal level in the liver of *Triplophysa dalaica* after the alkaline treatment [44]. In *E. sinensis*, the activity of T-AOC was significantly increased after the alkali treatment, while SOD showed no difference between the alkali-treated group and control group [37]. Alkali treatment stimulates the production of excessive free oxygen radicals in animals, and thus antioxidant enzymes are responsible for the elimination of the effects of these free oxygen radicals [45]. In the present study, the activities of all of the tested antioxidants showed no difference between 0 mmol/L and 4 mmol/L, indicating the alkaline concentration of 4 mmol/L did not result in changes in the antioxidative stress. In addition, alkali stress did not result in an increase in MDA, GSH, or GSH-PX levels, while the levels of SOD, CAT, and T-AOC were increased, indicating SOD, CAT, and T-AOC play essential roles in the response of *M. nipponense* to acute alkali stress. However, the role of the antioxidative defense system in the adaptive mechanism to alkali stress needs to be further investigated in *M. nipponense* through chronic exposure experiments.

Metabolic pathways, biosynthesis of secondary metabolites, biosynthesis of amino acids, and microbial metabolism in diverse environments have been identified as the main enriched metabolic pathways of DEMs when environmental stress occurs in plants and aquatic animals [46–49], which is consistent with the results of the present study. Secondary metabolites are natural products which show a restricted taxonomic distribution. Biosynthesis of secondary metabolites has been a hot research topic recently because they have positive effects on health [50,51]. Amino acids are essential substrates for the synthesis of many biologically active substances, playing essential roles in the maintenance of normal physiological and nutritional status in animals [52]. The present study predicted that biosynthesis of secondary metabolites and biosynthesis of amino acids significantly regulated the response to alkali stress in *M. nipponense*.

In the present study, only 184 and 149 genes were differentially expressed between 0 mmol/L and 4 mmol/L and between 0 mmol/L and 8 mmol/L, respectively. This indicated that a low concentration of alkali treatment did not result in significant changes in gene expression. A total of 3949 genes were identified to be differentially expressed between 0 mmol/L and 12 mmol/L, and endocytosis, RNA transport, protein processing in endoplasmic reticulum, lysosome, ubiquitin mediated proteolysis, ribosome, mTOR signaling pathway, and oxidative phosphorylation were the most enriched metabolic pathways of DEGs.

Endocytosis is a cellular process which has been reported to be involved in the regulation of cell signaling and the mediation of receptor internalization and nutrient uptake. The endocytic vesicle usually fuses with the early endosome after endocytosis, which accepts newly endocytosed material, serving as a sorting station that directs incoming proteins and lipids to their final destination [53]. TNF receptor-associated factor 6 (TRAF6) is a kind of ubiquitin-ligase, playing an important role in inflammation and immune response. TRAF6 has been identified as a transduction factor, involved in the activation of receptor activator of nuclear factor κB ligand (RANKL), RANK, NFATcl, and lipopolysaccharide signaling [54,55]. Lysosomes mediate a broad range of fundamental processes, including plasma membrane repair, signaling, secretion, and energy metabolism, which has significant implications for health and disease [56,57]. NPC intracellular cholesterol transporter (NPC) is an essential gene in lysosomes, which has been identified to be involved in mitochondrial dysfunction and mTOR suppression [58,59]. Ubiquitin-mediated protein degradation is one of the important mechanisms of protein degradation in cells, playing essential roles in the regulation of various cellular biological processes, including cell cycle, signal transduction, DNA repair, and immune response [60,61]. Ubiquitin E3 ligases (E3) have functions in the reorganization of the target protein, playing essential roles in the mediation of the covalent linkage between target and ubiquitin moieties. These ligases promote target specificity and uniqueness in the process of ubiquitination [62,63]. In the present study, endocytosis, lysosome, and ubiquitin-mediated proteolysis are significantly changed after the alkalinity exposure, mainly functioning in the recognition and digestion of damaged or aged cells caused by the exposure to alkalinity. The alkali concentration of 12 mmol/L significantly stimulated the expressions of TRAF6, NPC2, and E3 FANCL, indicating these genes are involved in the regulation of alkali tolerance in this species.

The endoplasmic reticulum (ER) is an organelle, and proteins are folded with the help of lumenal chaperones in the ER. Newly synthesized peptides are glycosylated in the ER. Correctly folded proteins are packaged into transport vesicles and transferred to the Golgi complex. Misfolded proteins are retained within the ER lumen and finally degraded [64,65]. Heat shock protein 90 (*HSP90*) proteins regulate the process of protein folding, signal transduction, protein degradation, and morphologic evolution. *HSP90* plays essential roles in folding newly synthesized proteins or stabilizing and refolding denatured proteins after stress [66,67]. Eukaryotic translation initiation factor 2 (*eIF2*) is a key protein involved in translation initiation of eukaryotic cells. It plays essential roles in the conversion of eIF2-GDP (inactive state of eIF2) into eIF2-GTP (active state of eIF2) during the process of translation initiation [68,69]. Ribosomes regulate the process of RNA translation into protein and can obtain the genetic information from messenger RNA and convert it into amino acid sequences to synthesize proteins [70,71]. Ribosomal proteins (RPs) are used to synthesize the ribosome. RPs are highly conserved proteins involved in translational control and cellular homeostasis [72]. Thus, protein processing in endoplasmic reticulum and ribosomes were suggested to participate in the regulation of alkali tolerance through ensuring the accuracy of protein synthesis in *M. nipponense* after the exposure to alkalinity. The significantly up-regulated genes from these two metabolic pathways, including *39S-RPL32*, *39S-RPL33*, *60S-RPL19*, *HSP90*, and *eIF2*, possibly promoted protein processing, which contributed to the adaptation to alkali stress in *M. nipponense*.

Oxidative phosphorylation is the main reaction to produce ATP in wild organisms [73]. Cellular respiration is an important process to produce energy in most eukaryotic organisms [74–76]. The cytochrome bc1 complex (Cbc) is an essential component of cellular respiration, promoting the generation of ATP [77]. Adenosine triphosphate (ATP) synthase promotes the production of ATP in cells. ATP synthase-coupling factor 6 (*ATP-CF6*) is released from the vascular endothelial cells and was considered as a cardiovascular therapeutic target through inhibiting prostacyclin synthesis and promoting nitric oxide (NO) synthesis [78]. In addition, ATP synthase-coupling factor 6 was identified to inhibit the JAK1-STAT6 signaling pathway and thus suppress male-predominant HCC [79]. Thus, the changes in oxidative phosphorylation in the present study were predicted to regulate the process of alkali tolerance through providing ATP in *M. nipponense*. Furthermore, *Cbc-7*, *Cbc-10*, and *ATP-CF6* were significantly up-regulated under alkali exposure in *M. nipponense*, which showed a positive response to the alkali stress.

Three genes were differentially expressed among all three comparisons, predicting these three genes play essential roles in the mechanism of alkali tolerance of *M. nipponense*, including hypothetical protein JAY84_18770, Ras-like GTP-binding protein, and *DMRT1-a*. Previous study identified that bacterial GTP-binding proteins are a key factor in the regulation of protein biosynthesis and protein secretion [80]. The member of the ras superfamily of GTP-binding proteins act as molecular binary switches, which were identified to be involved in the various cellular processes of an organism, especially for cell growth [81,82]. *DMRT1-a* is a transcription factor which was identified to regulate the process of male sex determination and differentiation. The main functions for *DMRT1-a* included the controlling of testis development and germ cell proliferation, which can act both as a transcription repressor and activator [83,84].

The qPCR verification of DEGs was generally consistent with those of RNA-Seq, indicating the accuracy of RNA-Seq. qPCR analyses revealed that the expression of four DEGs was sensitive to the changes in alkali concentrations, especially that of *RaG*, of which the expression was increased with the increase in alkali concentration, indicating these four genes play essential roles in the protection of the body from the damage caused by alkali treatment. In addition, the other tested DEGs showed the highest expressions at the alkali concentration of 12 mmol/L, and slightly changed between 0 mmol/L, 4 mmol/L, and 8 mmol/L, indicating only a high alkali concentration can stimulate significant changes in gene expressions and these genes are involved in the process of alkali tolerance in *M. nipponense*.

5. Conclusions

In conclusion, the results of the present study indicated that the death rate of adult wild *M. nipponense* was increased with the increase in alkali concentration. Low-concentration of alkali treatment (<4 mmol/L) did not result in changes in histology and antioxidant enzymes, while higher alkali concentrations stimulated the activities of SOD, CAT, and T-AOC, indicating these enzymes play essential roles in the protection of the body from the damage of alkali treatment. Furthermore, only the alkali concentration of 12 mmol/L led to significant changes in gene expressions, and endocytosis, RNA transport, protein processing in endoplasmic reticulum, lysosome, ubiquitin-mediated proteolysis, ribosome, mTOR signaling pathway, and oxidative phosphorylation represented the most enriched metabolic pathways. Endocytosis, lysosome, and ubiquitin-mediated proteolysis are immune-related metabolic pathways, which protect the body from the damage of alkali treatment and degrade damaged or aged cells. Protein processing in the endoplasmic reticulum and ribosome promoted protein synthesis. Oxidative phosphorylation produces ATP to support the adaptation to alkali treatment in this species. Interestingly, qPCR analyses revealed that four genes were differentially expressed among all three comparisons, predicting these genes were sensitive to the changes in alkali concentration, including *HP-JAY84*, *RaG*, *Dmrt1-a*, and *InR*. The present study identified the changes in antioxidant status, morphology, metabolites, and genes in the hepatopancreas of *M. nipponense* caused by the alkalinity exposure, providing valuable evidence to find out the mechanism of alkali adaptation in this species.

Author Contributions: Conceptualization, S.J. (Shubo Jin) methodology, S.J. (Shubo Jin) and M.X.; software, H.Q.; validation, S.J. (Sufei Jiang) and Y.X.; formal analysis, W.Z.; investigation, Y.W. and H.F.; resources, Y.X.; data curation, H.Q. and X.G.; writing—original draft preparation, S.J. (Shubo Jin); writing—review and editing, M.X. and H.F.; funding acquisition, H.F. All authors have read and agreed to the published version of the manuscript.

Funding: This research was supported by grants from Central Public-Interest Scientific Institution Basal Research Fund, CAFS (2023JBFM04, 2023TD39); the Seed Industry Revitalization Project of Jiangsu Province (JBGS [2021] 118); Jiangsu Agricultural Industry Technology System; the earmarked fund for CARS-48-07; the New Cultivar Breeding Major Project of Jiangsu Province (PZCZ201745); the Natural Science Foundation of Jiangsu Province (BK20221207).

Institutional Review Board Statement: Permissions for the experiments involved in the present study were obtained from the Institutional Animal Care and Use Ethics Committee of the Freshwater Fisheries Research Center, Chinese Academy of Fishery Sciences (Wuxi, China) (Authorization NO. 20210716139, 12 July 2023).

Informed Consent Statement: Not applicable.

Data Availability Statement: The raw data of the present study have been submitted to NCBI with the accession numbers SRX22243687–SRX22243698 and MetaboLights with the accession number MTBLS8831. All other data are contained within the main manuscript.

Acknowledgments: Thanks to the Jiangsu Province Platform for the Conservation and Utilization of Agricultural Germplasm.

Conflicts of Interest: The authors declare no conflicts of interest.

References

1. Fu, H.T.; Jiang, S.F.; Xiong, Y.W. Current Status and Prospects of Farming the Giant River Prawn (*Macrobrachium Rosenbergii*) and the oriental River Prawn (*Macrobrachium Nipponense*) in china. *Aquac. Res.* **2012**, *43*, 993–998.
2. Zhang, X.L.; Cui, L.F.; Li, S.M.; Liu, X.Z.; Han, X.; Jiang, K.Y. Bureau of Fisheries, Ministry of Agriculture, P.R.C. Fisheries Economic Statistics. In *China Fishery Yearbook*; Beijing China Agricultural Press: Beijing, China, 2020; p. 24.
3. Chi, B.J.; Liang, L.Q.; Liu, C.L.; Chang, Y.M.; Wang, S.; Han, Q.X.; Gao, G.Q. Adaptability of *Tribolodon brandti* (Dybowski)to NaCI concentration and alkalinity. *J. Fish. China* **2011**, *18*, 689–694.
4. Lei, Y.Z.; Dong, S.L.; Shen, C.G. Study on the toxicity of carbonate-alkaline to fishes. *J. Fish. Sci.* **1985**, *9*, 171–183.
5. Wang, Z.; Yao, Z.L.; Lin, T.T.; Shi, J.Q.; Zhou, K.; Wang, H.; Qi, H.F.; Lai, Q.F. Effects of carbonate alkalinity stress on SOD, ACP, and AKP activities in the liver and kidney of juvenile *Gymnocypris przewalskii*. *J. Fish. China* **2013**, *20*, 1212–1218. [CrossRef]
6. Fang, W.H.; Wang, H.; Lai, Q.F. Toxicity of carbonate-alkalinity and pH to larval *Penaeus chinensis*. *J. Fish. China* **2000**, *4*, 78–81.
7. Yang, Y.F.; Li, X.J.; Yang, X.Q.; Sun, L.M. Adapt ability of *Litopenaeus vannamei* to carbonate saline–alkaline waters in north east China. *Mar. Sci.* **2008**, *1*, 41–44.
8. Liu, F. Effects of Carbonate Alkalinity Stress on the Survival, Growth, Reproduction, and Immune Enzyme Activities of Exopalaemon carinicauda. Master's Thesis, Shanghai Ocean University, Shanghai, China, 2016.
9. Ren, S.S.; Sun, B.; Luo, L.; Zhang, L.M.; Chang, Y.M.; Liang, L.Q. Tolerance of Freshwater Shrimp (*Macrobrachium nipponense*) to Alkalinity and Low Temperature in Northeast China. *Chin. J. Fish.* **2020**, *33*, 24–28.
10. Xu, J.; Li, Q.; Xu, L.M.; Wang, S.L.; Jiang, Y.L.; Zhao, Z.X.; Zhang, Y.; Li, J.T.; Dong, C.J.; Xu, P.; et al. Gene expression changes leading extreme alkaline tolerance in Amur ide (*Leuciscus waleckii*) inhabiting soda lake. *BMC Genom.* **2013**, *14*, 628. [CrossRef]
11. Wang, L.B.; Pan, M.J.; Wang, M.Y.; Wang, R.Z.; Li, L.; Dong, S.L.; Li, W.D.; Tian, X.L. Kidney Transcriptomic Response of *Lateolabrax maculatus* to Long-Term Alkalinity Stressing. *Period. Ocean Univ. China* **2023**, *2*, 32–43.
12. Shang, X.C.; Geng, L.W.; Yang, J.; Zhang, Y.T.; Xu, W. Transcriptome analysis reveals the mechanism of alkalinity exposure on spleen oxidative stress, inflammation and immune function of *Luciobarbus capito*. *Ecotox. Environ. Saf.* **2021**, *225*, 112748. [CrossRef]

13. Chang, Y.M.; Tang, R.; Dou, X.J.; Tao, R.; Sun, X.W.; Liang, L.Q. Transcriptome and expression profiling analysis of *Leuciscus waleckii*: An exploration of the alkali-adapted mechanisms of a freshwater teleost. *Mol. Biosyst.* **2014**, *10*, 491–504. [CrossRef] [PubMed]
14. *SC/T 9406-2012*; Water quality for aquaculture in saline-alkali land, Ministry of Agriculture. China Agricultural Press: Beijing, China, 2012.
15. Ma, X.K.; Liu, X.Z.; Wen, H.S.; Xu, Y. Histological observation on gonadal sex differentiation in *Cynoglossus semilaevis* Günther. *Mar. Fish. Res.* **2006**, *27*, 55–61.
16. ShangGuan, B.M.; Liu, Z.Z.; Li, S.Q. Histological studies on ovarian development in *Scylla serrata*. *J. Fish. China* **1991**, *15*, 96–103.
17. Ortiz-Villanueva, E.; Navarro-Martín, L.; Jaumot, J.; Benavente, F.; Sanz-Nebot, V. Metabolic disruption of zebrafish (*Danio rerio*) embryos by bisphenol A. An integrated metabolomic and transcriptomic approach. *Environ. Pollut.* **2017**, *231 Pt 1*, 22–36. [CrossRef] [PubMed]
18. Jin, S.B.; Hu, Y.N.; Fu, H.T.; Sun, S.M.; Jiang, S.F.; Xiong, Y.W.; Qiao, H.; Zhang, W.Y.; Gong, Y.S.; Wu, Y. Analysis of testis metabolome and transcriptome from the oriental river prawn (*Macrobrachium nipponense*) in response to different temperatures and illumination times. *Comp. Biochem. Physiol. Part D* **2020**, *34*, 100662. [CrossRef] [PubMed]
19. Jin, S.B.; Fu, H.T.; Zhou, Q.; Sun, S.M.; Jiang, S.F.; Xiong, Y.W.; Gong, Y.S.; Qiao, H.; Zhang, W.Y. Transcriptome analysis of androgenic gland for discovery of novel genes from the oriental river prawn, Macrobrachium nipponense, using Illumina Hiseq 2000. *PLoS ONE* **2013**, *8*, e76840. [CrossRef]
20. Jin, S.B.; Fu, Y.; Hu, Y.N.; Fu, H.T.; Jiang, S.F.; Xiong, Y.W.; Qiao, H.; Zhang, W.Y.; Gong, Y.S.; Wu, Y. Identification of candidate genes from androgenic gland in *Macrobrachium nipponense* regulated by eyestalk ablation. *Sci. Rep.* **2021**, *11*, 1985. [CrossRef]
21. Chen, S.; Zhou, Y.; Chen, Y.; Gu, J. fastp: An ultra-fast all-in-one FASTQ preprocessor. *Bioinformatics* **2018**, *34*, i884–i890. [CrossRef]
22. Kim, D.; Paggi, J.M.; Park, C.; Bennett, C.; Salzberg, S.L. Graph-based genome alignment and genotyping with HISAT2 and HISAT-genotype. *Nat. Biotechnol.* **2019**, *37*, 907–915. [CrossRef]
23. Ashburner, M.; Ball, C.A.; Blake, J.A.; Botstein, D.; Butler, H.; Cherry, J.M.; Davis, A.P.; Dolinski, K.; Dwight, S.S.; Eppig, J.T.; et al. Gene ontology: Tool for the unification of biology. *Nat. Genet.* **2000**, *25*, 25–29. [CrossRef]
24. Tatusov, R.L.; Fedorova, N.D.; Jackson, J.D.; Jacobs, A.R.; Kiryutin, B.; Koonin, E.V.; Krylov, D.M.; Mazumder, R.; Mekhedov, S.L.; Nikolskaya, A.N.; et al. The COG database: An updated version includes eukaryotes. *BMC Bioinformat.* **2003**, *4*, 41. [CrossRef] [PubMed]
25. Kanehisa, M.; Araki, M.; Goto, S.; Hattori, M.; Itoh, M. KEGG for linking genomes to life and the environment. *Nucleic Acids Res.* **2008**, *36*, 480–484. [CrossRef] [PubMed]
26. Anders, S.; Pyl, P.T.; Huber, W. HTSeq-A Python framework to work with high-throughput sequencing data. *Bioinformatics* **2015**, *31*, 166–169. [CrossRef] [PubMed]
27. Love, M.I.; Huber, W.; Anders, S. Moderated estimation of fold change and dispersion for RNA-seq data with DESeq2. *Genome Biol.* **2014**, *15*, 550. [CrossRef] [PubMed]
28. Ferreira, J.A.; Zwinderman, A.H. On the benjamini–hochberg method. *Ann. Statist.* **2006**, *34*, 1827–1849. [CrossRef]
29. Jin, S.B.; Hu, Y.N.; Fu, H.T.; Jiang, S.F.; Xiong, Y.W.; Qiao, H.; Zhang, W.Y.; Gong, Y.S.; Wu, Y. Potential functions of Gem-associated protein 2-like isoform X1 in the oriental river prawn *Macrobrachium nipponense*: Cloning, qPCR, in situ hybridization, and RNAi analysis. *Int. J. Mol. Sci.* **2019**, *20*, 3995. [CrossRef]
30. Jin, S.B.; Hu, Y.N.; Fu, H.T.; Jiang, S.F.; Xiong, Y.W.; Qiao, H.; Zhang, W.Y.; Gong, Y.S.; Wu, Y. Identification and characterization of the succinate dehydrogenase complex iron sulfur subunit B gene in the oriental river prawn *Macrobrachium nipponense*. *Front. Genet.* **2021**, *12*, 698318. [CrossRef]
31. Hu, Y.N.; Fu, H.T.; Qiao, H.; Sun, S.M.; Zhang, W.Y.; Jin, S.B.; Jiang, S.F.; Gong, Y.S.; Xiong, Y.W.; Wu, Y. Validation and evaluation of reference genes for Quantitative real-time PCR in *Macrobrachium nipponense*. *Int. J. Mol. Sci.* **2018**, *19*, 2258. [CrossRef]
32. Livak, K.J.; Schmittgen, T.D. Analysis of relative gene expression data using real-time quantitative PCR and the $2^{-\Delta\Delta CT}$ method. *Methods* **2001**, *25*, 402–408. [CrossRef]
33. Gao, S.; Chang, Y.M.; Zhao, X.F.; Sun, B.; Zhang, L.M.; Liang, L.Q.; Dong, Z.G. The effect of different bicarbonate alkalinity of the gill structure of Amur ide (*Leuciscus waleckii*). *Acta Hydrobiol. Sin.* **2020**, *44*, 736–743.
34. Matey, V.; Richards, J.; Wang, Y.; Wood, C.M.; Rogers, J.; Davies, R.; Murray, B.W.; Chen, X.Q.; Du, J.; Brauner, C.J. The effect of hypoxia on gill morphology and ionoregulatory status in the Lake Qinghai scaleless carp, *Gymnocypris przewalskii*. *J. Exp. Biol.* **2008**, *211*, 1063–1074. [CrossRef] [PubMed]
35. Qin, G.X.; Wei, Q.; Yu, J.Q. Histological characterization muscular and gill of *Gymnocypris przewalskii*. *J. Qinghai Univ.* **2010**, *28*, 4–7.
36. Zhang, R.Y.; Li, G.G.; Zhang, C.F.; Tang, Y.T.; Zhao, K. Morphological differentiations of the gills of two *Gymnocypris przewalskii* subspecies in different habitats and their functional adaptations. *Zool. Res.* **2013**, *34*, 387–391.
37. Zhang, J.B.; Cui, G.T.; Cai, C.F.; Ren, S.J.; Ni, Q.; Wang, C.R.; Li, W.J.; Ge, Y.Y.; Ding, H.M.; Zhang, C. Effects of short-term extreme pH stress on physiology and growth performance of *Eriocheir sinensis*. *Freshw. Fish.* **2020**, *50*, 99–106.
38. Cagol, L.; Baldisserotto, B.; Becker, A.G.; Souza, C.D.F.; Ballester, E.L.C. Essential oil of Lippia alba in the diet of *Macrobrachium rosenbergii*: Effects on antioxidant enzymes and growth parameters. *Aquac. Res.* **2020**, *51*, 2243–2251. [CrossRef]
39. Kong, Y.Q.; Ding, Z.L.; Zhang, Y.X.; Ye, J.Y.; Du, Z.Y. Dietary selenium requirement of juvenile oriental river prawn *Macrobrachium nipponense*. *Aquaculture* **2017**, *476*, 72–78. [CrossRef]

40. Li, Y.; Liu, B.; Peng, Y.; Liu, C.; Li, C. Exogenous GABA alleviates alkaline stress in *Malus hupehensis* by regulating the accumulation of organic acids. *Sci. Hortic.* **2020**, *261*, 108982. [CrossRef]
41. Sriramachandrasekharan, M.V.; Gokula, P.N.; Manivannan, R. Ameliorative Role of Silicon on Osmoprotectants, Antioxidant Enzymes and Growth of Maize Grown Under Alkaline Stress. *Silicon* **2022**, *14*, 6577–6585. [CrossRef]
42. Sun, Y.L.; Hong, S.K. Exogenous proline mitigates the detrimental effects of saline and alkaline stresses in *Leymus chinensis* (Trin.). *J. Plant Biotechnol.* **2010**, *37*, 529–538. [CrossRef]
43. Han, C.Y.; Zheng, Q.M.; Sun, Z.T. Gene Expression and Activities of Antioxidant Enzymes in Liver of Hybrid Tilapia, *Oreochromis niloticus* × *Oreochromis aureus*, Under Acute pH Stress. *J. World Aquacult. Soc.* **2016**, *47*, 260–267. [CrossRef]
44. Wu, P.F. Study on Saline-Alkali Adaptability of Loach in the Dali Lake Plateau. Master's Thesis, Dalian Ocean University, Dalian, China, 2017.
45. Wang, Z.; Cai, C.; Cao, X.; Zhu, J.; Jie, H.; Ping, W.; Ye, Y. Supplementation of dietary astaxanthin alleviated oxidative damage induced by chronic high pH stress, and enhanced carapace astaxanthin concentration of Chinese mitten crab *Eriocheir sinensis*. *Aquaculture* **2018**, *483*, 230–237. [CrossRef]
46. Basanta, K.D.; Chakraborty, H.J.; Rout, A.K.; Behera, B.K. De novo whole transcriptome profiling of *Edwardsiella tarda* isolated from infected fish (*Labeo catla*). *Gene* **2019**, *701*, 152–160.
47. Fu, J.F.; Zhang, J.; Zhang, Y.J.; Yang, C.; Cao, G.X.; Zong, G.L. Analysis of genome sequence and natamycin biosynthetic gene cluster on high producing strain *Streptomyces gilvosporeus* F607. *Microbiol. China* **2019**, *46*, 2312–2325.
48. Yin, M.H.; Deng, H.G.; Jiang, Y.; Wan, L.; Wu, L.X.; Ling, F.; Wang, J.H. GC/MS Metabonomics Analysis of Dioscorea bulbifera L. Microtubers Conserved in vitro at Low Temperature. *Bull. Bot. Res.* **2018**, *38*, 238–246.
49. Zhao, W.S.; Guo, Q.G.; Dong, L.H.; Wang, P.P.; Su, Z.H.; Zhang, X.Y.; Lu, X.Y.; Li, S.Z.; Ma, P. Transcriptome and Proteome Analysis of Bacillus subtilis NCD-2 Response to L-proline from Cotton Root Exudates. *Sci. Agric. Sin.* **2021**, *21*, 4585–4600.
50. Asadollahei, M.V.; Yousefifard, M.; Tabatabaeian, J.; Nekonam, M.S.; Mahdavi, S.M.E. Effect of elicitors on secondary metabolites biosynthesis in *Zataria multiflora* Boiss. *Ind. Crop. Prod.* **2022**, *181*, 114789. [CrossRef]
51. Gantait, S.; Das, A.; Mitra, M.; Chen, J.T. Secondary metabolites in orchids: Biosynthesis, medicinal uses, and biotechnology. *S. Afr. J. Bot.* **2021**, *139*, 338–351. [CrossRef]
52. Rezaei, R.; Wang, W.; Wu, Z.; Dai, Z.; Wang, J.; Wu, G. Biochemical and physiological bases for utilization of dietary amino acids by young pigs. *J. Anim. Sci. Biotechnol.* **2013**, *4*, 7. [CrossRef]
53. Doherty, G.J.; McMahon, H.T. Mechanisms of endocytosis. *Annu. Rev. Biochem.* **2009**, *78*, 857–902. [CrossRef]
54. Hull, C.; Mclean, G.; Wong, F.; Duriez, P.J.; Karsan, A. Lipopolysaccharide Signals an Endothelial Apoptosis Pathway Through TNF Receptor-Associated Factor 6-Mediated Activation of c-Jun NH2-Terminal Kinase. *J. Immunol.* **2002**, *169*, 2611–2618. [CrossRef]
55. Zhang, P.; Yue, S.; Jiang, Y.; Zhang, X.; Liu, Y. TNF receptor-associated factor 6 regulates proliferation, apoptosis, and invasion of glioma cells. *Mol. Cell. Biochem.* **2013**, *377*, 87–96.
56. Kornfeld, S.; Mellman, I. The Biogenesis of Lysosomes. *Annu. Rev. Cell Biol.* **1989**, *5*, 483–525. [CrossRef] [PubMed]
57. Woo, S.M.; Kwon, T.K. The Functional Role of Lysosomes as Drug Resistance in Cancer. *J. Life Sci.* **2021**, *31*, 527–535.
58. Sharom, F.J. Lipid transporters and binding proteins; MsbA and NPC1. *FASEB J.* **2010**, *24*, 408.401. [CrossRef]
59. O'Neill, K. Triple Negative Breast Cancer is Dependent on the Lysosomal Cholesterol Transporter NPC1. *J. Endocr. Soc.* **2021**, *5*, A1034. [CrossRef]
60. Moran-Crusio, K.; Reavie, L.B.; Aifantis, I. Regulation of hematopoietic stem cell fate by the ubiquitin proteasome system. *Trends Immunol.* **2012**, *33*, 357–363. [CrossRef] [PubMed]
61. Strikoudis, A.; Guillamot, M.; Aifantis, I. Regulation of stem cell function by protein ubiquitylation. *EMBO Rep.* **2014**, *15*, 365–382. [CrossRef]
62. Metzger, M.B.; Hristova, V.A.; Weissman, A.M. HECT and RING finger families of E3 ubiquitin ligases at a glance. *J. Cell Sci.* **2012**, *125*, 531–537. [CrossRef]
63. Mohammed, Z.; Timothy, A.H.; Mark, T.D.; Donald, L.C. A Francisella tularensis DNA clone complements Escherichia coli defective for the production of Era, an essential Ras-like GTP-binding protein. *Gene* **1997**, *189*, 31–34.
64. Lawson, W.E.; Crossno, P.F.; Polosukhin, V.V.; Roldan, J.; Cheng, D.S.; Lane, K.B.; Blackwell, T.R.; Xu, C.; Markin, C.; Ware, L.B. Endoplasmic reticulum stress in alveolar epithelial cells is prominent in IPF: Association with altered surfactant protein processing and herpesvirus infection. *Am. J. Physiol. Lung Cell Mol. Physiol.* **2008**, *294*, L1119–L1126. [CrossRef]
65. Wood, P.; Elliott, T. Glycan-regulated Antigen Processing of a Protein in the Endoplasmic Reticulum Can Uncover Cryptic Cytotoxic T Cell Epitopes. *J. Exp. Med.* **1998**, *188*, 773–778. [CrossRef] [PubMed]
66. Ding, H.R.; Qian, J.; Tong, J.; Tang, J.N.; Lin, H.; Chu, J.P.; Zhu, G.Q.; Chen, F.; Liu, X.B. HSP90 pathway in intermediate mononuclear cells causes plaque erosion via induction of neutrophil hyper-responsiveness. *Eur. Heart J.* **2021**, *42*, ehab724.1301. [CrossRef]
67. Ghosh, A.; Garee, G.; Sweeny, E.A.; Nakamura, Y.; Stuehr, D.J. Hsp90 chaperones hemoglobin maturation in erythroid and nonerythroid cells. *Proc. Natl. Acad. Sci. USA* **2018**, *22*, E1117–E1126. [CrossRef]
68. Miyasaka, H.; Endo, S.; Shimizu, H. Eukaryotic translation initiation factor 1 (eIF1), the inspector of good AUG context for translation initiation, has an extremely bad AUG context. *J. Biosci. Bioeng.* **2010**, *109*, 635–637. [CrossRef] [PubMed]

69. Sun, Y.L.; Gong, H.S.; Jin, C.W.; Hong, S.K. Molecular cloning and identification of eukaryotic translation initiation factor 1 family genes (eIF1, eIF1A and eIF1B) in Leymus chinensis (Trin.). *Biotechnol. Biotec. Eq.* **2015**, *29*, 609–616. [CrossRef]
70. Stan, A.; Mayer, C. Tethered Ribosomes: Toward the Synthesis of Nonproteinogenic Polymers in Bacteria. *Chembiochem Eur. J. Chem. Boil.* **2023**, *24*, e202200578. [CrossRef]
71. Wang, A.; Hassan, A.H.; Freitas, F.C.; Singh, V.; Amunts, A.; Whitford, P. Understanding the energetics of translation in bacterial and eukaryotic ribosomes. *Biophys. J.* **2023**, *122*, 317a. [CrossRef]
72. Homann, H.E.; Nierhaus, K.H. Ribosomal Proteins. *Eur. J. Biochem.* **1971**, *20*, 249–257. [CrossRef]
73. Dimroth, P.; Kaim, G.; Matthey, U. Crucial role of the membrane potential for ATP synthesis by F(1) F(o) ATP synthases. *J. Exp. Biol.* **2000**, *203*, 51–59. [CrossRef]
74. Althoff, T.; Mills, D.J.; Popot, J.L.; Kuhlbrandt, W. Arrangement of electron transport chain components in bovine mitochondrial supercomplex I1III2IV1. *EMBO J.* **2011**, *30*, 4652–4664. [CrossRef]
75. Dudkina, N.V.; Kudryashev, M.; Stahlberg, H.; Boekema, E.J. Interaction of complexes I, III, and IV within the bovine respirasome by single particle cryoelectron tomography. *Proc. Natl. Acad. Sci. USA* **2011**, *108*, 15196–15200. [CrossRef] [PubMed]
76. Schagger, H.; Pfeiffer, K. Supercomplexes in the respiratory chains of yeast and mammalian mitochondria. *EMBO J.* **2000**, *19*, 1777–1783. [CrossRef]
77. Yang, W.C.; Li, H.; Wang, F.; Zhu, X.L.; Yang, G.F. Rieske Iron–Sulfur Protein of the Cytochrome bc1 Complex: A Potential Target for Fungicide Discovery. *ChemBioChem* **2012**, *13*, 1542–1551. [CrossRef] [PubMed]
78. Yi, F.; Zhu, Y. Ectopic ATP Synthase in Endothelial Cells: A Novel Cardiovascular Therapeutic Target. *South China J. Cardiovasc. Dis.* **2011**, *S1*, 35–36.
79. Yang, W.; Yan, L.; Xu, Y.; Xu, L.; Wei, Z.; Wu, Y.; Long, L.; Shen, P. Estrogen Represses Hepatocellular Carcinoma (HCC) Growthvia Inhibiting Alternative Activation of Tumor-associated Macrophages (TAMs). *J. Biol. Chem.* **2012**, *287*, 40140–40149. [CrossRef] [PubMed]
80. Kaziro, Y.; Itoh, H.; Kozasa, T.; Nakafuku, M.; Satoh, T. Structure and function of signal-transducing GTP-binding proteins. *Annu. Rev. Biochem.* **1991**, *60*, 349–400. [CrossRef] [PubMed]
81. Hoshino, M.; Nakamura, S. The Ras-like small GTP-binding protein Rin is activated by growth factor stimulation. *Biochem. Biophys. Res. Commun.* **2002**, *295*, 651–656. [CrossRef]
82. Morreale, F.E.; Walden, H. Types of ubiquitin ligases. *Cell* **2016**, *165*, 248. [CrossRef]
83. Yoshimoto, S.; Ikeda, N.; Izutsu, Y.; Shiba, T.; Takamatsu, N.; Ito, M. Opposite roles of DMRT1 and its W-linked paralogue, DM-W, in sexual dimorphism of Xenopus laevis: Implications of a ZZ/ZW-type sex-determining system. *Development* **2010**, *137*, 2519–2526. [CrossRef]
84. Yoshimoto, S.; Okada, E.; Oishi, T.; Numagami, R.; Umemoto, H.; Tamura, K.; Kanda, H.; Shiba, T.; Takamatsu, N.; Ito, M. Expression and promoter analysis of Xenopus DMRT1 and functional characterization of the transactivation property of its protein. *Dev. Growth Differ.* **2006**, *48*, 597–603. [CrossRef]

antioxidants

Article

A High-Fat-Diet-Induced Microbiota Imbalance Correlates with Oxidative Stress and the Inflammatory Response in the Gut of Freshwater Drum (*Aplodinotus grunniens*)

Miaomiao Xue [1], Pao Xu [1,2], Haibo Wen [1,2], Jianxiang Chen [1,2], Qingyong Wang [1], Jiyan He [1], Changchang He [1], Changxin Kong [1], Xiaowei Li [1], Hongxia Li [1,2,*] and Changyou Song [1,2,*]

[1] Wuxi Fisheries College, Nanjing Agricultural University, Wuxi 214081, China
[2] Key Laboratory of Freshwater Fisheries and Germplasm Resources Utilization, Ministry of Agriculture and Rural Affairs, Freshwater Fisheries Research Center, Chinese Academy of Fishery Sciences, Wuxi 214081, China
* Correspondence: lihx@ffrc.cn (H.L.); songchangyou@ffrc.cn (C.S.)

Abstract: Lipids are critical nutrients for aquatic animals, and excessive or insufficient lipid intake can lead to physiological disorders, which further affect fish growth and health. In aquatic animals, the gut microbiota has an important regulatory role in lipid metabolism. However, the effects of a high-fat diet on physical health and microbiota diversity in the gut of freshwater drum (*Aplodinotus grunniens*) are unclear. Therefore, in the present study, a control group (Con, 6%) and a high-fat diet group (HFD, 12%) were established for a 16-week feeding experiment in freshwater drum to explore the physiological changes in the gut and the potential regulatory mechanisms of bacteria. The results indicated that a high-fat diet inhibited antioxidant and immune capacity while increasing inflammation, apoptosis and autophagy in gut cells. Transcriptome analysis revealed significant enrichment in immune-related, apoptosis-related and disease-related pathways. Through 16S rRNA analysis, a total of 31 genus-level differentially abundant bacterial taxa were identified. In addition, a high-fat diet reduced gut microbial diversity and disrupted the ecological balance of the gut microbiota (Ace, Chao, Shannon and Simpson indices). Integrated analysis of the gut microbiota combined with physiological indicators and the transcriptome revealed that gut microbial disorders were associated with intestinal antioxidants, immune and inflammatory responses, cell apoptosis and autophagy. Specifically, genus-level bacterial taxa in Proteobacteria (*Plesiomonas*, *Arenimonas*, *Erythrobacter* and *Aquabacteriumb*) could serve as potential targets controlling the response to high-fat-diet stimulation.

Keywords: high-fat diet; gut microbiota; physiological homeostasis; *Aplodinotus grunniens*

check for updates

Citation: Xue, M.; Xu, P.; Wen, H.; Chen, J.; Wang, Q.; He, J.; He, C.; Kong, C.; Li, X.; Li, H.; et al. A High-Fat-Diet-Induced Microbiota Imbalance Correlates with Oxidative Stress and the Inflammatory Response in the Gut of Freshwater Drum (*Aplodinotus grunniens*). *Antioxidants* **2024**, *13*, 363. https://doi.org/10.3390/antiox13030363

Academic Editors: Evangelos Zoidis and Erchao Li

Received: 11 December 2023
Revised: 29 February 2024
Accepted: 4 March 2024
Published: 18 March 2024

1. Introduction

Carbohydrates, lipids and proteins are the main nutrients for fish growth. However, the production of fish meal, the optimal protein resource for aquatic animals, is increasingly scarce worldwide, which strictly limits the development of aquaculture. Therefore, lipids and carbohydrates are widely applied as non-protein energy sources in diets to partially replace the protein content [1]. However, fish have a low tolerance to carbohydrates, and their ability to utilize energy from digestible carbohydrates is limited [2]. Therefore, lipids are considered to be the main non-protein energy source for fish [3], providing energy and essential fatty acids and acting as carriers of nutrients such as the fat-soluble vitamins A, D and K [4,5]. With the fast development of aquaculture, high-fat diets are widely utilized because they promote growth and protein retention [6]. However, excessive lipids can cause various metabolic diseases [7,8] such as metabolic disorders, fat accumulation, inflammation and endoplasmic reticulum stress [9–11], which can affect fish health.

The gut is not only a major site for digestion and nutrient absorption but also a key part of host defense [1,12] The fish gut is a complex ecosystem in which the gut microbiota is an important component of intestinal environment and plays a crucial role in enhancing host immunity, nutrient metabolism and digestion, as well as healthy growth and reproduction [13–15]. Many factors, such as species, growth stage and environment, can regulate the gut microbial community [16]. It is universally recognized that diet plays a key role in determining the composition of the gut microbiota and that dietary lipid levels influence the metabolic capacity of the gut microbial community [11,17]. However, as the freshwater drum is a newly domesticated aquatic animal, the effects of a high-fat diet on its gut and microbiota have not been studied. In this study, we hypothesized that a high-fat diet may induce immune and inflammatory responses in the gut and that alterations in the gut microbiota are in response to an imbalance in the physiological homeostasis of the gut.

It is well known that alterations in the lipid content of aquaculture feeds may affect multiple rather than individual genes and signaling pathways in fish [18–20], and transcriptomics provides an effective and rapid technique to study the molecular mechanisms of tissues and organs under different conditions [21]. In addition, 16S rRNA gene sequencing has become a major technique for studying bacterial communities [22]. Therefore, transcriptome and gut microbiota analyses can be used to understand changes in gut molecular regulatory networks and microbiota in animals on high-fat diets. The combined analysis of the two can further identify the role of key gut bacteria in the regulation of functional genes.

The freshwater drum (*Aplodinotus grunniens*) is native to North and Central America and is the only freshwater species in the genus *Aplodinotus* [23]. Freshwater drum have no intermuscular bones and possess a higher proportion of edible mass than most fish, with pleasant-tasting and nutritious flesh rich in proteins, amino acids and fatty acids, especially the unsaturated fatty acids DHA and EPA. The inclusion of freshwater drum can also improve the fish quality and processing of aquatic products [24]. Therefore, we introduced freshwater drum and conducted a great deal of related research. We conducted feed domestication and discovered potential regulatory mechanisms in freshwater drum [25], explored the mechanisms of intestinal adaptation to hypothermia [26] and identified the role of *miR-1/AMPK* in the regulation of glycolipid metabolism [24] as well as the role of the PPAR signaling pathway in the hypothermic regulation of lipid and amino acid metabolic homeostasis [27]. The above studies found that freshwater drum are suitable for nationwide culture and promotion. However, in recent aquaculture, we have discovered that a high-fat diet can adversely affect freshwater drums. Based on this finding, we evaluated the effects of high-fat diet on growth performance (condition factor (CF), viscerosomatic index (VSI) and hepatosomatic index (HSI)) and physiological homeostasis, and we explored the potential regulatory mechanisms in the liver of freshwater drum under lipid deposition [28]. However, limited information is known about the mechanisms that regulate the gut and the microbiota of freshwater drum. Therefore, in the present study, we assessed the effects of a high-fat diet on the gut antioxidant capacity, immunocompetence and inflammatory response as well as apoptosis and autophagy in freshwater drum. The interactions between gut bacteria and differentially expressed genes were also investigated. These results can reveal the effects of a high-fat diet on gut health and physiological homeostasis as well as the response mechanisms of the gut microbiota, which is important for the sustainability of freshwater drum as a resource.

2. Materials and Methods

2.1. Ethics Statement

The care and use of the animals followed the guidelines of the Animal Care and Use Committee of the Nanjing Agricultural University, China, and was approved under those guidelines (WXFC 2021-0006). All animal procedures were performed in accordance with the Guideline for the Care and Use of Laboratory Animals in China. The ethical situation in this experiment was identical to that in our previously published paper [26].

2.2. Experimental Animals and Experimental Design

Experiments were performed at the Freshwater Fisheries Research Center, Chinese Academy of Fishery Sciences. About 40,000 healthy freshwater drum with an average of 20.88 ± 2.75 g were transported randomly to two outdoor fish ponds (pond size 667 m^2, about 20,000 fish per pond) for the experiment. During the experiment, the control group (Con, 6% fat) and high-fat diet group (HFD, 12% fat) were fed compounded diets four times a day for four months during the experimental period (Table S1). The feeding amount was 3~5% of body weight every day. The water was taken from an underground source and then fully aerated. The water quality was maintained as follows: DO > 6 mg/L, pH 7.2~7.8, NO_2^- < 0.02 mg/L and NH_3 < 0.05 mg/L.

2.3. Sample Collection

After 16 weeks of rearing experiments, fish were starved for 24 h to evacuate the residual feed in the intestine. Freshwater drum from the control (Con) and high-fat diet (HFD) groups were randomly selected (27 per group) from each pond and anesthetized with MS-222 (100 mg/L, tricaine methanesulfonate, Sigma, St. Louis, MO, USA) to collect samples. The sampled fish were dissected on ice to collect the gut tissue. The whole gut and contents were collected, immediately frozen in liquid nitrogen and stored at −80 °C for sequencing and RT-PCR analysis.

2.4. Biochemical Index Analysis

According to the manufacturer's instructions, enzyme activity levels were measured with 10% gut homogenate supernatant. Specifically, a total of nine gut samples were selected and measured in duplicate for each group. A 0.1~0.2 g portion of each gut sample was isolated and homogenized in ninefold saline (*w*/*v*). Centrifugation (2500 rpm, 4 °C) was performed for 10 min to collect the supernatant for further measurement.

The analyzed antioxidant parameters included glutathione peroxidase (GPx), glutathione (GSH), catalase (CAT) and malondialdehyde (MDA). In detail, GPx was measured by the colorimetric method (Category No. A005-1-2), GSH by the microplate method (Category No. A006-2-1), CAT by the ammonium molybdate method (Category No. A007–1-1) and MDA by the thiobarbituric acid (TBA) method (Category No. A003-1-1). All kits were purchased from Nanjing Jiancheng Bioengineering Institute, Nanjing, China.

2.5. Enzyme-Linked Immunosorbent Assay Analysis

According to the manufacturer's instructions, nine gut samples were selected from each group and rinsed with pre-cooled PBS (0.01 M, pH = 7.4) to remove residual contents. The gut samples were separated 0.1~0.2 g, combined with ice-cold PBS to form a 10% (*w*/*v*) tissue homogenate and centrifuged at 5000× *g* and 4 °C for 10 min to collect the supernatant. Specifically, we analyzed tumor necrosis factor alpha (TNF-α, Category No. ml07310), interleukin 6 (IL-6, Category No. ml0257510), immunoglobulin M (IgM, Category No. ml025819) and immunoglobulin G (IgG, Category No. ml823766) content by the double antibody sandwich ELISA method using commercial kits (Shanghai Enzyme-linked Biotechnology Co., Ltd., Shanghai, China).

2.6. RNA Extraction and De Novo High-Throughput Sequencing

According to the method described by Chen [24], three gut samples were selected from each group for transcriptome sequencing. Specifically, total RNA was isolated from gut tissues with a Trizol kit (Invitrogen, Carlsbad, CA, USA). Agilent 2100 and Nanodrop apparatuses (ThermoFisher Ltd., Waltham, MA, USA) were used to examine RNA quality, and high-quality RNA (1.8 < OD260/280 < 2.0, RNA integrity number (RIN) ≥ 1.8, 28S/18S ≥ 1.0) was processed with oligo (dT) to enrich mRNA. Next, the mRNA was randomly split into small fragments of about 300 bp using random primers. Then, cDNA was synthesized using commercial kits (NEB7530, New England Biolabs, Ipswich, MA, USA). Finally, de novo high-throughput sequencing was performed with an Illumina NovaSeq6000 (Majorbio

Bio-pharm Technology Co., Ltd., Shanghai, China). The raw data were filtered with fastp (version 0.18.0) to obtain high-quality reads. Sequences were read using Trinity assembly and splicing as well as functionally annotated against the NR, Swiss Port, Pfam, KOG and GO databases. Transcripts with $|\log_2\text{fold change}| > 1$ and $p < 0.05$ were considered differentially expressed genes (DEGs). The enrichment analysis was performed with the Gene Ontology (GO) and Kyoto Encyclopedia of Genes and Genomes (KEGG) databases.

2.7. Microbial DNA Extraction and 16S Sequencing

In order to explore the diversity and composition of gut bacteria under the condition of a high-fat diet, six samples of gut contents were randomly selected for microbiome analysis according to the method described by Song [25]. First, an E.Z.N.A.® Soil DNA kit (Omega Bio-Tek, Norcross, GA, USA) was used to extract microbial DNA [29]. The integrity of the obtained DNA was measured using 1% agarose gel electrophoresis and a Nanodrop ND2000 spectrophotometer (Thermo Scientific, Waltham, MA, USA). Isolated DNA concentrations were measured using a Quant-iT PicoGreen dsDNA Assay Kit (Invitrogen, Eugene, OR, USA) and a fluorometer and diluted to 20 ng/μL. Full-length 16S rRNA was amplified using primers 338 F (5′-ACTCCTACGGGAGGCAGCAG-3′) and 806 R (5′-GGACTACHVGGGTWTCTAAT-3′) by PCR. The PCR products from the same samples were mixed and detected using 2% gel electrophoresis, and the purified PCR products were recycled using an AxyPrepDNA Gel Extraction Kit (AXYGEN, Hangzhou, China). PCR products were quantified using a QuantiFluor™-ST Blue Fluorescence Quantification System (Promega, Madison, WI, USA). The purified PCR products were sequenced using the MiSeq platform (Illumina, San Diego, CA, USA).

PE reads obtained from Illumina sequencing were first spliced based on overlap relationships, while quality control and filtering of sequence quality were performed to arrive at the final valid data (fastq files). The RDP classifier Bayesian algorithm was used to analyze the taxonomy of the OTUs' representative sequences at a 97% similarity level and to count the community species composition of each sample at the domain, kingdom, phylum, class, order, family, genus and species levels, respectively. Alpha diversity indices (Ace, Chao, Shannon and Simpson indices) of the samples were analyzed using Mothur 1.30.2 software. Beta diversity (weighted UniFrac and unweighted UniFrac distances) analysis was performed using QIIME software (version 1.9.1). The species abundance of the samples was counted at both the phylum and genus taxonomic levels, and the composition of the bacteria was analyzed.

2.8. Integrated Analysis between Key Genes and Bacteria

Differentially abundant bacteria were analyzed using Pearson correlation with gut-health-related genes (relevant to antioxidants, immunity, inflammation, autophagy and apoptosis) and DEGs in the transcriptome. In addition, DEGs and bacteria with $p < 0.05$ and $p < 0.01$ correlation were integrated for network interaction analysis in Cytoscape (version 3.9.0).

2.9. Transcriptional Expression Analysis by Real-Time PCR

Total RNA from nine gut samples in each group was extracted with RNAiso Plus reagent (Takara Co., Ltd., Dalian, China) and incubated with RNase-free DNase (Takara Co., Ltd., Dalian, China) to eliminate contaminating genomic DNA. The cDNA was synthesized from 1 μg of high-quality RNA (1.8 < OD260/280 < 2.0) with a PrimeScript™ RT reagent kit (TaKaRa, Dalian, China). Finally, RT-PCR was performed using a TB Green Premix Ex Taq™ II (Tli RNase Plus) kit (TaKaRa, Dalian, China). The relative expression levels of genes were normalized to β-actin and further calculated using the $2^{-\Delta\Delta CT}$ method. All the genes involved in this experiment were designed with Premier 5.0 based on the mRNA sequences obtained from the *A. grunniens* genome database in our laboratory and synthesized by Shanghai Generay Biotechnology Co., Ltd., Shanghai, China (Table S2).

2.10. Statistical Analysis

In this study, all data were analyzed using SPSS software (version 26.0) and presented as the mean ± standard error of the mean (SEM). Antioxidant parameters and ELISA data were analyzed in SPSS 26.0 with the independent-samples *t*-test. The $2^{-\Delta\Delta CT}$ method was used to calculate genes' relative expression levels, followed by an independent-samples *t*-test. In data analysis, an independent-samples *t*-test was used when data were normally distributed and homoscedastic; otherwise, two independent-samples nonparametric tests (Mann-Whitney U) were used. Similarly, Pearson correlation analysis of genes and bacteria was performed using SPSS. Alpha diversity indices (Ace, Chao, Shannon and Simpson indices) were analyzed by one-way analysis of variance (ANOVA) followed by Tukey's post hoc test using SPSS 26.0. $p < 0.05$ indicated a significant difference or correlation.

3. Results

3.1. A High-Fat Diet Inhibits Antioxidant Capacity in the Gut of A. grunniens

To explore the effects of a high-fat diet on the physiological homeostasis of the freshwater drum's gut, we first assessed the antioxidant capacity. Results showed that the expression levels of GPx ($p = 0.025$), GSH ($p = 0.024$) and CAT ($p = 0.0006$) were significantly decreased in the HFD group compared to the Con group (Figure 1A,C). In contrast, the content of MDA was remarkably increased ($p = 0.026$, Figure 1D). In addition, RT-PCR results further revealed that the HFD reduced the expression of the antioxidant-related genes Glutamic-Oxaloacetic Transaminase 1 (*GOT1*, $p = 0.045$), glutathione peroxidase-1 (*GPX1*, $p = 0.015$), glutathione peroxidase-4 (*GPX4*, $p = 0.009$) and heme oxygenase 1 (*HO-1*, $p = 0.029$) compared to the Con. However, the HFD activated the transcription level of nuclear erythroid-related factor 2 (*Nrf2*, $p = 0.044$) (Figure 1E). These results indicate that an HFD inhibits gut antioxidant capacity, leading to physiological disorders in freshwater drum.

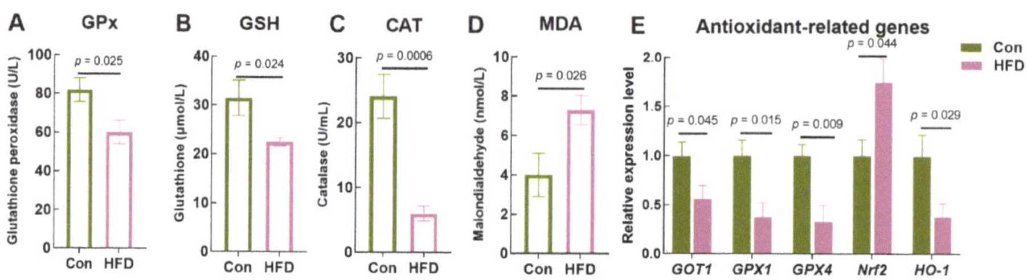

Figure 1. HFD inhibits antioxidant capacity in the gut of *A. grunniens*. (**A**), Glutathione peroxidase, GPx; (**B**), Glutathione, GSH; (**C**), Catalase, CAT; (**D**), Malondialdehyde, MDA; (**E**), Transcriptional expression of antioxidant-related genes. Results are indicated as the mean ± SEM, $n = 9$.

3.2. A High-Fat Diet Suppresses Immunocompetence and Induces Cellular Inflammation, Apoptosis and Autophagy

To determine whether physiologic disturbances resulting from high-fat diets caused immune and inflammatory responses, we evaluated the levels of immune and inflammatory factors. Enzyme-linked immunosorbent assays (ELISAs) showed decreasing trends in IgG ($p > 0.05$) and IgM ($p = 0.04$) in the gut; however, there were no significant differences in IgG after HFD feeding. Unexpectedly, *TNF-α* ($p = 0.006$) and *IL-6* ($p = 0.016$) levels were markedly increased after HFD feeding (Figure 2A). Meanwhile, RT-PCR results demonstrated that the immunity-related genes heat shock protein 70 (*HSP70*, $p = 0.009$) and heat shock protein 90 (*HSP90*, $p = 0.017$) were dramatically downregulated in freshwater drum receiving an HFD compared to the control group (Figure 2B). In addition, a high-fat diet increased the transcript levels of the inflammation-related genes toll-like receptor 1 (*TLR1*, $p = 0.034$), toll-like receptor 2 (*TLR2*, $p = 1.95 \times 10^{-5}$), myeloid differentiation protein-88 (*MyD88*, $p = 0.018$), interleukin-1 (*IL-1β*, $p = 8.23 \times 10^{-6}$) and interleukin-6 (*IL-6*,

$p = 2.23 \times 10^{-4}$) compared to the Con. However, there was no significant difference in nuclear factor kappa B p65 (*P65*, $p > 0.05$) (Figure 2C).

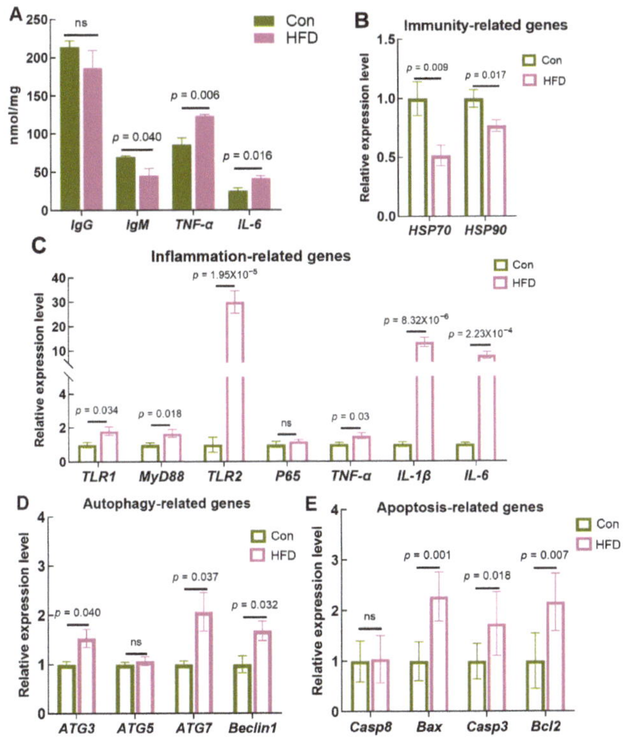

Figure 2. HFD suppresses immunity and induces cellular inflammation, apoptosis and autophagy in the gut of *A. grunniens*. (**A**), Enzyme-linked immunosorbent assay (ELISA) indicators; (**B**), Transcriptional expression of immunity-related genes; (**C**), Transcriptional expression of inflammation-related genes; (**D**), Transcriptional expression of autophagy-related genes; (**E**), Transcriptional expression of apoptosis-related genes. Results were indicated as mean ± SEM, $n = 9$. "ns" indicated no significant difference.

Immune and inflammatory responses usually lead to cellular autophagy and apoptosis. Therefore, we examined the transcriptional expression of apoptosis- and autophagy-related genes. The results showed that HFD feeding significantly increased the mRNA expression of autophagy-related genes 3 and 7 (*ATG3*, $p = 0.040$; *ATG7*, $p = 0.037$) and *Beclin1* ($p = 0.032$) (Figure 2D). In addition, the expression levels of the apoptosis-related genes bcl2-associated X (*Bax*, $p = 0.001$), caspase 3 (*Casp3*, $p = 0.018$) and B-cell lymphoma-2 (*Bcl2*, $p = 0.007$) were significantly increased (Figure 2E). The above results indicate that a high-fat diet disrupts the physiological homeostasis of the gut, which further induces gut immune and inflammatory responses and increases cellular autophagy and apoptosis.

3.3. Gut Transcriptome Analysis of A. grunniens on a High-Fat Diet

To reveal the potential regulatory mechanisms through which HFD feeding acts on the gut, we performed transcriptome analysis using high-throughput sequencing. Principal component analysis (PCA) showed that the samples in the Con and HFD groups were clustered into separate clusters, revealing clear differences between the transcriptome profiles of the Con and HFD groups (Figure 3A). A total of 151 differentially expressed genes (DEGs) were identified ($|\log_2\text{fold change}| > 1$, p-value < 0.05) compared with the Con group, of which 81 were significantly upregulated and 70 were downregulated (Figure 3B). Based on the expression levels, these DEGs were clustered into different subgroups on a heatmap (Figure 3C).

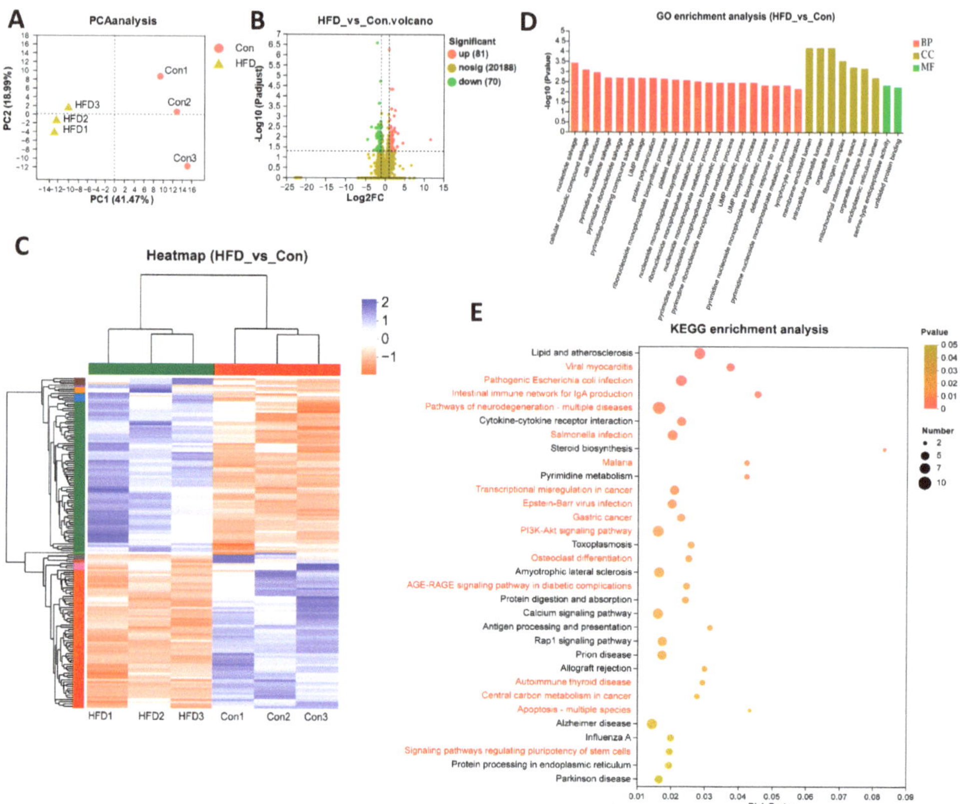

Figure 3. The transcriptome reveals that HFD dysregulates immunity and inflammation in the gut of *A. grunniens*. (**A**), PCA analysis between samples; (**B**), UniGene volcano plot; (**C**), Heatmap clustering of DEGs; (**D**), GO annotation of DEGs; (**E**), KEGG enrichment of DEGs (HFD vs. Con).

These DEGs were analyzed for GO enrichment, with a total of 135 GO items enriched, among which the top 30 items included 7 cellular components (CC), 21 biological processes (BP) and 2 molecular functions (MF) (Figure 3D, Table S3). Specifically, membrane-enclosed lumen (GO:0031974), intracellular organelle lumen (GO:0070013) and organelle lumen (GO:0043233) were the most significantly enriched GO items, suggesting that an HFD might affect gut organelles' normal structure. KEGG enrichment was also utilized for DEG analysis (Figure 3E, Table S4). The results showed that DEGs were mainly enriched in the immune system (intestinal immune network for IgA production, antigen processing and presentation), disease and cancer (such as pathogenic Escherichia coli infection, salmonella infection, malaria, transcriptional misregulation in cancer, AGE-RAGE signaling pathway in diabetic complications, prion disease, autoimmune thyroid disease and central carbon metabolism in cancer), apoptosis in multiple species, signal transduction (PI3K-Akt signaling pathway, signaling pathways regulating pluripotency of stem cells) and metabolism (steroid biosynthesis and pyrimidine metabolism). The transcriptome results suggest that an HFD disrupts the immune system, generates inflammation and apoptosis and affects signaling transduction and metabolic systems in the gut.

3.4. Gut Microbiota Alternation of *A. grunniens* Fed with a High-Fat Diet

Gut health is tightly correlated with microbiota composition; therefore, we determined the composition of the microbial community by 16S rRNA gene amplicon sequencing. A total of 2014 operational taxonomic units (OTUs) were identified, corresponding to 41 phyla, 109 classes, 246 orders, 415 families, 798 genera and 1332 species (Table S5). Among them, a total of 1532 OTUs were identified in the Con group and 1095 in the HFD group, with co-enrichment of 613 OTUs (Figure 4A). Meanwhile, PLS-DA analysis of the samples revealed that the two groups clustered into different subgroups, indicating that HFD feeding affected the composition of OTUs in the gut (Figure 4B). The results of the beta diversity indices (weighted UniFrac and unweighted UniFrac) analyses showed that the differences in the diversity of the gut bacterial community between the HFD and Con groups were not significant (Figure 4C,D, $p > 0.05$). Ace, Chao, Shannon and Simpson's analysis methods were applied to evaluate the alpha diversity of OTUs, and the results indicated that the OTUs exhibited a significant difference between the Con and HFD groups (Figure 4E, $p < 0.05$). In addition, gut microbiota composition was assessed by hierarchical clustering, and bacteria of the taxa Proteobacteria, Actinobacteriota, Firmicutes, Fusobacteriota, Bacteroidota and unclassified_k__norank__d__Bacteria were found to exhibit the richest diversity at the phylum level (Figure 4F).

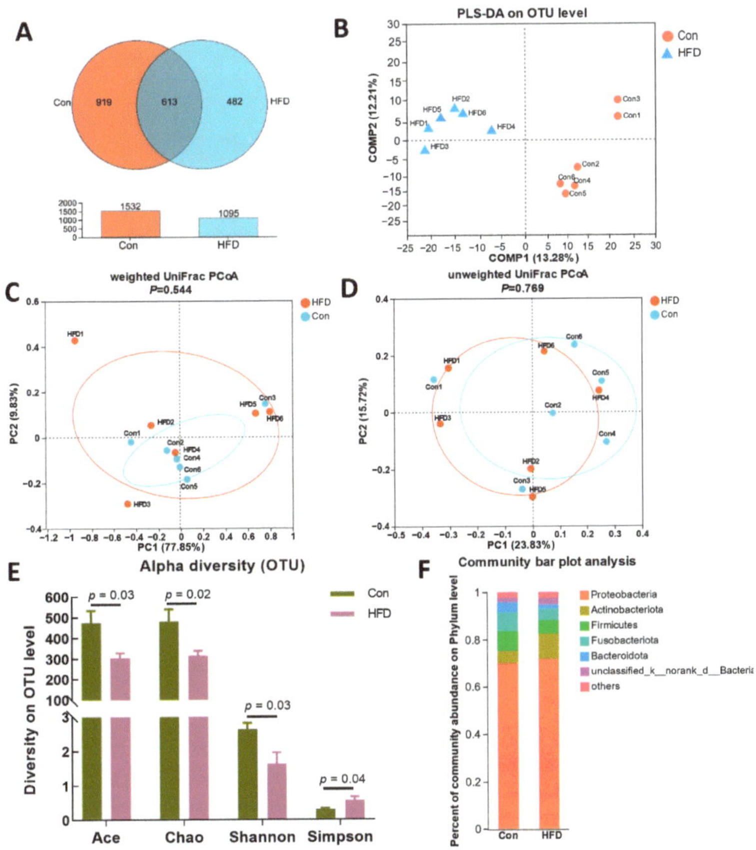

Figure 4. HFD-induced gut microbiome diversity alterations in *A. grunniens*. (**A**), Venn diagram analysis; (**B**), Partial least-squares discriminant analysis (PLS-DA) on the OTU level; (**C**), Weighted UniFrac index; (**D**), Unweighted UniFrac index; (**E**), Alpha diversity of the operational taxonomic units (OTUs); (**F**), Community bar plot analysis on the phylum level.

3.5. Dominant Microbe Distribution in the Gut of *A. grunniens* Fed with a High-Fat Diet

To further explore microbiota variation, we compared the distribution and proportions of major bacteria at the phylum and genus levels in all gut samples from freshwater drum. At the phylum level, a total of eight dominant phyla were identified, 6 of which had an average abundance >1%, including Proteobacteria (64.35%), Fusobacteriota (15.14%), Actinobacteriota (7.16%), Firmicutes (6.46%), Bacteroidota (2.62%) and unclassified_k__norank_d__Bacteria (2.08%) (Figure 5A,B). At the genus level, a total of 31 dominant genera were identified, 11 of which had an average abundance >1%, including *Achromobacter* (50.59%), *Cetobacterium* (15.41%), *norank_f__Rhizobiales_Incertae_Sedis* (4.18%), *norank_f__norank_o__PeM15* (2.48%), *unclassified_k__norank_d__Bacteria* (2.12%), *Zoogloea* (1.96%), *Enterococcus* (1.60%), *Prevotella* (1.46%), *Aurantimicrobium* (1.28%), *Acinetobacter* (1.24%) and *Mycobacterium* (1.06%) (Figure 5C,D). The above data indicate that Proteobacteria and *Achromobacter* are the most abundant colonizing bacteria at the phylum and genus levels in the gut of freshwater drums under HFD feeding.

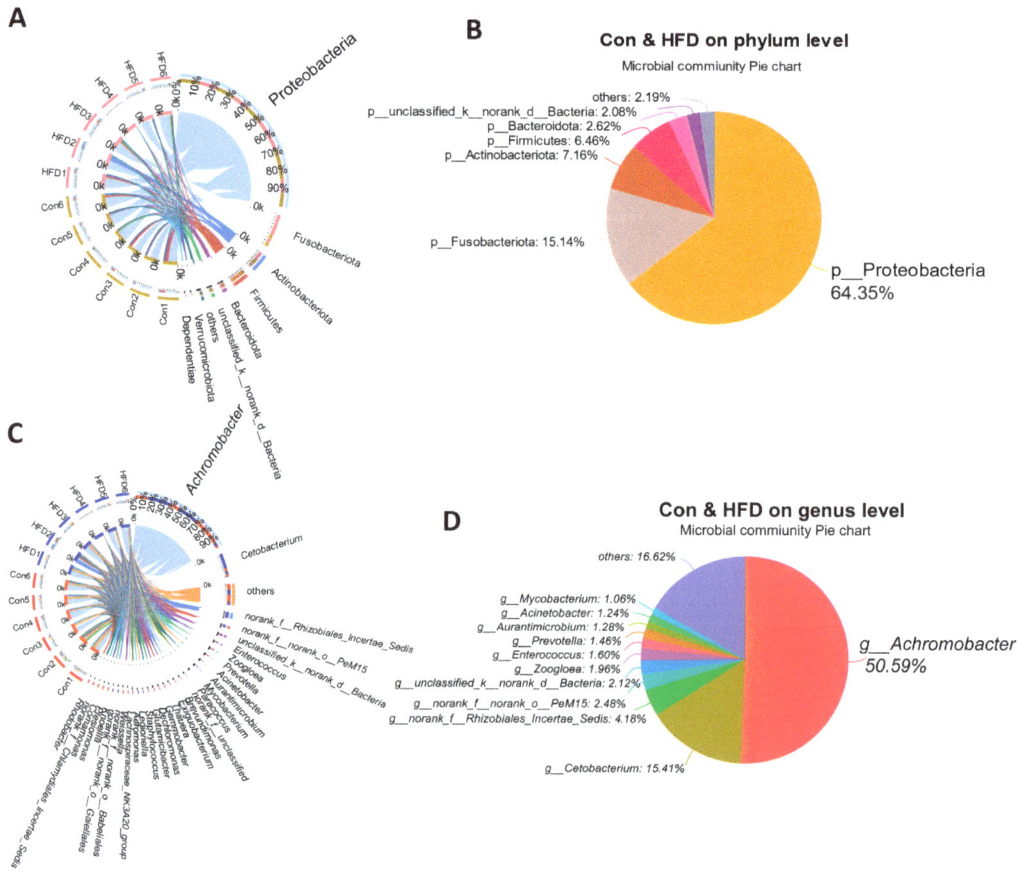

Figure 5. Dominant microbe distribution in the gut of *A. grunniens* fed with an HFD. (**A**), Circos samples and species relationships at the phylum level; (**B**), Species distribution pie chart at the phylum level; (**C**), Circos samples and species relationships at the genus level; (**D**), Species distribution pie chart at the genus level.

3.6. Microbial Comparison and Phylogenetic Tree Analysis

Based on the above findings, we further analyzed the bacteria with differential abundance between the two groups. At the genus level, a total of 31 differentially abundant bacteria were identified, and all of them exhibited a significant difference in the HFD group ($p < 0.05$, Figure 6A). With the classification of these bacteria, we found that they belonged to nine phyla, including Firmicutes (five different bacteria), Proteobacteria (eleven), Actinobacteriota (three), Deinococcota (one), Verrucomicrobiota (two), Bacteroidota (four), Gemmatimonadota (one), Chloroflexi (two) and Desulfobacterota (two). Interestingly, the most enriched phylum was Proteobacteria (Figure 6C). In addition, phylogenetic analyses of these dominant bacteria were conducted, which indicated that Proteobacteria are evolutionarily distant from other bacteria at the phylum level (Figure 6B).

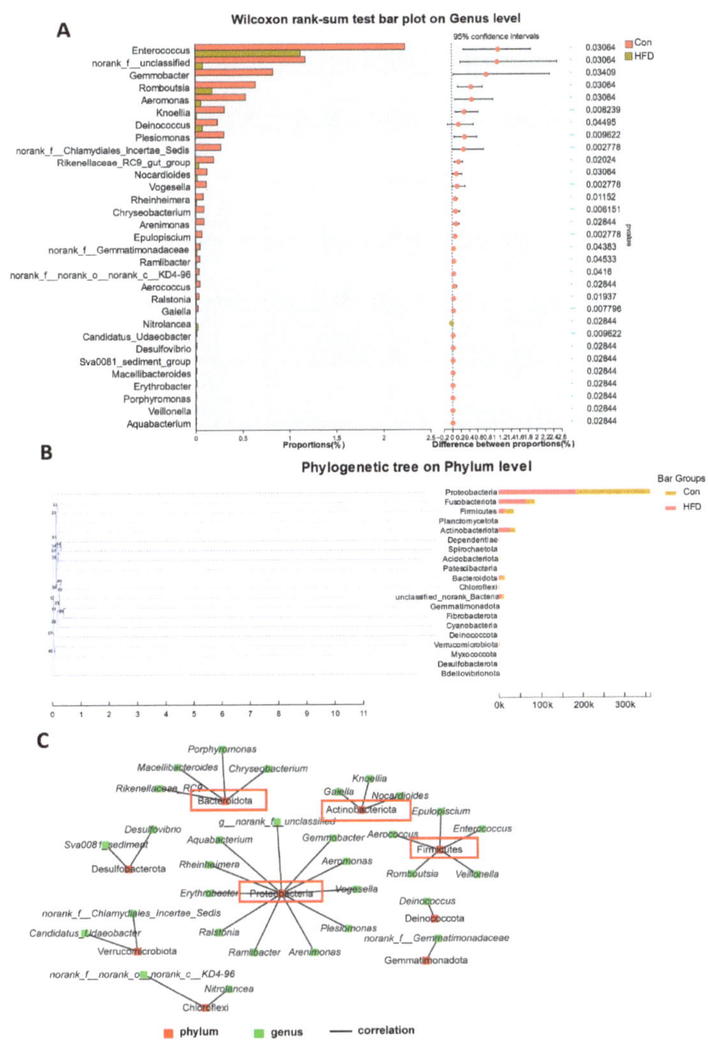

Figure 6. Microbial comparison and phylogenetic analysis in the gut of *A. grunniens* fed with an HFD. (**A**), Differentially abundant colonizing microbial communities on the genus level; (**B**), Classification of genus-level differentially abundant microorganisms and corresponding phyla; (**C**), Phylogenetic analysis on the phylum level. Red boxes represent phylum with more classification at the genus level.

3.7. Correlation Analysis of Gut Antioxidant, Immune, Inflammatory, Autophagy and Apoptosis Genes with Gut Bacteria

To investigate whether gut bacteria are involved in the regulation of gut health and physiological homeostasis, we subjected 31 differential abundant genus-level bacteria to Pearson correlation analysis with antioxidant, immune, inflammatory, autophagy and apoptosis genes as shown in Figure 7. We found that some gut bacteria are involved in the regulation of gut health after consumption of a high-fat diet. *Enterococcus* was significantly associated with antioxidant gene (*Nrf2*, $p < 0.05$). *Desulfovibrio* was significantly correlated with an immune gene (*HSP70*, $p < 0.05$). *Chlamydiales_Incertae_Sedis*, *Arenimonas*, *Veillonella*, *Norank*, *Unclassified*, *Gemmobacter*, *Gemmobacter*, *Rheinheimera*, *Macellibacteroides*, *Nocardioides* and *Porphyromonas* were significantly related to inflammatory genes (*TLR1*, *MyD88*, *TLR2*, *IL-6*, $p < 0.05$), respectively. *Nocardioides* and *Chryseobacterium* were significantly correlated with autophagy genes (*ATG3*, *ATG7*, $p < 0.05$). *Romboutsia*, *Arenimonas*, *Candidatus_Udaeobacter*, *Aquabacterium*, *Aeromonas*, *Knoellia* and *Rikenellaceae_RC9_gut_group* were significantly associated with apoptotic genes (*Casp8*, *Bcl2*, $p < 0.05$), respectively. The above results demonstrate that gut microbial dysbiosis on a high-fat diet is closely related to gut health and physiological homeostasis.

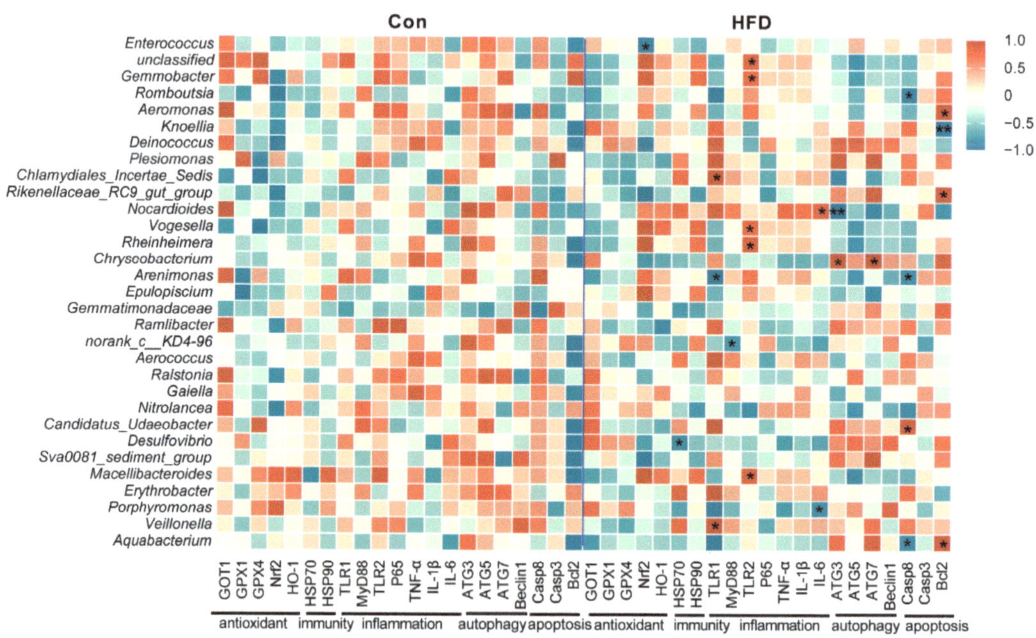

Figure 7. Correlation analysis between gut microbe abundance and gut health factors. Data were analyzed using Pearson correlation analysis with SPSS 26.0. * represents a statistical difference (*, $p < 0.05$; **, $p < 0.01$).

3.8. Integrated Analysis between DEGs and Bacteria of *A. grunniens* on a High-Fat Diet

The integrated network analysis between DEGs and bacteria revealed potential mechanisms for gut regulation through HFD feeding. Pearson correlation analysis combining DEGs in 16 key pathways (immune, inflammatory, apoptotic and metabolic disease-related pathways) and 31 genus-level differentially abundant bacteria revealed significant correlations between 20 bacteria and 16 DEGs ($p < 0.05$, Figure 8A,B). Furthermore, we found extremely significant correlations between ten bacteria and eleven DEGs ($p < 0.01$, Figure 8C), among which four belonged to Proteobacteria, including *Plesiomonas*, *Areni-*

monas, *Erythrobacter* and *Aquabacterium*. The above findings demonstrate that bacteria in the phylum Proteobacteria may be key factors in regulating gut health.

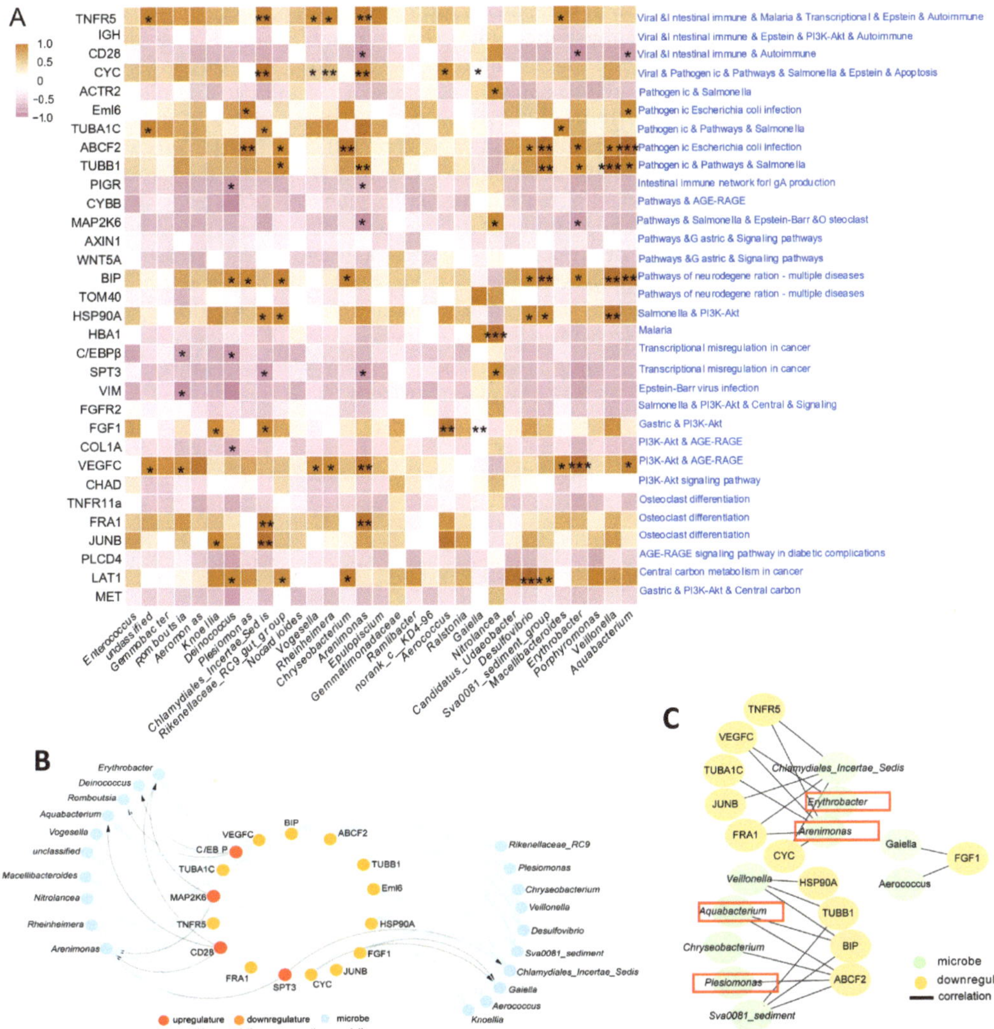

Figure 8. Integrated analysis of DEGs and bacteria on an HFD in *A. grunniens*. (**A**), Pearson correlation analysis between key bacteria at the genus level and enriched genes; (**B**), Interacting regulatory network between bacteria and genes with correlative differences ($p < 0.05$). In the network, red nodes represent upregulated DEGs, orange nodes represent downregulated DEGs, blue nodes represent bacteria, blue lines indicate positive correlations and grey lines indicate negative correlations. (**C**), Interacting regulatory network between bacteria and genes with correlative differences ($p < 0.01$). In the network, yellow nodes represent downregulated DEGs, green nodes represent bacteria and all lines indicate positive correlations. Red boxes represent bacteria at the level of the genus Proteobacteria. * Refers to a significant correlation between bacteria and genes (*, $p < 0.05$; **, $p < 0.01$; ***, $p < 0.001$).

3.9. Expression Validation of Key Genes

Next, we assessed the reliability of the transcriptome sequencing results using RT-PCR. Based on the integrated analysis between DEGs and bacteria, 12 DEGs were identified as having significant correlations with gut bacteria. These genes belong to four pathways, including protein digestion and absorption (Figure 9A, Fos-related antigen (*FRA1*, $p = 0.020$)), transcription factor jun-B (*JUNB*, $p = 0.020$)), transcriptional dysregulation in cancer cells (Figure 9B, tumor necrosis factor receptor superfamily member 5 (*TNFR5*, $p = 0.010$)), transcription initiation protein SPT3 homolog (*SPT3*, $p = 0.048$)), the PI3K-Akt signaling pathway (Figure 9C, Putative fibroblast growth factor 1 (*FGF1*, $p = 0.011$)), vascular endothelial growth factor C (*VEGFC*, $p = 0.019$), pathogenic Escherichia coli infection (Figure 9D, ATP-binding cassette sub-family F member 2 (*ABCF2*, $p = 0.033$)) and pathways of neurodegeneration—multiple diseases (endoplasmic reticulum chaperone BiP isoform (*BIP*, $p = 0.001$), dual specificity mitogen-activated protein kinase kinase 6 isoform X1 (*MAP2K6*, $p = 0.020$), cytochrome c (*CYC*, $p = 0.020$), tubulin alpha-1C chain isoform X2 (*TUBA1C*, $p = 0.044$), tubulin beta-1 chain-like (*TUBB1*, $p = 0.018$)) (Figure 9E). These data not only demonstrate the reliability of the results of differential gene expression analyses based on transcriptome sequencing data but also demonstrate that HFD affects gut health, cellular differentiation and signaling functions.

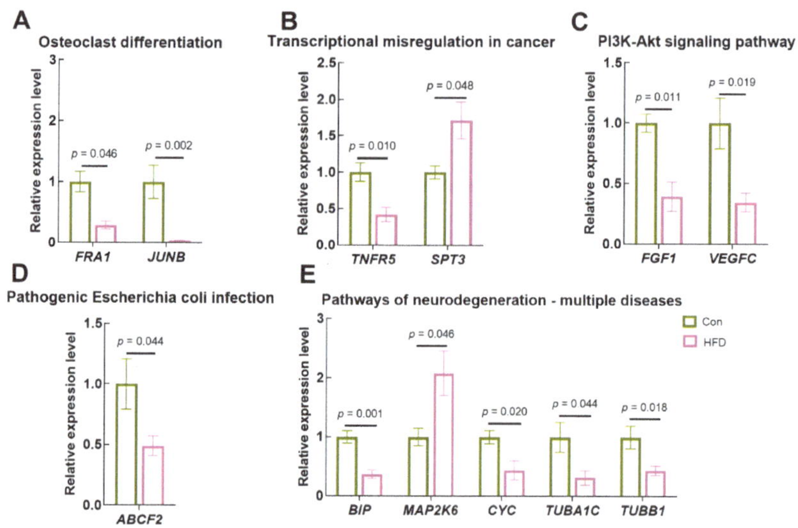

Figure 9. Transcriptional expression of key genes under HFD stimulation. (**A**), Transcriptional expression of genes related to protein digestion and absorption; (**B**), Transcriptional expression of genes related to transcriptional misregulation in cancer; (**C**), Transcriptional expression of genes related to the PI3K-Akt signaling pathway; (**D**), Transcriptional expression of genes related to pathogenic Escherichia coli infection; (**E**), Transcriptional expression of genes related to pathways of neurodegeneration—multiple diseases. Results are indicated as mean ± SEM, $n = 9$.

4. Discussion

Excessive lipid accumulation can have a serious impact on fish health and is one of the greatest challenges for aquaculture [17]. Our previous studies have shown that while an HFD promotes the growth of freshwater drum to some extent, excessive fat content produces more harm than good [28]. For example, HFD feeding led to liver lipid deposition in freshwater drum, along with decreased antioxidant capacity, and resulted in lipid metabolism disorders in the liver [28]. In fish, several studies have focused on microbial regulation and the contribution of the gut microbiota to host growth, immunomodulation and metabolism [30]. Therefore, based on the above studies, we further investigated the effects of an HFD on gut health and physiological homeostasis as well as on the ecological balance of microbial communities.

Like all aerobic organisms, fish are susceptible to reactive oxygen species (ROS). In addition to the organism's inherent capacity for antioxidant defense, fish antioxidant defense is also dependent on feeding behavior and nutritional factors [31]. It has been shown that fish fed with a high-fat diet produce excessive amounts of ROS [32], while excessive levels of ROS can lead to lipid peroxidation, resulting in the production of large amounts of MDA [33]. MDA is a biomarker for assessing lipid peroxidation and can indirectly reflect the extent of free radical attack in fish [34,35]. In the current study, MDA content was markedly increased with HFD feeding, indicating that a high-fat diet leads to lipid peroxidation in the gut. In addition, organisms rely on antioxidant protection systems to prevent oxidative damage from HFD-induced stress [36]. GPx converts superoxide into harmless hydroxyl compounds and water, preventing their oxidation and the formation of dangerous free radicals [37]. GSH plays the role of a scavenger by oxidizing to GSSG [38]. CAT is the first line of the enzymatic antioxidant system [39]. The antioxidant system protects freshwater drum from oxidative stress. However, the production of oxidized substances that cannot be completely eliminated by the antioxidant system leads to the inhibition of antioxidant effects. The inhibition of the antioxidant enzyme system comprising CAT activity, GPx and GSH content further supports the conclusion that an HFD induces oxidative stress in the gut of freshwater drum. The same results of reduced antioxidant enzyme activity in aquatic animals fed a high-fat diet have also been reported in previous studies [40–42]. Importantly, an HFD also suppressed the transcript levels of antioxidant-related genes (*GOT1*, *GPX1* and *GPX4*). *GOT1* is essential for redox homeostasis and procreation in cells [43]. Activation of *GPX1* and *GPX4* mediates antioxidant processes [44,45]. Importantly, HFD-induced stress activates *Nrf2* signaling. As a key regulator of the antioxidant response, *Nrf2* protects cells from apoptosis when the organism is subjected to oxidative stress [46–48]. We found that an HFD activated *Nrf2* to enhance gut oxidative stress resistance. In particular, *HO-1* exerts potent antioxidant and anti-apoptotic effects [49]. Therefore, *Nrf2* may play a regulatory role with key genes of lipid peroxidation at the transcriptional level to alleviate the side effects of lipid peroxidation to some extent.

The ongoing development of oxidative stress and a decline in antioxidant capacity will inevitably affect immune capacity and exacerbate the inflammatory response. Immunoglobulins play an essential role in combating microbial invasion in fish [50]. The results showed that HFD feeding reduced immunoglobulin (IgG and IgM) levels. In addition, heat shock proteins (HSPs) are involved in cellular antioxidant and innate immune responses [26]. Transcript levels of *HSP70* and *HSP90* were dramatically decreased under HFD feeding, which may be attributable to oxidative stress caused by HFD as well as suppression of antioxidant capacity, which further compromises the gut immune system. To explore the possible effects of an HFD on gut inflammation, typical inflammatory parameters (*TNF-α* and *IL-6*) were used to assess gut inflammation. *TNF-α* is an inflammatory cytokine that has been extensively studied in fish [51]. *IL-6* plays a central role in host defense, which can alleviate inflammation and trigger an immune response under acute inflammation [52]. Both ELISA and RT-PCR results showed that *TNF-α* and *IL-6* increased significantly after HFD feeding. Specifically, the expression levels of other pro-inflammatory cytokines such as *TLR1*, *TLR2*, *MyD88*, *IL-1β* and *NF-κB* were all dramatically increased. These results indicate that excessive lipid intake induces a gut inflammatory response in freshwater drum. Autophagy has an important role in cellular self-protection and maintenance of cellular physiological homeostasis [53]. Additionally, autophagy is regulated by ROS signaling and oxidative stress [54]. In this study, the autophagy-related genes *ATG3*, *ATG5*, *ATG7* and *Beclin1* (a biomarker of autophagy) showed upregulation, indicating that HFD may disrupt the physiological homeostasis of the gut. This may further support the fact that excessive lipid intake causes oxidative stress through increased ROS levels which in turn leads to oxidative damage. To maintain normal physiological functions, the organism removes unnecessary or abnormal cells through apoptosis [26]. We discovered that an HFD activates apoptotic genes (*Casp8*, *Casp3*, *Bax*, *Bcl2*) in the gut, demonstrating that apoptosis is a key response mechanism for maintaining homeostasis in the body under a high-fat diet. Moreover, powerful evidence shows that inhibition of antioxidant capacity also promotes apoptotic responses [55,56].

Based on the above studies, we further revealed the effects of a high-fat diet on intestinal molecular regulation and microbial homeostasis using transcriptome and 16S rRNA gene sequencing. The transcriptome sequencing results showed that DEGs were mainly enriched in pathways associated with intestinal metabolic disorders, inflammation and even disease, suggesting that an HFD disrupts the immune system, generates inflammation and apoptosis and affects signaling and metabolic systems in the gut. This further confirms the negative impact of a high-fat diet on the gut. In addition, diet can influence interactions between the host and the microbial community, leading to changes in the structure of the gut microbiota [57]. The gut microbiota plays an important function in fish health by stimulating the immune system and aiding nutrient absorption [1]. It is well known that the external environment, such as the living temperature, various external stress, dietary intake and bacterial or viral infection, can disrupt the abundance and diversity of the gut bacteria in fish, thereby strongly affecting the health of the host [58]. In the present study, although the HFD group showed some degree of similarity in beta diversity and no non-significant differences compared to Con, the alpha diversity index was significantly lower, suggesting that HFD feeding has an impact on gut microbial abundance and diversity. Analysis of the dominant bacteria revealed that Proteobacteria and *Achromobacter* were the most abundant at the phylum and genus levels, respectively. Through Pearson correlation analysis, we found that some of the 31 different genus-level bacteria were involved in the regulation of gut antioxidant function, immunity, inflammation, autophagy and apoptosis. In addition, analysis of genus-level differentially abundant bacteria and interaction analyses with DEGs in 16 pathways indicated that *Plesiomonas*, *Arenimonas*, *Erythrobacter* and *Aquabacteriumb* under Proteobacteria were significantly associated with gut disease, cell differentiation and signaling DEGs. Based on the above findings, we speculate that gut bacteria, especially key genera in the phylum Proteobacteria, are collectively involved in the regulation of gut inflammation and metabolism under a high-fat diet in freshwater drum.

Studies have shown that the phylum Proteobacteria is one of the dominant phyla of the gut microbiota [59,60], associated with intestinal microbiota instability, and an increase in its abundance is one of the potential diagnostic criteria for ecological disorders and diseases [61,62]. Proteobacteria have also been used in previous studies to act as a "microbiological signature" of disease [63]. We found that although overall and genus-level microbial abundance tended to decrease under HFD feeding, the abundance of bacteria in the particular phylum Proteobacteria tended to increase.

Studies have demonstrated that *Plesiomonas* is one of the core gut bacteria in zebrafish [64]. In addition, as a pathogenic bacterium, *Plesiomonas* has been shown to be associated with diseases in fish [65,66]. In the present study, HFD feeding was found to reduce the abundance of *Plesiomonas*. It was hypothesized that a high-fat diet would disturb the balance of the microbial community and may further induce gut diseases, which is similar to the findings in HFD-fed zebrafish [67]. Studies have shown that *Aquabacterium* has been detected in the insect gut and the water column [68–70], with especially large quantities being present in the water column [71]. Recent studies have identified *Aquabacterium* as the dominant microbe in the gill mucosa of *Salmo salar* [72]. Other studies have revealed that it may be a commensal bacterium that plays an important role in SVCV infections [59], but its exact function is not yet known. We hypothesize that the reduced abundance of *Aquabacterium* after HFD feeding may be associated with immune and inflammatory responses in the gut, but more studies are needed to provide evidence. Interestingly, previous studies have shown *Arenimonas* to be an environmentally relevant microbe, frequently isolated in iron ore and sewage, and have not described adverse effects on fish [73,74]. However, a reduced abundance of *Arenimonas* in the gut of zebrafish on a high-fat diet with added sea buckthorn was recently found [75]. Unusually, the opposite result was found in our study. Thus, the function of *Arenimonas* and the role it plays in relation to the fish gut deserve further investigation. Much of the research on *Erythrobacter* has focused on its utilization of light energy [76]. In addition to this, *Erythrobacter* was found to be present as a dominant microbial genus in the water column of *Penaeus vannamei* aquaculture [77], and the relative abundance of *Erythrobacter* was found to be increased in *Pandalus platyceros* on a high-fishmeal diet [78]. In the present study, the abundance of *Erythrobacter* was reduced. The inconsistency of the results of various studies indicates that diet-induced changes in the gut microbiota may depend on species and the health conditions of the fish itself. It has been shown that the functions of the gut microbiota are dynamic [79], which means that the same function can be performed by several bacteria. For the above bacteria whose specific functions are unknown, more research is needed to determine whether there is a symbiotic interaction between them to regulate the health of the gut. However, it is certain that important bacteria in the phylum Proteobacteria are important microbiota for the response of freshwater drum to metabolic disorders, inflammation and disease caused by a high-fat diet.

5. Conclusions

The present study indicates that oxidative stress induced by a high-fat diet in freshwater drum (20.88 $\pm$ 2.75 g) suppresses the antioxidant capacity of the gut, causing immune and inflammatory responses, as well as cellular autophagy and apoptosis. A high-fat diet leads to reduced abundance in the overall abundance of gut microbiota, some of which are strongly associated with intestinal immunity and the inflammatory response, cell apoptosis and autophagy. Importantly, genus-level bacterial taxa in Proteobacteria are involved in the regulation of gut health and physiological homeostasis.

Supplementary Materials: The following supporting information can be downloaded at: https://www.mdpi.com/article/10.3390/antiox13030363/s1, Table S1: Formulation and proximate composition of experimental diets; Table S2: Primers and sequences referred to in the experiment; Table S3: Transcriptome GO enrichment analysis statistics for the high-fat diet experiment; Table S4: Transcriptome KEGG enrichment analysis statistics for the high-fat diet experiment; Table S5: Gut microbiota composition of freshwater drum on a high-fat diet.

Author Contributions: M.X., conceptualization, data curation, methodology, software, validation, writing—original draft; P.X., funding acquisition, investigation, resources, supervision; H.W., funding acquisition, project administration, resources, supervision; J.C., conceptualization, software, validation; Q.W., formal analysis, visualization; J.H., methodology, validation, software; C.H., formal analysis, visualization; C.K., methodology, data curation; X.L., conceptualization, software; C.S., conceptualization, data curation, investigation, project administration, validation, writing—review and editing; H.L., data curation, funding acquisition, project administration, resources, validation, writing—review and editing. All authors have read and agreed to the published version of the manuscript.

Funding: This work was supported by the Jiangsu Agriculture Science and Technology Independent Innovation Fund [CX(20)2025]; the Central Public-Interest Scientific Institution Basal Research Fund, CAFS [2020TD62]; and the National Nonprofit Institute Research Grant of Freshwater Fisheries Research Center, Chinese Academy of Fishery Sciences [2021JBFM13].

Institutional Review Board Statement: This study was approved by the Animal Care and Use Committee of Nanjing Agricultural University (Nanjing, China) (WXFC 2021-0006). All animal procedures were performed according to the Guideline for the Care and Use of Laboratory Animals in China.

Informed Consent Statement: Not applicable.

Data Availability Statement: The data presented in this study are available in the article and Supplementary Materials.

Acknowledgments: The authors gratefully acknowledge the help given by the students of Wuxi Fisheries College of Nanjing Agricultural University during the experiment.

Conflicts of Interest: The authors declare no conflicts of interest.

References

1. Yu, C.; Zhang, J.; Qin, Q.; Liu, J.; Xu, J.; Xu, W. Berberine Improved Intestinal Barrier Function by Modulating the Intestinal Microbiota in Blunt Snout Bream (*Megalobrama amblycephala*) under Dietary High-Fat and High-Carbohydrate Stress. *Fish Shellfish Immunol.* **2020**, *102*, 336–349. [CrossRef]
2. Kamalam, B.S.; Medale, F.; Panserat, S. Utilisation of Dietary Carbohydrates in Farmed Fishes: New Insights on Influencing Factors, Biological Limitations and Future Strategies. *Aquaculture* **2017**, *467*, 3–27. [CrossRef]
3. Abimorad, E.G.; Carneiro, D.J. Digestibility and Performance of Pacu (*Piaractus mesopotamicus*) Juveniles? Fed Diets Containing Different Protein, Lipid and Carbohydrate Levels. *Aquac. Nutr.* **2007**, *13*, 1–9. [CrossRef]
4. Watanabe, T. Lipid Nutrition in Fish. *Comp. Biochem. Physiol. Part B Comp. Biochem.* **1982**, *73*, 3–15. [CrossRef]
5. Xie, R.; Amenyogbe, E.; Chen, G.; Huang, J. Effects of Feed Fat Level on Growth Performance, Body Composition and Serum Biochemical Indices of Hybrid Grouper (*Epinephelus fuscoguttatus* × *Epinephelus polyphekadion*). *Aquaculture* **2021**, *530*, 735813. [CrossRef]
6. Zhou, W.; Rahimnejad, S.; Lu, K.; Wang, L.; Liu, W. Effects of Berberine on Growth, Liver Histology, and Expression of Lipid-Related Genes in Blunt Snout Bream (*Megalobrama amblycephala*) Fed High-Fat Diets. *Fish Physiol. Beachem.* **2018**, *45*, 83–91. [CrossRef] [PubMed]
7. Chatzifotis, S.; Panagiotidou, M.; Papaioannou, N.; Pavlidis, M.; Nengas, I.; Mylonas, C.C. Effect of Dietary Lipid Levels on Growth, Feed Utilization, Body Composition and Serum Metabolites of Meagre (*Argyrosomus regius*) Juveniles. *Aquaculture* **2010**, *307*, 65–70. [CrossRef]
8. Du, Z.-Y.; Clouet, P.; Zheng, W.-H.; Degrace, P.; Tian, L.-X.; Liu, Y.-J. Biochemical Hepatic Alterations and Body Lipid Composition in the Herbivorous Grass Carp (*Ctenopharyngodon idella*) Fed High-Fat Diets. *Br. J. Nutr.* **2006**, *95*, 905–915. [CrossRef] [PubMed]
9. Xie, S.; Lin, Y.; Wu, T.; Tian, L.; Liang, J.; Tan, B. Dietary Lipid Levels Affected Growth Performance, Lipid Accumulation, Inflammatory Response and Apoptosis of Japanese Seabass (*Lateolabrax japonicus*). *Aquac. Nutr.* **2021**, *27*, 807–816. [CrossRef]
10. Wang, A.; Meng, D.; Hao, Q.; Xia, R.; Zhang, Q.; Ran, C.; Yang, Y.; Li, D.; Liu, W.; Zhang, Z.; et al. Effect of Supplementation of Solid-State Fermentation Product of Bacillus Subtilis HGcc-1 to High-Fat Diet on Growth, Hepatic Lipid Metabolism, Epidermal Mucus, Gut and Liver Health and Gut Microbiota of Zebrafish. *Aquaculture* **2022**, *560*, 738542. [CrossRef]
11. Naiel, M.A.E.; Negm, S.S.; Ghazanfar, S.; Shukry, M.; Abdelnour, S.A. The Risk Assessment of High-fat Diet in Farmed Fish and Its Mitigation Approaches: A Review. *J. Anim. Physiol. Anim. Nutr.* **2022**, *107*, 948–969. [CrossRef]
12. Hamlin, H.J.; Von Herbing, I.H.; Kling, L.J. Histological and Morphological Evaluations of the Digestive Tract and Associated Organs of Haddock throughout Post-hatching Ontogeny. *J. Fish Biol.* **2000**, *57*, 716–732. [CrossRef]
13. Nayak, S.K. Role of Gastrointestinal Microbiota in Fish. *Aquacult. Res.* **2010**, *41*, 1553–1573. [CrossRef]
14. Ghanbari, M.; Kneifel, W.; Domig, K.J. A New View of the Fish Gut Microbiome: Advances from next-Generation Sequencing. *Aquaculture* **2015**, *448*, 464–475. [CrossRef]
15. Yin, P.; Xie, S.; Zhuang, Z.; He, X.; Tang, X.; Tian, L.; Liu, Y.; Niu, J. Dietary Supplementation of Bile Acid Attenuate Adverse Effects of High-Fat Diet on Growth Performance, Antioxidant Ability, Lipid Accumulation and Intestinal Health in Juvenile Largemouth Bass (*Micropterus salmoides*). *Aquaculture* **2021**, *531*, 735864. [CrossRef]
16. Sullam, K.E.; Essinger, S.D.; Lozupone, C.A.; O'connor, M.P.; Rosen, G.L.; Knight, R.; Kilham, S.S.; Russell, J.A. Environmental and Ecological Factors That Shape the Gut Bacterial Communities of Fish: A Meta-analysis. *Mol. Ecol.* **2012**, *21*, 3363–3378. [CrossRef] [PubMed]
17. Wu, S.; Pan, M.; Zan, Z.; Jakovlić, I.; Zhao, W.; Zou, H.; Ringø, E.; Wang, G. Regulation of Lipid Metabolism by Gut Microbiota in Aquatic Animals. *Rev. Aquac.* **2023**, *16*, 34–46. [CrossRef]
18. Kortner, T.M.; Skugor, S.; Penn, M.H.; Mydland, L.; Djordjevic, B.; Hillestad, M.; Krasnov, A.; Krogdahl, Å. Dietary Soyasaponin Supplementation to Pea Protein Concentrate Reveals Nutrigenomic Interactions Underlying Enteropathy in Atlantic Salmon (*Salmo salar*). *BMC Vet. Res.* **2012**, *8*, 101. [CrossRef]
19. Tacchi, L.; Bickerdike, R.; Douglas, A.; Secombes, C.J.; Martin, S.A.M. Transcriptomic Responses to Functional Feeds in Atlantic Salmon (*Salmo salar*). *Fish Shellfish Immunol.* **2011**, *31*, 704–715. [CrossRef]
20. Tacchi, L.; Secombes, C.J.; Bickerdike, R.; Adler, M.A.; Venegas, C.; Takle, H.; Martin, S.A. Transcriptomic and Physiological Responses to Fishmeal Substitution with Plant Proteins in Formulated Feed in Farmed Atlantic Salmon (*Salmo salar*). *BMC Genom.* **2012**, *13*, 363. [CrossRef]
21. Xu, W.-H.; Guo, H.-H.; Chen, S.-J.; Wang, Y.-Z.; Lin, Z.-H.; Huang, X.-D.; Tang, H.-J.; He, Y.-H.; Sun, J.-J.; Gan, L. Transcriptome Analysis Revealed Changes of Multiple Genes Involved in Muscle Hardness in Grass Carp (*Ctenopharyngodon idellus*) Fed with Faba Bean Meal. *Food Chem.* **2020**, *314*, 126205. [CrossRef]
22. Zhang, L.; Yu, Y.; Dong, L.; Gan, J.; Mao, T.; Liu, T.; Li, X.; He, L. Effects of Moderate Exercise on Hepatic Amino Acid and Fatty Acid Composition, Liver Transcriptome, and Intestinal Microbiota in Channel Catfish (*Ictalurus punctatus*). *Comp. Biochem. Phys. Part D Genom. Proteom.* **2021**, *40*, 100921. [CrossRef]

23. Hernández-Gómez, R.E.; Contreras-Sánchez, W.M.; Hernández-Franyutti, A.; Perera-García, M.A.; Torres-Martínez, A. Testicular Structure and Development of the Male Germinal Epithelium in the Freshwater Drum *Aplodinotus grunniens* (Perciformes: Sciaenidae) from the Usumacinta River, Southern Mexico. *Acta Zool.* **2021**, *103*, 414–432. [CrossRef]
24. Chen, J.; Song, C.; Wen, H.; Liu, G.; Wu, N.; Li, H.; Xue, M.; Xu, P. miR-1/AMPK-Mediated Glucose and Lipid Metabolism under Chronic Hypothermia in the Liver of Freshwater Drum, *Aplodinotus grunniens*. *Metabolites* **2022**, *12*, 697. [CrossRef]
25. Song, C.; Wen, H.; Liu, G.; Ma, X.; Lv, G.; Wu, N.; Chen, J.; Xue, M.; Li, H.; Xu, P. Gut Microbes Reveal Pseudomonas Medicates Ingestion Preference via Protein Utilization and Cellular Homeostasis Under Feed Domestication in Freshwater Drum, *Aplodinotus grunniens*. *Front. Microbiol.* **2022**, *13*, 861705. [CrossRef]
26. Chen, J.; Xu, P.; Wen, H.; Xue, M.; Wang, Q.; He, J.; He, C.; Su, S.; Li, J.; Yu, F.; et al. Hypothermia-Mediated Oxidative Stress Induces Immunosuppression, Morphological Impairment and Cell Fate Disorder in the Intestine of Freshwater Drum, *Aplodinotus grunniens*. *Aquaculture* **2023**, *575*, 739805. [CrossRef]
27. Wu, N.; Wen, H.; Xu, P.; Chen, J.; Xue, M.; Li, J.; Wang, M.; Song, C.; Li, H. PPAR Signaling Maintains Metabolic Homeostasis under Hypothermia in Freshwater Drum (*Aplodinotus grunniens*). *Metabolites* **2023**, *13*, 102. [CrossRef]
28. Xue, M.; Xu, P.; Wen, H.; Chen, J.; Wang, Q.; He, J.; He, C.; Kong, C.; Song, C.; Li, H. Peroxisome Proliferator-Activated Receptor Signaling-Mediated 13-S-Hydroxyoctadecenoic Acid Is Involved in Lipid Metabolic Disorder and Oxidative Stress in the Liver of Freshwater Drum, *Aplodinotus grunniens*. *Antioxidants* **2023**, *12*, 1615. [CrossRef]
29. Sun, Y.; Han, W.; Liu, J.; Huang, X.; Zhou, W.; Zhang, J.; Cheng, Y. Bacterial Community Compositions of Crab Intestine, Surrounding Water, and Sediment in Two Different Feeding Modes of Eriocheir Sinensis. *Aquac. Rep.* **2020**, *16*, 100236. [CrossRef]
30. Liu, X.; Shi, H.; He, Q.; Lin, F.; Wang, Q.; Xiao, S.; Dai, Y.; Zhang, Y.; Yang, H.; Zhao, H. Effect of Starvation and Refeeding on Growth, Gut Microbiota and Non-Specific Immunity in Hybrid Grouper (*Epinephelus fuscoguttatus*♀×*E. Lanceolatus*♂). *Fish Shellfish Immunol.* **2020**, *97*, 182–193. [CrossRef]
31. Martínez-Álvarez, R.M.; Morales, A.E.; Sanz, A. Antioxidant Defenses in Fish: Biotic and Abiotic Factors. *Rev. Fish Biol. Fish.* **2005**, *15*, 75–88. [CrossRef]
32. Sies, H.; Jones, D.P. Reactive Oxygen Species (ROS) as Pleiotropic Physiological Signalling Agents. *Nat. Rev. Mol. Cell Biol.* **2020**, *21*, 363–383. [CrossRef] [PubMed]
33. Lykkesfeldt, J.; Svendsen, O. Oxidants and Antioxidants in Disease: Oxidative Stress in Farm Animals. *Vet. J.* **2007**, *173*, 502–511. [CrossRef] [PubMed]
34. Tsikas, D. Assessment of Lipid Peroxidation by Measuring Malondialdehyde (MDA) and Relatives in Biological Samples: Analytical and Biological Challenges. *Anal. Biochem.* **2017**, *524*, 13–30. [CrossRef] [PubMed]
35. Wang, B.; Liu, Y.; Feng, L.; Jiang, W.-D.; Kuang, S.-Y.; Jiang, J.; Li, S.-H.; Tang, L.; Zhou, X.-Q. Effects of Dietary Arginine Supplementation on Growth Performance, Flesh Quality, Muscle Antioxidant Capacity and Antioxidant-Related Signalling Molecule Expression in Young Grass Carp (*Ctenopharyngodon idella*). *Food Chem.* **2015**, *167*, 91–99. [CrossRef]
36. Wang, B.; Sun, J.; Li, X.; Zhou, Q.; Bai, J.; Shi, Y.; Le, G. Resveratrol Prevents Suppression of Regulatory T-Cell Production, Oxidative Stress, and Inflammation of Mice Prone or Resistant to High-Fat Diet–Induced Obesity. *Nutr. Res.* **2013**, *33*, 971–981. [CrossRef]
37. Imai, H.; Nakagawa, Y. Biological Significance of Phospholipid Hydroperoxide Glutathione Peroxidase (*PHGPx*, *GPx4*) in Mammalian Cells. *Free Radical Biol. Med.* **2003**, *34*, 145–169. [CrossRef]
38. Kosower, N.S.; Kosower, E.M. The Glutathione Status of Cells. *Int. Rev. Cytol.* **1978**, *54*, 109–160.
39. Ighodaro, O.M.; Akinloye, O.A. First Line Defence Antioxidants-Superoxide Dismutase (SOD), Catalase (CAT) and Glutathione Peroxidase (GPX): Their Fundamental Role in the Entire Antioxidant Defence Grid. *Alexandria J. Med.* **2018**, *54*, 287–293. [CrossRef]
40. Ding, T.; Xu, N.; Liu, Y.; Du, J.; Xiang, X.; Xu, D.; Liu, Q.; Yin, Z.; Li, J.; Mai, K.; et al. Effect of Dietary Bile Acid (BA) on the Growth Performance, Body Composition, Antioxidant Responses and Expression of Lipid Metabolism-Related Genes of Juvenile Large Yellow Croaker (*Larimichthys crocea*) Fed High-Lipid Diets. *Aquaculture* **2020**, *518*, 734768. [CrossRef]
41. Jia, Y.; Jing, Q.; Niu, H.; Huang, B. Ameliorative Effect of Vitamin E on Hepatic Oxidative Stress and Hypoimmunity Induced by High-Fat Diet in Turbot (*Scophthalmus maximus*). *Fish Shellfish Immunol.* **2017**, *67*, 634–642. [CrossRef]
42. Liu, Y.; Zhou, X.; Liu, B.; Gao, Q.; Sun, C.; Zhou, Q.; Zheng, X.; Liu, B. Effects of High Fat in the Diet on Growth, Antioxidant, Immunity and Fat Deposition of *Macrobrachium rosenbergii* Post-Larvae. *Fish Shellfish Immunol.* **2022**, *129*, 13–21. [CrossRef]
43. Kremer, D.M.; Nelson, B.S.; Lin, L.; Yarosz, E.L.; Halbrook, C.J.; Kerk, S.A.; Sajjakulnukit, P.; Myers, A.; Thurston, G.; Hou, S.W.; et al. GOT1 Inhibition Promotes Pancreatic Cancer Cell Death by Ferroptosis. *Nat. Commun.* **2021**, *12*, 4860. [CrossRef] [PubMed]
44. Gong, D.; Chen, M.; Wang, Y.; Shi, J.; Hou, Y. Role of Ferroptosis on Tumor Progression and Immunotherapy. *Cell Death Discov.* **2022**, *8*, 427. [CrossRef]
45. de Haan, J.B.; Crack, P.J.; Flentjar, N.; Iannello, R.C.; Hertzog, P.J.; Kola, I. An Imbalance in Antioxidant Defense Affects Cellular Function: The Pathophysiological Consequences of a Reduction in Antioxidant Defense in the Glutathione Peroxidase-1 (*Gpx1*) Knockout Mouse. *Redox Rep.* **2003**, *8*, 69–79. [CrossRef]
46. Ishii, T.; Itoh, K.; Takahashi, S.; Sato, H.; Yanagawa, T.; Katoh, Y.; Bannai, S.; Yamamoto, M. Transcription Factor *Nrf2* Coordinately Regulates a Group of Oxidative Stress-Inducible Genes in Macrophages. *J. Biol. Chem.* **2000**, *275*, 16023–16029. [CrossRef]
47. Itoh, K.; Mimura, J.; Yamamoto, M. Discovery of the Negative Regulator of Nrf2, Keap1: A Historical Overview. *Antioxid. Redox Signaling* **2010**, *13*, 1665–1678. [CrossRef] [PubMed]
48. Sajadimajd, S.; Khazaei, M. Oxidative Stress and Cancer: The Role of Nrf2. *Curr. Cancer Drug Targets* **2018**, *18*, 538–557. [CrossRef]

49. Furfaro, A.L.; Traverso, N.; Domenicotti, C.; Piras, S.; Moretta, L.; Marinari, U.M.; Pronzato, M.A.; Nitti, M. The *Nrf2/HO-1* Axis in Cancer Cell Growth and Chemoresistance. *Oxid. Med. Cell Longev.* **2016**, *2016*, 1958174. [CrossRef]
50. Rombout, J.H.W.M.; Abelli, L.; Picchietti, S.; Scapigliati, G.; Kiron, V. Teleost Intestinal Immunology. *Fish Shellfish Immunol.* **2011**, *31*, 616–626. [CrossRef]
51. Castillo, J.; Teles, M.; Mackenzie, S.; Tort, L. Stress-Related Hormones Modulate Cytokine Expression in the Head Kidney of Gilthead Seabream (*Sparus aurata*). *Fish Shellfish Immunol.* **2009**, *27*, 493–499. [CrossRef] [PubMed]
52. Bird, S.; Zou, J.; Savan, R.; Kono, T.; Sakai, M.; Woo, J.; Secombes, C. Characterisation and Expression Analysis of an Interleukin 6 Homologue in the Japanese Pufferfish. *Dev. Comp. Immunol.* **2005**, *29*, 775–789. [CrossRef]
53. Xia, X.; Wang, X.; Qin, W.; Jiang, J.; Cheng, L. Emerging Regulatory Mechanisms and Functions of Autophagy in Fish. *Aquaculture* **2019**, *511*, 734212. [CrossRef]
54. Zhang, D.-G.; Zhao, T.; Hogstrand, C.; Ye, H.-M.; Xu, X.-J.; Luo, Z. Oxidized Fish Oils Increased Lipid Deposition via Oxidative Stress-Mediated Mitochondrial Dysfunction and the CREB1-Bcl2-Beclin1 Pathway in the Liver Tissues and Hepatocytes of Yellow Catfish. *Food Chem.* **2021**, *360*, 129814. [CrossRef]
55. Murakawa, M.; Jung, S.-K.; Iijima, K.; Yonehara, S. Apoptosis-Inducing Protein, AIP, from Parasite-Infected Fish Induces Apoptosis in Mammalian Cells by Two Different Molecular Mechanisms. *Cell Death Differ.* **2001**, *8*, 298–307. [CrossRef]
56. Woo, S.; Park, I.-C.; Park, M.-J.; Lee, H.-C.; Lee, S.-J.; Chun, Y.-J.; Lee, S.-H.; Hong, S.-I.; Rhee, C. Arsenic Trioxide Induces Apoptosis through a Reactive Oxygen Species-Dependent Pathway and Loss of Mitochondrial Membrane Potential in HeLa Cells. *Int. J. Oncol.* **2002**, *21*, 57–63. [CrossRef]
57. Tan, R.; Dong, H.; Chen, Z.; Jin, M.; Yin, J.; Li, H.; Shi, D.; Shao, Y.; Wang, H.; Chen, T.; et al. Intestinal Microbiota Mediates High-Fructose and High-Fat Diets to Induce Chronic Intestinal Inflammation. *Front. Cell Infect. Microbiol.* **2021**, *11*, 654074. [CrossRef]
58. Liu, C.; Zhao, L.-P.; Shen, Y.-Q. A Systematic Review of Advances in Intestinal Microflora of Fish. *Fish Physiol. Biochem.* **2021**, *47*, 2041–2053. [CrossRef]
59. Fiedler, A.W.; Drågen, M.K.R.; Lorentsen, E.D.; Vadstein, O.; Bakke, I. The Stability and Composition of the Gut and Skin Microbiota of Atlantic Salmon throughout the Yolk Sac Stage. *Front. Microbiol.* **2023**, *14*, 1177972. [CrossRef]
60. Borges, N.; Keller-Costa, T.; Sanches-Fernandes, G.M.M.; Louvado, A.; Gomes, N.C.M.; Costa, R. Bacteriome Structure, Function, and Probiotics in Fish Larviculture: The Good, the Bad, and the Gaps. *Annu. Rev. Anim. Biosci.* **2021**, *9*, 423–452. [CrossRef] [PubMed]
61. Tran, N.T.; Xiong, F.; Hao, Y.-T.; Zhang, J.; Wu, S.-G.; Wang, G.-T. Starvation Influences the Microbiota Assembly and Expression of Immunity-Related Genes in the Intestine of Grass Carp (*Ctenopharyngodon idellus*). *Aquaculture* **2018**, *489*, 121–129. [CrossRef]
62. Shin, N.-R.; Whon, T.W.; Bae, J.-W. Proteobacteria: Microbial Signature of Dysbiosis in Gut Microbiota. *Trends Biotechnol.* **2015**, *33*, 496–503. [CrossRef] [PubMed]
63. Meng, K.-F.; Ding, L.-G.; Wu, S.; Wu, Z.-B.; Cheng, G.-F.; Zhai, X.; Sun, R.-H.; Xu, Z. Interactions between Commensal Microbiota and Mucosal Immunity in Teleost Fish During Viral Infection With SVCV. *Front. Immunol.* **2021**, *12*, 654758. [CrossRef] [PubMed]
64. Roeselers, G.; Mittge, E.K.; Stephens, W.Z.; Parichy, D.M.; Cavanaugh, C.M.; Guillemin, K.; Rawls, J.F. Evidence for a Core Gut Microbiota in the Zebrafish. *ISME J.* **2011**, *5*, 1595–1608. [CrossRef]
65. Santos, J.A.; Rodríguez-Calleja, J.-M.; Otero, A.; García-López, M.-L. Plesiomonas. In *Molecular Medical Microbiology*; Elsevier: Amsterdam, The Netherlands, 2015; pp. 1111–1123. [CrossRef]
66. Pękala-Safińska, A. Contemporary Threats of Bacterial Infections in Freshwater Fish. *J. Vet. Res.* **2018**, *62*, 261–267. [CrossRef]
67. Liu, Y.; Zhu, D.; Liu, J.; Sun, X.; Gao, F.; Duan, H.; Dong, L.; Wang, X.; Wu, C. Pediococcus Pentosaceus PR-1 Modulates High-Fat-Died-Induced Alterations in Gut Microbiota, Inflammation, and Lipid Metabolism in Zebrafish. *Front. Nutr.* **2023**, *10*, 1087703. [CrossRef]
68. Kalmbach, S.; Manz, W.; Wecke, J.; Szewzyk, U. *Aquabacterium* Gen. Nov., with Description of *Aquabacterium citratiphilum* sp. nov., *Aquabacterium parvum* sp. Nov. and *Aquabacterium commune* sp. nov., Three in Situ Dominant Bacterial Species from the Berlin Drinking Water System. *Int. J. Syst. Evol. Microbiol.* **1999**, *49*, 769–777. [CrossRef]
69. Lin, M.-C.; Jiang, S.-R.; Chou, J.-H.; Arun, A.B.; Young, C.-C.; Chen, W.-M. *Aquabacterium fontiphilum* sp. nov., Isolated from Spring Water. *Int. J. Syst. Evol. Microbiol.* **2009**, *59*, 681–685. [CrossRef]
70. Shelomi, M.; Lo, W.-S.; Kimsey, L.S.; Kuo, C.-H. Analysis of the Gut Microbiota of Walking Sticks (*Phasmatodea*). *BMC Res. Notes* **2013**, *6*, 368. [CrossRef] [PubMed]
71. Li, Y.; Ding, Y.; Zhang, S.; Zhang, A.; Song, X.; Wang, L.; Li, H.; Chen, W. Effects of Culture Methods on the Nutrient Levels, Physiological Characteristics and Intestinal Microbiota of the Innkeeper Worm *Urechis unicinctus*. *Aquac. Res.* **2021**, *52*, 3843–3853. [CrossRef]
72. Slinger, J.; Adams, M.B.; Stratford, C.N.; Rigby, M.; Wynne, J.W. The Effect of Antimicrobial Treatment upon the Gill Bacteriome of Atlantic Salmon (*Salmo salar* L.) and Progression of Amoebic Gill Disease (AGD) In Vivo. *Microorganisms* **2021**, *9*, 987. [CrossRef]
73. Zhang, Y.; Liu, Y.; Ma, H.; Sun, M.; Wang, X.; Jin, S.; Yuan, X. Insufficient or Excessive Dietary Carbohydrates Affect Gut Health through Change in Gut Microbiota and Regulation of Gene Expression of Gut Epithelial Cells in Grass Carp (*Ctenopharyngodon idella*). *Fish Shellfish Immunol.* **2023**, *132*, 108442. [CrossRef]

74. Rocha, S.D.C.; Lei, P.; Morales-Lange, B.; Mydland, L.T.; Øverland, M. From a Cell Model to a Fish Trial: Immunomodulatory Effects of Heat-Killed *Lactiplantibacillus plantarum* as a Functional Ingredient in Aquafeeds for Salmonids. *Front. Immunol.* **2023**, *14*, 1125702. [CrossRef]
75. Lan, Y.; Wang, C.; Zhang, C.; Li, P.; Zhang, J.; Ji, H.; Yu, H. Dietary Sea Buckthorn Polysaccharide Reduced Lipid Accumulation, Alleviated Inflammation and Oxidative Stress, and Normalized Imbalance of Intestinal Microbiota That Was Induced by High-Fat Diet in Zebrafish (*Danio rerio*). *Fish Physiol. Biochem.* **2022**, *48*, 1717–1735. [CrossRef]
76. Zheng, Q.; Lin, W.; Liu, Y.; Chen, C.; Jiao, N. A Comparison of 14 Erythrobacter Genomes Provides Insights into the Genomic Divergence and Scattered Distribution of Phototrophs. *Front. Microbiol.* **2016**, *7*, 984. [CrossRef] [PubMed]
77. Hu, X.; Xu, Y.; Su, H.; Xu, W.; Wen, G.; Xu, C.; Yang, K.; Zhang, S.; Cao, Y. Effect of a Bacillus Probiotic Compound on Penaeus Vannamei Survival, Water Quality, and Microbial Communities. *Fishes* **2023**, *8*, 362. [CrossRef]
78. Xie, S.; Wei, D.; Tan, B.; Liu, Y.; Tian, L.; Niu, J. *Schizochytrium limacinum* Supplementation in a Low Fish-Meal Diet Improved Immune Response and Intestinal Health of Juvenile *Penaeus monodon*. *Front. Physiol.* **2020**, *11*, 613. [CrossRef]
79. Moya, A.; Ferrer, M. Functional Redundancy-Induced Stability of Gut Microbiota Subjected to Disturbance. *Trends Microbiol.* **2016**, *24*, 402–413. [CrossRef] [PubMed]

antioxidants

Brief Report

H_2O_2-Induced Oxidative Stress Responses in *Eriocheir sinensis*: Antioxidant Defense and Immune Gene Expression Dynamics

Qinghong He [1,†], Wenrong Feng [2,†], Xue Chen [2,†], Yuanfeng Xu [2], Jun Zhou [3], Jianlin Li [2], Pao Xu [2] and Yongkai Tang [1,2,*]

1 College of Fisheries and Life, Shanghai Ocean University, Shanghai 201306, China; hqh13990750864@163.com
2 Key Laboratory of Freshwater Fisheries and Germplasm Resources Utilization, Ministry of Agriculture and Rural Affairs, Freshwater Fisheries Research Center, Chinese Academy of Fishery Sciences, Wuxi 214081, China; fengwenrong@ffrc.cn (W.F.); chenxue@ffrc.cn (X.C.); xuyuanfeng@ffrc.cn (Y.X.); lijl@ffrc.cn (J.L.); xup@ffrc.cn (P.X.)
3 Freshwater Fisheries Research Institute of Jiangsu Province, Nanjing 210017, China; finedrizzle@163.com
* Correspondence: tangyk@ffrc.cn
† These authors contributed equally to this work.

Abstract: *Eriocheir sinensis*, a key species in China's freshwater aquaculture, is threatened by various diseases, which were verified to be closely associated with oxidative stress. This study aimed to investigate the response of *E. sinensis* to hydrogen peroxide (H_2O_2)-induced oxidative stress to understand the biological processes behind these diseases. Crabs were exposed to different concentrations of H_2O_2 and their antioxidant enzyme activities and gene expressions for defense and immunity were measured. Results showed that activities of antioxidant enzymes—specificallysuperoxide dismutase (SOD), catalase (CAT), total antioxidant capacity(T-AOC), glutathione (GSH), and glutathione peroxidase (GSH-Px)—varied with exposure concentration and duration, initially increasing then decreasing. Notably, SOD, GSH-Px, and T-AOC activities dropped below control levels at 96 h. Concurrently, oxidative damage markers, including malondialdehyde (MDA), H_2O_2, and 8-hydroxy-2′-deoxyguanosine (8-OHdG) levels, increased with exposure duration. The mRNA expression of *SOD*, *CAT*, and *GSH-Px* also showed an initial increase followed by a decrease, peaking at 72 h. The upregulation of *phenoloxidaseloxidase (proPO)* and *peroxinectin (PX)* was also detected, but *proPO* was suppressed under high levels of H_2O_2. *Heat shock protein 70 (HSP70)* expression gradually increased with higher H_2O_2 concentrations, whereas *induced nitrogen monoxide synthase (iNOS)* was upregulated but decreased at 96 h. These findings emphasize H_2O_2's significant impact on the crab's oxidative and immune responses, highlighting the importance of understanding cellular stress responses for disease prevention and therapy development.

Keywords: *Eriocheir sinensis*; oxidative stress; antioxidation; gene expression; H_2O_2

Citation: He, Q.; Feng, W.; Chen, X.; Xu, Y.; Zhou, J.; Li, J.; Xu, P.; Tang, Y. H_2O_2-Induced Oxidative Stress Responses in *Eriocheir sinensis*: Antioxidant Defense and Immune Gene Expression Dynamics. *Antioxidants* **2024**, *13*, 524. https://doi.org/10.3390/antiox13050524

Academic Editor: Alessandra Napolitano

Received: 5 March 2024
Revised: 4 April 2024
Accepted: 11 April 2024
Published: 26 April 2024

1. Introduction

Eriocheir sinensis holds the second highest rank in terms of production volume in the field of crustacean aquaculture in China. It is highly prized for its culinary attributes and economic value. *E. sinensis* has a long history of consumption in China. It has high content of protein, fats, and various vitamins [1]. The fatty paste and roe, that is, the gonads, represent luxury foodstuffs and are often featured in traditional Chinese cuisine. Over the years, there have been considerable advancements in the aquaculture techniques for *E. sinensis*. However, with an expansion in farm size and increased stocking densities, there has been a concomitant increase in the incidence of disease. This rise is attributable to combinations of environmental stressors and escalated pathogen load. The hepatopancreas of *E. sinensis* is not only an edible tissue but also serves multiple physiological functions. It is involved in digestion, absorption, and storage of nutrients, particularly during molting [2] and gonadal maturation [3]. Additionally, it plays a role in detoxification and metabolic regulation. Due to its important roles, the hepatopancreas is sensitive to changes in the external and internal environments. When *E. sinensis* is subjected to environmental stressors, its hepatopancreas is typically one of the first organs to be affected.

In aquaculture, *E. sinensis* is exposed to a variety of environmental stressors, such as hypoxia [4], elevated temperature [5], heavy metal ions [6], pesticides [7], and high levels of ammonia nitrogen [8]. These stressors trigger two types of stress responses: oxidative stress and nitrosative stress [9]. Oxidative stress induces the overproduction of reactive oxygen species (ROS), including superoxide anion radicals (O^{2-}), hydroxyl radicals (OH), and H_2O_2. Nitrosative stress induces the production and release of nitric oxide (NO), triggering a series of chain reactions that result in the formation of reactive nitrogen species (RNS). Research has demonstrated that many diseases in *E. sinensis* are frequently accompanied by strong oxidative stress. Infections caused by *Aeromonas hydrophora* lead to an upregulation of antioxidant defenses, including T-AOC), GSH, and GSH-PX, as well as lysozyme (LZM) and phenoloxidase (PO) activities [10]. When *E. sinensis* was subjected to acute salt stress, there was an elevation in antioxidant enzyme activities, such as CAT, SOD, T-AOC, GSH-PX, and MDA levels, alongside an upregulation of the heat shock protein 90 (*HSP90*) gene, which enhanced resistance [11]. Conversely, under ammonia nitrogen stress, there was a notable decrease in the antioxidant capacity indicators of T-AOC, T-SOD, and GSH-Px in the hemolymph, alongside a significant increase in MDA, marking reduced antioxidant capacity and increased oxidative damage of *E. sinensis* [12]. Saline-alkali stress exposure resulted in initial increases followed by decreases in SOD, CAT, and T-AOC activities in the hepatopancreas of *E. sinensis*; the decreases in antioxidant capacity were in correlation with hepatopancreatic damage [13]. Yang's work demonstrated that acute hepatopancreatic necrosis syndrome (AHPNS) in *E. sinensis* led to higher blood levels of aspartate aminotransferase (AST) and glutamic pyruvic transaminase (GPT) compared to healthy specimens; contrastingly, alkaline phosphatase (ALP) and acid phosphatase (ACP) activities in the hepatopancreas were notably lower, with a concurrent significant increase in MDA levels, indicating both oxidative stress and organ damage [14]. Thus, investigating the effects of oxidative stress on *E. sinensis* may offer new insights into exploring the pathogenesis of disease.

H_2O_2 is a widely prevalent ROS with a remarkable ability to penetrate cell membranes, leading to oxidative stress or triggering apoptosis within the cell. Consequently, H_2O_2 is frequently used as a standard reagent to experimentally induce oxidative stress in animals. In this study, one-year old juvenile *E. sinensis* were subjected to H_2O_2 stress tests to explore their physiological response to oxidative stress. The activity of antioxidase in the hepatopancreas and hemolymph, and the mRNA expression levels of antioxidative and immune-related genes in the hepatopancreas, were measured after 96 h of gradient H_2O_2 treatment. This research provides a scientific basis for the in-depth study of the oxidative stress response in *E. sinensis* and brings a novel perspective to the prevention and treatment strategies for diseases caused by oxidative damage.

2. Materials and Methods

2.1. Ethics Statement

The crabs were handled and the experimental procedures were performed in accordance with the guidelines for the care and use of animals for scientific purposes set by the Animal Ethics Committee of the Freshwater Fisheries Research Center (FFRC) Chinese Academy of Fishery Sciences, and the necessary ethical protocol code is LAECFFRC-2023-09-12. All operations were performed to minimize the suffering of the crabs.

2.2. Crabs and Rearing Conditions

Juvenile *E. sinensis* were obtained from Yangcheng Lake Shrimp and Crab Green Cultivation Base, Freshwater Fisheries Center, Chinese Academy of Fisheries Sciences. Juvenile crabs (13.34 ± 2.56 g) were acclimated to the aquatic environment in a laboratory aquarium (100 cm $\times$ 45 cm $\times$ 50 cm) for one week. During the acclimation period, continuous aeration was provided to maintain a dissolved oxygen concentration (DO) of $\geq$7.0 mg$\cdot$L^{-1}. The ambient water temperature was regulated at 25 $\pm$ 2 °C with a pH of 8.0 $\pm$ 0.2. Commercial feed was administered every morning, and one-third of the water volume was replaced every other day. Feeding was ceased 24 h prior to experimentation, and individuals in intermolt with healthy, intact appendages were selected for the study.

2.3. H_2O_2 Stress Treatment

Six treatment groups were set up in the experiment, with H_2O_2 concentrations set at 0 (control group), 3, 6, 9, 12, and 15 mmol$\cdot$L^{-1}. Each group containing 70 juvenile *E. sinensis* was raised separately in two tanks (100 cm $\times$ 45 cm $\times$ 50 cm) with the same conditions. During the experiment, the water was completely changed every 24 h, with the concentration of hydrogen peroxide being adjusted to meet the specified experimental requirements. During the experiment, the water quality parameters were maintained at a temperature of 25 $\pm$ 2 °C, DO $\geq$ 7.0 mg$\cdot$L^{-1}, pH = 8.0 $\pm$ 0.2, ammonia $\leq$ 0.02 mg$\cdot$L^{-1}, and nitrite $\leq$ 0.05 mg$\cdot$L^{-1}. Samples were collected at 0, 24, 48, 72, and 96 h of exposure. For each sample point, nine juvenile crabs were picked randomly and immediately anesthetized in an ice water bath. Hemolymph was extracted using a disposable sterile syringe from the basal membrane of the third walking leg, followed by dissection on ice for hepatopancreas sampling. After the hemolymph clotting at room temperature, it was centrifuged at 1000$\times$ *g* for 10 min to obtain serum. Hepatopancreas samples were flash-frozen in liquid nitrogen. Samples were stored at $-$80 °C for subsequent experiments.

2.4. Biochemical Analysis

Hepatopancreas tissues were immersed in physiological saline (with a weight/volume ratio of 1:9) and homogenized using a high throughput tissue grinder (SCIENTZ-48, Ningbo, China). After centrifuging at 5000$\times$ *g* for 15 min at 4 °C, the supernatant was collected for measurement. Serum was diluted using saline for enzyme activity determination. All parameters were determined using commercial assay kits according to the manufacturer's protocols provided by Nanjing Jiancheng Bioengineering Institute (Nanjing, China). Total protein content (TP) was determined by the Coomassie Brilliant Blue assay (A045-2). MDA levels were assessed using the thiobarbituric acid (TBA) method (A003-1). T-AOC was measured via the ABTS method (A015-2-1). SOD activity was quantified through the nitro blue tetrazolium (NBT) method (A001-1), while CAT activity was evaluated using the ammonium molybdate method (A007-1-1). GSH (A006-2-1) and GSH-Px (A005-1) activities and H_2O_2 (A064-1-1) content were determined by colorimetric assay. The concentration of 8-OHdG was measured using a Crab 8-hydroxydeoxyguanosine Elisa Kit (H165-1).

2.5. Quantitative Real-Time Fluorescent PCR (qPCR) Analysis

Total RNA was extracted from the hepatopancreas by the TRIzol method. RNA quality, including purity and concentration, was assessed by spectrophotometry (NanoPhotometer® N50, Implen, Munich, Germen) at 260/280 nm. The cDNA was synthesized from 2 µg of total RNA using the PrimeScript™ RT reagent kit with gDNA Eraser (Takara). Primers for *SOD*, *CAT*, *GPS-Px*, *iNOS*, *HSP70*, *PX*, *proPO*, and *β-actin* were designed by the Primer Premier 5.0 software (USA) based on known sequences from *E. sinensis*. The primer sequences and GenBank accession numbers are listed in Table 1. *β-actin* served as the internal reference gene. The qPCR was performed on a Thermal Cycler Dice® Real Time System TP800 and programmed as follows: an initial denaturation step at 95 °C for 30 s, followed by 40 cycles of denaturation at 95 °C for 5 s and annealing at 60 °C for 30 s. The melting curve analysis was performed with the following temperatures and times: 95 °C for 15 s, 60 °C for 30 s, 95 °C for 5 s. Three replicates were performed for each sample. Each sample was subjected to three repetitions, and the data were converted to cycle/threshold (Ct) values after each reaction. The relative gene expression levels were calculated by the $2^{-\Delta\Delta Ct}$ method.

Table 1. Sequences of primers used in qPCR.

Gene	Primer Sequence (5′-3′)	Product Length (bp)	GenBank Accession Number
iNOS	TTGCCAGAGCCGTCAAGTT GCGCCTCGTGTTCTATGTTG	201	XM_050876720.1
GSH-Px	ATCCTGTACCCTGCAACCAC CTCTGGGAACAGCTTCTTGG	174	FJ617305.1
SOD	TGGACTGACGGAAGGGCTGC TGGCGTTAGGGGCGGAGTG	128	FJ617306.1
CAT	CCTGCTCGCAGGAATCGGTG GTCCAAGGAGGTGGCGGTCA	159	MH178391.1
HSP70	GGCAAGGCAGCGAAGGTCATC CGGCATTGGTGACAGACTGACG	127	KC493625.1
peroxinectin	CAGCAACGACTACAACCCGA TCCTTGCACCAGGGAATGAC	91	GU353176.1
Prophenoloxidase	CCATGTCATCATTGCAGCGG TGTACTTGTGCCAGCGGTAG	119	EF493829.1
β-actin	TGGGTATGGAATCCGTTGGC AGACAGAACGTTGTTGGCGA	101	KM244725.1

2.6. Data Analysis

The results are expressed as mean ± standard error (mean ± SE). Data analysis was conducted using SPSS Statistic 23.0 software (IBM, Armonk, NY, USA), with one-way analysis of variance (ANOVA) employed to evaluate differences among groups, and the Tukey test was used for post hoc comparisons to assess the significance of differences between groups ($p < 0.05$). Tests for homogeneity of variance were utilized to verify the assumption of normal distribution of the data. Graphical representations were generated using GraphPad Prism 8.0.

3. Results

3.1. Effect of H_2O_2 Stress on Antioxidant Response in Hepatopancreas

Following exposure to H_2O_2 stress, the SOD activity in hepatopancreas showed an initially increasing and subsequently declining response over time. Notably, the activities peaked at 72 h and were significantly lower than in the control at 96 h of stress ($p < 0.05$, Figure 1a) in all the treated groups. The CAT activity displayed a similar trend, showed a rise and subsequent fall over the course of the experiment, reaching peak levels at 48 h for concentrations of 3 mmol·L^{-1} and 15 mmol·L^{-1}, and at 72 h for concentrations of 6 mmol·L^{-1}, 9 mmol·L^{-1}, and 12 mmol·L^{-1}. Notably, CAT activities at 12 mmol·L^{-1} and 15 mmol·L^{-1} significantly diminished compared to those of the control at 96 h ($p < 0.05$, Figure 1b). T-AOC within the 6 and 9 mmol·L^{-1} treatment groups showed an initial increase, peaking at 48 h, while the 12 and 15 mmol·L^{-1} groups peaked at 24 h before exhibiting a downward trend. At 96 h of stress, T-AOC levels in all treatment groups were significantly reduced compared to those of the control group ($p < 0.05$, Figure 1c). GSH levels in treated groups also rose and then fell, with the greatest levels observed at 72 h. Notably, the 15 mmol·L^{-1} treatment group showed a significantly lower GSH activity than the control at 96 h ($p < 0.05$, Figure 1d). GSH-PX activity in the treated groups displayed an initial rise followed by a decline, with activities substantially lower than those of the control group at 96 h of stress ($p < 0.05$, Figure 1e). MDA, 8-OHdG, and H_2O_2 levels all exhibited a consistent upward trend in response to both increased experimental duration and elevated stress concentrations (Figure 1f–h).

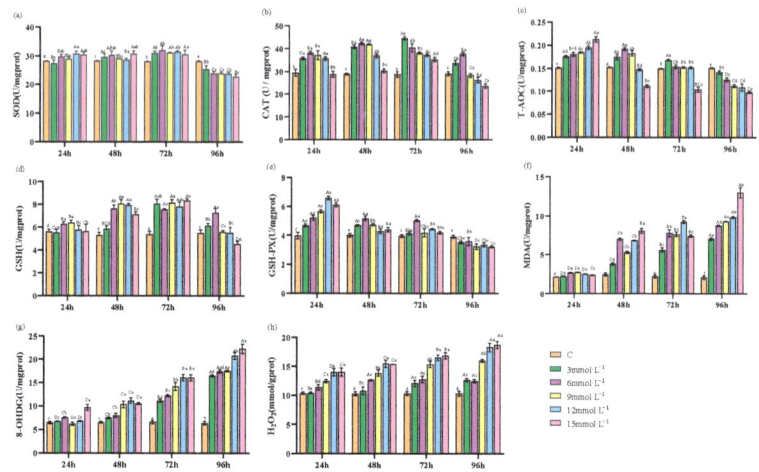

Figure 1. Effects of H_2O_2 stress on the antioxidant parameters of hepatopancreas. Distinct lowercase letters indicate significant differences at the same time point ($p < 0.05$), and distinct uppercase letters indicate significant differences at different time points within the same treatment group ($p < 0.05$).

3.2. Effect of H_2O_2 Stress on Antioxidant Response in Hemolymph

The activities of SOD, T-AOC, CAT, GSH, and GSH-PX in all treatment groups initially increased and subsequently decreased over the duration of the experiment. Specifically, SOD activity in the 3 and 12 mmol·L^{-1} H_2O_2 treatment groups reached a maximum at 72 h, while peak activity in the other concentration groups occurred at 48 h (Figure 2a). T-AOC, CAT, and GSH activities reached their respective maxima at 72 h (Figure 2b–d). GSH-PX activity showed a peak at 48 h in the 12 and 15 mmol·L^{-1} H_2O_2 treatment groups, and at 72 h in the lower concentration groups of 3, 6, and 9 mmol·L^{-1} (Figure 2e). Conversely, the concentrations of MDA and H_2O_2 in the hemolymph showed an increasing trend with experimental time. The concentrations of MDA and H_2O_2 in the hemolymph exhibited a progressively increasing trend as time continued (Figure 2f,g).

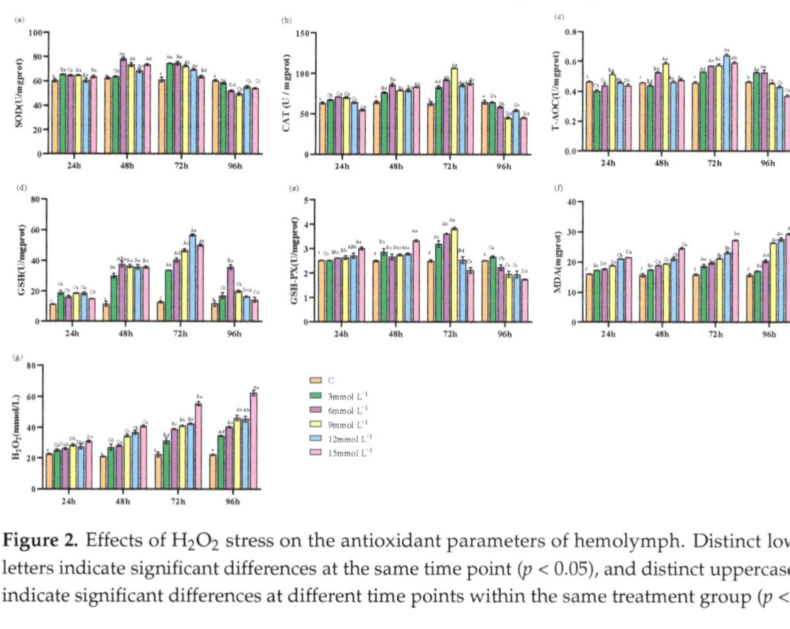

Figure 2. Effects of H_2O_2 stress on the antioxidant parameters of hemolymph. Distinct lowercase letters indicate significant differences at the same time point ($p < 0.05$), and distinct uppercase letters indicate significant differences at different time points within the same treatment group ($p < 0.05$).

3.3. Effect of H_2O_2 Stress on Genes Expression in Hepatopancreas

During H_2O_2-induced stress, the mRNA expression levels of *SOD*, *CAT*, *GSH-Px*, and *iNOS* in each treatment group exhibited a tendency to increase and then decrease over time. These expression levels of genes peaked at 72 h post-treatment. By 96 h, the mRNA expression across all stressed groups was significantly increasing compared to the control ($p < 0.05$, Figure 3a–c,g). Concurrently, the mRNA expression of *proPO* in the 3, 6, 9, and 12 mmol·L^{-1} H_2O_2 concentrations of different treatment groups also increased and then decreased, reaching the highest value at 72 h. Notably, in the 15 mmol·L^{-1} H_2O_2 concentration group, the mRNA expression of *proPO* demonstrated a consistent decline over time and was significantly reduced compared to that of the control group at 48 h ($p < 0.05$, Figure 3d). Additionally, the mRNA expression levels of *HSP70* and *PX* showed a steady increase as the duration of time extended (Figure 3e,f).

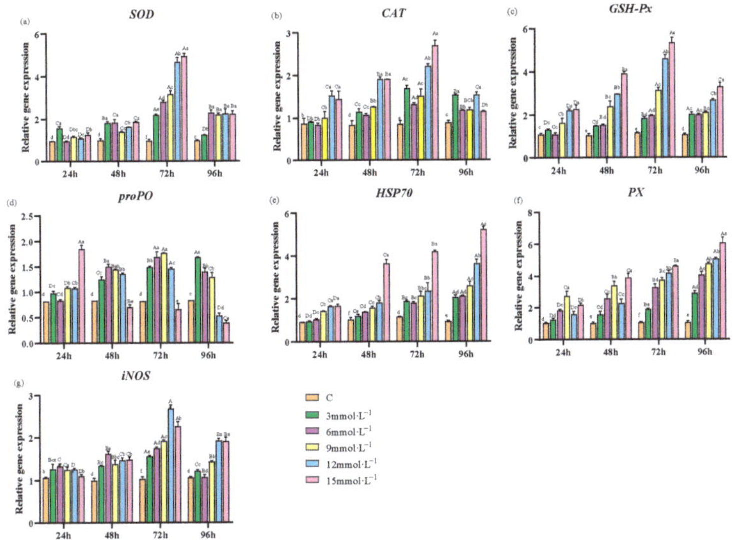

Figure 3. Effects of H_2O_2 stress on the expression of antioxidant-related genes in hepatopancreas. Distinct lowercase letters indicate significant differences at the same time point ($p < 0.05$), and distinct uppercase letters indicate significant differences at different time points within the same treatment group ($p < 0.05$).

4. Discussion

Extensive studies have demonstrated that variations in salinity, alkalinity, dissolved oxygen, temperature, and ammonia nitrogen within the aquatic environment can induce defense responses in organisms, including oxidative stress responses [15–18]. Under such stress conditions, the continuous production of ROS can disrupt the balance between the oxidative and antioxidant system, inflicting oxidative damage on lipids, proteins, DNA, and carbohydrates. When subjected to external stressors, the antioxidant system responds swiftly, enhancing its antioxidative capacity and modulating the expression of relevant genes to mitigate the stresses of oxidative challenge.

4.1. Effect of H_2O_2 Stress on Antioxidative Enzyme Activities in Hepatopancreas of E. sinensis

Under normal physiological conditions, organisms generate ROS as a byproduct of metabolism. However, an excessive accumulation of ROS can negatively impact the organism's physiological state [19]. Antioxidants, which organisms intrinsically possess, can promptly and effectively remove ROS, thereby preventing oxidative stress. Hepatopancreas of *E. sinensis* plays a key role in eliminating excessive ROS [20]. Under stress conditions,

SOD and CAT are critical antioxidant enzymes and function as the primary line of defense against the overproduction of ROS, mitigating potential adverse effects [21]. SOD removes the conversion of superoxide radicals ($\cdot O^{2-}$) into H_2O_2 and O_2, while CAT further decomposes H_2O_2 into water (H_2O) and oxygen (O_2). Their combined activity effectively eliminates oxidative damage from superoxide radicals, thus preserving the organism's internal homeostasis [22]. T-AOC is the cumulative antioxidant potential of tissue and represents the organism's overall capacity to scavenge ROS [23]. Studies showed that air exposure caused oxidative stress in *E. sinensis*, with the activities of SOD and CAT in the hepatopancreas initially increasing and then decreasing as the duration of exposure extended [24]. In the Pacific white shrimp (*Litopenaeus vannamei*), SOD and CAT activities in the hepatopancreas were found to be elevated by hypoxia treatment; however, post-reoxygenation, the activities first rose and then diminished [25]. Furthermore, after recovery from cold-shock treatment in Pacific white shrimp, activities of SOD, CAT, and T-AOC showed an initial increase followed by a gradual decrease [26]. The results of our study indicated that during H_2O_2 stress, SOD, CAT, and T-AOC levels in the hepatopancreas initially increased and then decreased. This suggests that the antioxidative capacity of the organism was rapidly enhanced in response to H_2O_2 stress. Firstly, the presence of H_2O_2 directly enhanced CAT activity. Additionally, oxidative stress elevated $\cdot O^{2-}$ levels, which increased SOD activity. SOD converted $\cdot O^{2-}$ into H_2O_2, which in turn boosted CAT activity to eliminate excess H_2O_2. This resulted in an elevated activity of T-AOC. However, with the persistence of oxidative stress, the antioxidant system exceeded its reductive limit, leading to oxidative damage. This occurred when the cell failed to counterbalance the damage or the synthesis of new enzymes became impaired, as Sohal, R.S. indicated [27], which subsequently led to a decrease in antioxidant activities.

GSH possesses the capability to scavenge ROS, including free radicals, peroxides, and lipid peroxides, thereby playing a crucial role in cellular antioxidative defense mechanisms [28]. GSH-Px is an important peroxidolytic enzyme, catalyzing the specific reduction of ROS by oxidizing reduced GSH to its oxidized form against lipid peroxidation [29]. Wang et al. found that the administration of aflatoxin B1 to *L. vannamei* significantly increased the activities of CAT, SOD, and GSH-PX in the hepatopancreas compared to controls, with a tendency to increase and then decrease [30]. Duan et al. studied the oxidative stress response of *Penaeus monodon* to *Vibrio parahaemolyticus* infection, noting that GSH-Px and SOD activities in the hepatopancreas initially increased and then decreased, while the MDA content persistently rose [31]. In our study, when exposed to H_2O_2, the activities of GSH and GSH-Px both exhibited a trend of initial increase followed by a decrease. Specifically, the activity of GSH-PX peaked at 24 h post-stress and then progressively decreased from 48 h to 96 h. Meanwhile, GSH activity reached a higher level at 48 h and 72 h. This pattern may be due to the role of GSH as the substrate for GSH-Px. Increased activity of GSH-Px led to the consumption of GSH. Notably, the activity of GSH-Px decreased after 48 h, which consequently led to a continued rise in GSH levels [32]. At a H_2O_2 concentration of 15 mmol$\cdot L^{-1}$, GSH and GSH-Px activities showed a significant reduction compared to those of the control. This reduction may be attributed to exacerbated lipid peroxidation, resulting in hepatopancreatic damage compared to the control.

H_2O_2 is a significant byproduct of oxidative stress that belongs to ROS. MDA, a typical product of ROS-induced lipid peroxidation, serves as a crucial indicator of oxidative stress, reflects the rate and intensity of lipid peroxidation, and indirectly indicates the degree of tissue peroxidative damage [33,34]. When ROS attack DNA molecules, 8-OHdG is formed as an oxidative adduct. It is widely recognized as a sensitive biomarker for oxidative DNA damage [35]. Lin et al. found that when Cd stress was applied to *E. sinensis*, the activities of SOD, CAT, and GPx followed a trend of initial increase and then decrease, coinciding with increased MDA and H_2O_2 content, which led to tissue damage and apoptosis [36]. When *Charybdis japonica* was exposed to sulfide, MDA content had an ascending trend [37]. Additionally, the hepatopancreatic cells of *E. sinensis* showed an increase in 8-OHdG content after in vitro stimulation with abamectin, indicating DNA damage [38]. In red swamp crayfish (*Procambarus clarkii*), there was a significant increase in 8-OHdG levels in response to the pesticide deltamethrin [39]. The results of our experiment showed that under H_2O_2 stress, the MDA and H_2O_2 levels in the hepatopancreas of *E. sinensis* showed a gradual increase. This indicates that the production and accumulation of ROS in the hepatopancreas led to aggravative lipid peroxidation. Additionally, the significant elevation in the levels of 8-OHdG observed after 48 h of exposure highlights a time-dependent aggravation of oxidative DNA damage.

4.2. Effects of H_2O_2 Stress on the Antioxidant Enzyme Activities of Hemolymph in E. sinensis

Crustaceans depend on the innate immune defense system to combat infections. The hemolymph serves as the primary vehicle for immunological defense and is vital in mediating the host's defensive reactions [40]. Singaram et al. observed that in the mud crab (*Scylla serrata*), antioxidant parameters such as SOD, CAT, and GPx in the hemolymph increased initially and then decreased when exposed to mercury stress [41]. Similarly, *E. sinensis* exhibited a comparable response under thiamethoxam stress, with activities of SOD, CAT, T-AOC, and GSH-Px in the hemolymph showing an initial rise followed by a decline [42]. Furthermore, when *E. sinensis* was subjected to acute ammonia-N stress, there was a significant decrease in hemolymph antioxidants T-AOC and T-SOD, while levels of GSH-Px and MDA were concomitantly elevated [12]. In our study, the levels of SOD, CAT, and T-AOC in the hemolymph showed an initial increase followed by a subsequent decrease under H_2O_2-induced oxidative stress. This indicates that the crab initially upregulated SOD and CAT activity to counteract the accumulation of ROS. However, as the duration of stress extended, the activities of these antioxidant enzymes became suppressed. This could be due to an excessive accumulation of ROS exceeding the detoxification capacity of SOD and CAT, and subsequently leading to a reduction in T-AOC activity. Moreover, GSH and GSH-Px activities showed a tendency of initial increase and then decrease. This pattern triggered by the initial accumulation of peroxides in hemolymph. But, theree activities decreased when oxidative stress was overwhelmed at a particular ROS threshold. Furthermore, upon exposure to deltamethrin, *E. sinensis* exhibited a significant elevation in oxidative stress markers H_2O_2 and MDA in the hemolymph [43]. Similarly, our findings revealed a consistent increase in MDA and H_2O_2 levels in the hemolymph of *E. sinensis* under H_2O_2 stress. The results indicated that H_2O_2, acting as inducer, can lead to significant accumulation in both ROS and lipid peroxidation products in the hemolymph.

4.3. Effects of H_2O_2 Stress on the Expression of Antioxidant- and Immune-Related Genes in Hepatopancreas

In *E. sinensis*, the innate immune system is principally composed of the antioxidant systems, prophenoloxidase (proPO) system, and multiple immune factors [44]. Among them, key antioxidant enzymes such as SOD, CAT, and GPx are the first line of defense against external invasions [45]. Studies have shown that the mRNA expression of *SOD*, *CAT*, and *GPx* in the hepatopancreas of *Portunus trituberculatus* initially increased and then decreased when infected by *Plasmodium trituberculatus* [46]. Similarly, the kuruma prawn (*Marsupenaeus japonicas*) exhibited an initial increase and subsequent decrease in these mRNA expressions when subjected to nitrite stress [47]. In *L. vannamei*, a comparable trend in the mRNA expression of *CAT* and *GPx* was observed after recovery from cold shock [26]. In this study, the expression of *SOD*, *CAT*, and *GSH-Px* first increased and then decreased under H_2O_2 stress, reaching the peak at 72 h, consistent with the activity profiles of these enzymes. This indicates that oxidative stress triggered the gene expression of antioxidant enzymes in the hepatopancreas. However, as the stress intensified, the levels of gene expression decreased.

The proPO system, significantly implicated in the melanization process, is integral to crustacean immune responses and participates in the acute reaction to pathogenic challenges [48]. It is a complex cascade consisting of proPO, PO, pattern recognition proteins (PRPs), and multiple serine proteases. Upon invasion by external pathogens, PRPs initiate a cascade of reactions that activates proPO into its active form, PO. Within this system, an important immune factor known as PX is also activated alongside proPO, thereby acquiring biological activity [49]. Studies have revealed distinct responses of genes in crustaceans' proPO system following various challenges. In *P. clarkii*, infection with *Aeromonas veronii* led to an initial upregulation followed by a downregulation of *proPO* expression [50]. Similarly, infection with *Aeromonas astaci* in Japanese water shrimp (*Macrobrachium nipponense*) caused a significant upregulation of *proPO* mRNA levels [51]. Additionally, the immunostimulant β-glucan was found to induce an upregulation of *PX* in the Indian white shrimp (*Fenneropenaeus indicus*) [52]. In the current study, consistent elevation of *proPO* mRNA was observed at the H_2O_2 concentration of 3 mmol·L^{-1}. Conversely, H_2O_2 concentrations of 6, 9, 12, and 15 mmol·L^{-1} induced an initial increase and then a decrease in *proPO* mRNA levels. This indicates that while low concentrations of H_2O_2 activated the proPO system, excessively high levels may disrupt it, causing a decrease in *proPO* mRNA expression. Moreover, the relative expression of *PX* continuously increased over time with rising H_2O_2 concentrations, suggesting that *PX* expression increased in accordance with oxidative stress, thereby enhancing the immune system and disease resistance of *E. sinensis*.

Heat shock proteins (HSPs) are ubiquitously distributed within the cells of both eukaryotes and prokaryotes with a highly conserved evolutionary process. HSPs perform multiple biomolecular functions, including as molecular chaperones, antioxidants, regulators of apoptotic, and mediators of immune responses [53]. As sensitive biomarkers of environmental stress, HSPs can provide indications to diverse stressors, such as water environmental factors, salinity, air exposure, and pesticides—all of which can elicit an increase in *HSP* expression levels [54–57]. When ridgetail white shrimp (*Exopalaemon carinicauda*) were exposed to *Prorocentrum minimum*, an increase in *HSP70* gene expression was observed in hemocytes and the hepatopancreas [54]. In *Macrophthalmus japonicus*, the mRNA expression of *HSP70* and *HSP90* was significantly upregulated in the hepatopancreas under salinity or bisphenol A (BPA) stress [55]. Under conditions of air exposure, mud crab (*Scylla paramamosain*) exhibited raised levels of *HSP90* and *HSP70* mRNAs in the hepatopancreas [56]. Moreover, in the black tiger prawn (*Penaeus maculatus*), *HSP70* in the muscle was significantly increased under the stress of the pesticides endosulfan and deltamethrin [57]. However, *E. sinensis* showed an initial increase followed by a decrease in *HSP70* gene expression when exposed to glyphosate [58]. In the current study, the expression of *HSP70* in the hepatopancreas of *E. sinensis* showed a continuous increase under H_2O_2 stress, correlating with both the duration of exposure and rising H_2O_2 concentrations. Notably, at a higher concentration of 15 mmol·L^{-1}, a substantial upsurge in expression was observed. The study demonstrated that the elevated expression of *HSP70* may play a crucial role in mitigating oxidative damage.

iNOS is a vital component of the innate immune system, possessing antiviral, antibacterial, and antiparasitic properties. iNOS exerts these effects by directly or indirectly targeting the bases and chains of DNA, proteins, and membrane lipids, thereby inflicting damage to the DNA, enzymes, and membranes of pathogens [59]. Post-infection with the White Spot Syndrome Virus (WSSV), the expression of *iNOS* in Chinese white shrimp (*Fenneropenaeus chinensis*) and *M. japonicas* showed an initial upregulation followed by a subsequent reduction [60]. Similarly, *S. paramamosain* showed a significant increase at the mRNA levels of *NOS* within the intestine, hepatopancreas, and hemocytes upon pathogens infection, suggesting a correlation between *NOS* activity and immune system functionality [61]. In this study, the expression of iNOS in *E. sinensis* during H_2O_2 stress also followed a trend of first increasing and then decreasing, indicating the role of *iNOS* in modulating the immune response of *E. sinensis*. Additionally, iNOS has the ability to generate NO, which may lead to an increase in RNS, and thereby intensify damage in *E. sinensis* [62].

4.4. Effects of Stressors on the Antioxidant Capacity of Crustaceans

The crustacean antioxidant enzyme system plays a pivotal role in combating oxidative stress, representing an intricate mechanism by which these organisms maintain physiological homeostasis amidst environmental perturbations. We compared the trends in oxidative stress markers in crustaceans under various stressors to gain deeper insight into their physiological responses to oxidative stress over time (Table S1). We observed that different stress treatments elicited varying antioxidant responses across different crustaceans. For crustaceans, the primary external stressors include environmental pressures such as temperature, hypoxia, salinity, ammonia, and desiccation; and anthropogenic stressors such as heavy metals (copper, cadmium), toxic substances, and pesticides (aflatoxin, bisphenol, abamectin, deltamethrin). Furthermore, pathogenic microorganisms (bacteria and viruses), compound the oxidative burden. In response to oxidative stress, the activity of antioxidant enzymes in crustaceans can exhibit three distinct trends: an increase, a decrease, or an initial increase followed by a decrease. Generally, upon exposure to stressors, there is an upregulation of antioxidant markers to neutralize the surge in ROS. However, a decrease in certain antioxidant markers may occur due to depletion in response to excessive ROS or as a result of tissue damage. Additionally, antioxidant markers initially rise due to the pro-oxidant characteristics; if the stress is prolonged or excessive, the antioxidant system may become depleted or damaged, leading to a decrease in the activity of antioxidant enzymes. This inability to effectively clear ROS aggravates cellular damage. The elevation of lipid peroxidation products (such as MDA) and DNA damage markers (such as 8-OHdG) are also significant indicators of oxidative stress, signaling damage to cell membranes and genetic material.

5. Conclusions

In summary, we exposed *E. sinensis* to various concentrations of H_2O_2 and monitored physiological and biochemical markers of oxidative stress. We also measured expression levels of genes associated with the antioxidant response and immune function. The conveyed data support the dynamic and biphasic nature of the oxidative stress response in *E. sinensis*. The findings demonstrate that an organism's initial response to oxidative stress is to enhance its antioxidative defenses. However, if the intensity or duration of the stress surpasses a certain threshold, the protective mechanisms become overwhelmed, resulting in oxidative damage. These findings have significant implications for comprehending the stress responses at the cellular and molecular levels and can be critical for devising strategies to shield organisms from oxidative harm.

Supplementary Materials: The following supporting information can be downloaded at: https://www.mdpi.com/article/10.3390/antiox13050524/s1. References [63–74] are cited in the Supplementary Materials.

Author Contributions: Conceptualization, Q.H. and W.F.; methodology, W.F.; validation, Y.T. and W.F.; formal analysis, X.C.; investigation, Q.H.; resources, J.Z. and J.L.; data curation, Y.X.; writing—original draft preparation, Q.H. and W.F.; writing—review and editing, Y.T. and W.F.; visualization, J.L.; supervision, W.F.; project administration, J.L. and P.X.; funding acquisition, Y.T., W.F., J.Z. and P.X. All authors have read and agreed to the published version of the manuscript.

Funding: This research was funded by National Natural Science Foundation of China (grant number 32202966); the Key Project for Jiangsu Agricultural New Variety Innovation (grant number PZCZ201749); the Key Research and Development Program of Jiangsu Province (grant number BE2022360); and the Jiangsu Revitalization of Seed Industry (grant number JBGS[2021]031).

Institutional Review Board Statement: The study was approved by the Ethics Committee of FFRC (protocol code LAECFFRC-2023-09-12).

Informed Consent Statement: Informed consent was obtained from all subjects involved in the study.

Data Availability Statement: The original contributions presented in the study are included in the article/Supplementary Materials; further inquiries can be directed to the corresponding authors.

Conflicts of Interest: The authors declare no conflicts of interest.

References

1. Wang, Q.; Wu, X.; Long, X.; Zhu, W.; Ma, T.; Cheng, Y. Nutritional quality of different grades of adult male chinese mitten crab, *Eriocheir sinensis*. *J. Food Sci. Technol.* **2018**, *55*, 944–955. [CrossRef] [PubMed]
2. Huang, S.; Wang, J.; Yue, W.; Chen, J.; Gaughan, S.; Lu, W.; Lu, G.; Wang, C. Transcriptomic variation of hepatopancreas reveals the energy metabolism and biological processes associated with molting in Chinese mitten crab, *Eriocheir sinensis*. *Sci. Rep.* **2015**, *5*, 14015. [CrossRef] [PubMed]
3. Jiang, S.; Zhang, W.; Xiong, Y.; Cheng, D.; Wang, J.; Jin, S.; Gong, Y.; Wu, Y.; Qiao, H.; Fu, H. Hepatopancreas transcriptome analyses provide new insights into the molecular regulatory mechanism of fast ovary maturation in *Macrobrachium nipponense*. *BMC Genom.* **2022**, *23*, 625. [CrossRef] [PubMed]
4. Qiu, R.Q.R.; Cheng, Y.C.Y.; Huang, X.H.X.; Wu, X.W.X.; Yang, X.Y.X.; Tong, R.T.R. Effect of hypoxia on immunological, physiological response, and hepatopancreatic metabolism of juvenile Chinese mitten crab *Eriocheir sinensis*. *Aquac. Int.* **2011**, *19*, 283–299. [CrossRef]
5. Li, Z.; Zhao, Z.; Luo, L.; Wang, S.; Zhang, R.; Guo, K.; Yang, Y. Immune and intestinal microbiota responses to heat stress in Chinese mitten crab (*Eriocheir sinensis*). *Aquaculture* **2023**, *563*, 738965. [CrossRef]
6. Bu, X.; Song, Y.; Pan, J.; Wang, X.; Qin, C.; Jia, Y.; Du, Z.; Qin, J.; Chen, L. Toxicity of chronic copper exposure on Chinese mitten crab (*Eriocheir sinensis*) and mitigation of its adverse impact by myo-inositol. *Aquaculture* **2022**, *547*, 737511. [CrossRef]
7. Yang, X.; Song, Y.; Zhang, C.; Pang, Y.; Song, X.; Wu, M.; Cheng, Y. Effects of the glyphosate-based herbicide roundup on the survival, immune response, digestive activities and gut microbiota of the Chinese mitten crab, *Eriocheir sinensis*. *Aquat. Toxicol.* **2019**, *214*, 105243. [CrossRef] [PubMed]
8. Wang, T.; Yang, C.; Zhang, S.; Rong, L.; Yang, X.; Wu, Z.; Sun, W. Metabolic changes and stress damage induced by ammonia exposure in juvenile *Eriocheir sinensis*. *Ecotoxicol. Environ. Saf.* **2021**, *223*, 112608. [CrossRef] [PubMed]
9. Kurutas, E.B. The importance of antioxidants which play the role in cellular response against oxidative/nitrosative stress: Current state. *Nutr. J.* **2016**, *15*, 71. [CrossRef] [PubMed]
10. Zheng, N.; Wang, N.; Wang, Z.; Abdallah, G.; Zhang, B.; Wang, S.; Yao, Q.; Chen, Y.; Wang, Q.; Zhang, D. Effect of infection with *Aeromonas hydrophila* on antioxidant capacity, inflammation response, and apoptosis proteins in Chinese mitten crab (*Eriocheir sinensis*). *Comp. Biochem. Physiol. Part C Toxicol. Pharmacol.* **2022**, *252*, 109220. [CrossRef] [PubMed]
11. Wang, X.; Yao, Q.; Zhang, D.; Lei, X.; Wang, S.; Wan, J.; Liu, H.; Chen, Y.; Zhao, Y.; Wang, G. Effects of acute salinity stress on osmoregulation, antioxidant capacity and physiological metabolism of female Chinese mitten crabs (*Eriocheir sinensis*). *Aquaculture* **2022**, *552*, 737989. [CrossRef]
12. Yang, X.; Shi, A.; Song, Y.; Niu, C.; Yu, X.; Shi, X.; Pang, Y.; Ma, X.; Cheng, Y. The effects of ammonia-N stress on immune parameters, antioxidant capacity, digestive function, and intestinal microflora of Chinese mitten crab, *Eriocheir sinensis*, and the protective effect of dietary supplement of melatonin. *Comp. Biochem. Physiol. Part C Toxicol. Pharmacol.* **2021**, *250*, 109127. [CrossRef] [PubMed]
13. Zhang, R.; Zhao, Z.; Li, M.; Luo, L.; Wang, S.; Guo, K.; Xu, W. Effects of saline-alkali stress on the tissue structure, antioxidation, immunocompetence and metabolomics of *Eriocheir sinensis*. *Sci. Total Environ.* **2023**, *871*, 162109. [CrossRef] [PubMed]
14. Yang, Z.Y. *Preliminary Studies on the Etiology and Pathogenic Mechanism of Hepatopancreatic Necrosis Syndrome in Chinese Mitten Crab Eriocheir Sinensis*; Shanghai Ocean University: Shanghai, China, 2018; (In Chinese with English abstract)
15. Chang, C.; Mayer, M.; Rivera-Ingraham, G.; Blondeau-Bidet, E.; Wu, W.; Lorin-Nebel, C.; Lee, T. Effects of temperature and salinity on antioxidant responses in livers of temperate (*Dicentrarchus labrax*) and tropical (*Chanos chanos*) marine euryhaline fish. *J. Therm. Biol.* **2021**, *99*, 103016. [CrossRef] [PubMed]
16. Jiang, X.; Dong, S.; Liu, R.; Huang, M.; Dong, K.; Ge, J.; Gao, Q.; Zhou, Y. Effects of temperature, dissolved oxygen, and their interaction on the growth performance and condition of rainbow trout (*Oncorhynchus mykiss*). *J. Therm. Biol.* **2021**, *98*, 102928. [CrossRef] [PubMed]
17. Fan, Z.; Wu, D.; Li, J.; Li, C.; Zheng, X.; Wang, L. Phosphorus Nutrition in Songpu Mirror Carp (*Cyprinus carpio* Songpu) During Chronic Carbonate Alkalinity Stress: Effects on Growth, Intestinal Immunity, Physical Barrier Function, and Intestinal Microflora. *Front. Immunol.* **2022**, *13*, 900793. [CrossRef] [PubMed]
18. Singh, S.P.; Ahmad, T.; Sharma, J.; Chakrabarti, R. Effect of temperature on food consumption, immune system, antioxidant enzymes, and heat shock protein 70 of Channa punctata (Bloch, 1793). *Fish Physiol. Biochem.* **2021**, *47*, 79–91. [CrossRef] [PubMed]
19. Wang, Z.; Cai, C.; Cao, X.; Zhu, J.; He, J.; Wu, P.; Ye, Y. Supplementation of dietary astaxanthin alleviated oxidative damage induced by chronic high pH stress, and enhanced carapace astaxanthin concentration of Chinese mitten crab *Eriocheir sinensis*. *Aquaculture* **2018**, *483*, 230–237. [CrossRef]
20. Schieber, M.; Chandel, N.S. ROS function in redox signaling and oxidative stress. *Curr. Biol.* **2014**, *24*, R453–R462. [CrossRef] [PubMed]
21. Moniruzzaman, M.; Kumar, S.; Das, D.; Sarbajna, A.; Chakraborty, S.B. Enzymatic, non enzymatic antioxidants and glucose metabolism enzymes response differently against metal stress in muscles of three fish species depending on different feeding niche. *Ecotoxicol. Environ. Saf.* **2020**, *202*, 110954. [CrossRef] [PubMed]
22. Jia, X.; Zhang, D.; Wang, F.; Dong, S. Immune responses of *Litopenaeus vannamei* to non-ionic ammonia stress: A comparative study on shrimps in freshwater and seawater conditions. *Aquac. Res.* **2017**, *48*, 177–188. [CrossRef]

23. Liu, T.; Zhang, G.; Feng, Y.; Kong, C.; Ayisi, C.L.; Huang, X.; Hua, X. Dietary soybean antigen impairs growth and health through stress-induced non-specific immune responses in Pacific white shrimp, *Litopenaeus vannamei*. *Fish Shellfish Immunol.* **2019**, *84*, 124–129. [CrossRef] [PubMed]
24. Bao, J.; Li, X.; Xing, Y.; Feng, C.; Jiang, H. Respiratory metabolism and antioxidant response in Chinese mitten crab *Eriocheir sinensis* during air exposure and subsequent reimmersion. *Front. Physiol.* **2019**, *10*, 907. [CrossRef] [PubMed]
25. Parrilla-Taylor, D.P.; Zenteno-Savín, T. Antioxidant enzyme activities in Pacific white shrimp (*Litopenaeus vannamei*) in response to environmental hypoxia and reoxygenation. *Aquaculture* **2011**, *318*, 379–383. [CrossRef]
26. Xu, Z.; Regenstein, J.M.; Xie, D.; Lu, W.; Ren, X.; Yuan, J.; Mao, L. The oxidative stress and antioxidant responses of *Litopenaeus vannamei* to low temperature and air exposure. *Fish Shellfish Immunol.* **2018**, *72*, 564–571. [CrossRef] [PubMed]
27. Sohal, R.S.; Allen, R.G. Oxidative stress as a causal factor in differentiation and aging: A unifying hypothesis. *Exp. Gerontol.* **1990**, *25*, 499–522. [CrossRef] [PubMed]
28. Ueno, Y.; Kizaki, M.; Nakagiri, R.; Kamiya, T.; Sumi, H.; Osawa, T. Dietary glutathione protects rats from diabetic nephropathy and neuropathy. *J. Nutr.* **2002**, *132*, 897–900. [CrossRef]
29. Ye, J.; Han, Y.; Zhao, J.; Lu, T.; Liu, H.; Yang, Y. Effects of dietary olaquindox on antioxidant enzymes system in hepatopancreas of *Cyprinus carpio*. *Shuichan Xuebao* **2004**, *28*, 231–235.
30. Wang, Y.; Wang, B.; Liu, M.; Jiang, K.; Wang, M.; Wang, L. Aflatoxin B1 (AFB1) induced dysregulation of intestinal microbiota and damage of antioxidant system in pacific white shrimp (*Litopenaeus vannamei*). *Aquaculture* **2018**, *495*, 940–947. [CrossRef]
31. Duan, Y.; Zhang, J.; Dong, H.; Wang, Y.; Liu, Q.; Li, H. Oxidative stress response of the black tiger shrimp Penaeus monodon to Vibrio parahaemolyticus challenge. *Fish Shellfish Immunol.* **2015**, *46*, 354–365. [CrossRef]
32. Wang, J.; Sun, L.; Li, X.; Tao, S.; Wang, F.; Shi, Y.; Guan, H.; Yang, Y.; Zhao, Z. Alkali exposure induces autophagy through activation of the MAPKpathway by ROS and inhibition of mTOR in *Eriocheir sinensis*. *Aquat. Toxicol.* **2023**, *258*, 106481. [CrossRef] [PubMed]
33. Yao, J.; Wang, J.; Liu, L.; Li, Y.; Xun, A.; Zeng, W.; Jia, C.; Wei, X.; Feng, J.; Zhao, L. Anti-oxidant effects of resveratrol on mice with DSS-induced ulcerative colitis. *Arch. Med. Res.* **2010**, *41*, 288–294. [CrossRef] [PubMed]
34. Hong, Y.; Huang, Y.; Yan, G.; Huang, Z. Effects of deltamethrin on the antioxidant defense and heat shock protein expression in Chinese mitten crab, *Eriocheir sinensis*. *Environ. Toxicol. Pharmacol.* **2019**, *66*, 1–6. [CrossRef] [PubMed]
35. Pilger, A.; Rüdiger, H.W. 8-Hydroxy-2′-deoxyguanosine as a marker of oxidative DNA damage related to occupational and environmental exposures. *Int. Arch. Occup. Environ. Health* **2006**, *80*, 1–15. [CrossRef] [PubMed]
36. Lin, Y.; Huang, J.; Dahms, H.; Zhen, J.; Ying, X. Cell damage and apoptosis in the hepatopancreas of *Eriocheir sinensis* induced by cadmium. *Aquat. Toxicol.* **2017**, *190*, 190–198. [CrossRef] [PubMed]
37. Xu, X.H.; Zhang, Y.Q.; Yan, B.L.; Xu, J.T.; Tang, Y.; Du, D.D. Immunological and histological responses to sulfide in the crab *Charybdis japonica*. *Aquat. Toxicol.* **2014**, *150*, 144–150. [CrossRef] [PubMed]
38. Hong, Y.; Huang, Y.; Dong, Y.; Xu, D.; Huang, Q.; Huang, Z. Cytotoxicity induced by abamectin in hepatopancreas cells of Chinese mitten crab, *Eriocheir sinensis*: An in vitro assay. *Ecotoxicol. Environ. Saf.* **2023**, *262*, 115198. [CrossRef] [PubMed]
39. Hong, Y.; Huang, Y.; Yan, G.; Yin, H.; Huang, Z. DNA damage, immunotoxicity, and neurotoxicity induced by deltamethrin on the freshwater crayfish, *Procambarus clarkii*. *Environ. Toxicol.* **2021**, *36*, 16–23. [CrossRef] [PubMed]
40. Jia, Z.; Wang, L.; Jiang, S.; Sun, M.; Wang, M.; Yi, Q.; Song, L. Functional characterization of hemocytes from Chinese mitten crab *Eriocheir sinensis* by flow cytometry. *Fish Shellfish Immunol.* **2017**, *69*, 15–25. [CrossRef]
41. Singaram, G.; Harikrishnan, T.; Chen, F.; Bo, J.; Giesy, J.P. Modulation of immune-associated parameters and antioxidant responses in the crab (*Scylla serrata*) exposed to mercury. *Chemosphere* **2013**, *90*, 917–928. [CrossRef] [PubMed]
42. Sun, Y.; Yuan, C.; Cui, Q. Acute toxic effects of thiamethoxam on Chinese mitten crab *Eriocheir sinensis*. *Environ. Sci. Pollut. Res. Int.* **2022**, *29*, 63512–63519. [CrossRef] [PubMed]
43. Hong, Y.; Yang, X.; Huang, Y.; Yan, G.; Cheng, Y. Oxidative stress and genotoxic effect of deltamethrin exposure on the Chinese mitten crab, *Eriocheir sinensis*. *Comp. Biochem. Physiol. Part C Toxicol. Pharmacol.* **2018**, *212*, 25–33. [CrossRef] [PubMed]
44. Lee, S.Y.; Söderhäll, K. Early events in crustacean innate immunity. *Fish Shellfish Immunol.* **2002**, *12*, 421–437. [PubMed]
45. Czapski, G.; Goldstein, S. The uniqueness both of superoxide toxicity and of the protective role of superoxide dismutase. In *Oxygen Radicals in Biology and Medicine*; Springer: Boston, MA, USA, 1988; pp. 43–46.
46. Ren, X.; Lv, J.; Gao, B.; Li, J.; Liu, P. Immune response and antioxidant status of *Portunus trituberculatus* inoculated with pathogens. *Fish Shellfish Immunol.* **2017**, *63*, 322–333. [CrossRef] [PubMed]
47. Zheng, J.; Mao, Y.; Su, Y.; Wang, J. Effects of nitrite stress on mRNA expression of antioxidant enzymes, immune-related genes and apoptosis-related proteins in *Marsupenaeus japonicus*. *Fish Shellfish Immunol.* **2016**, *58*, 239–252. [CrossRef] [PubMed]
48. Gai, Y.; Zhao, J.; Song, L.; Li, C.; Zheng, P.; Qiu, L.; Ni, D. A prophenoloxidase from the Chinese mitten crab *Eriocheir sinensis*: Gene cloning, expression and activity analysis. *Fish Shellfish Immunol.* **2008**, *24*, 156–167. [CrossRef] [PubMed]
49. Goudru, H.G.; Kumar, S.; Jayalakshmi, S.K.; Ballal, C.R.; Sharma, H.C.; Sreeramulu, K. Purification and characterization of prophenoloxidase from cotton bollworm, *Helicoverpa armigera*. *Entomol. Res.* **2013**, *43*, 55–62. [CrossRef]

50. Zhu, L.; Wang, X.; Hou, L.; Jiang, X.; Li, C.; Zhang, J.; Pei, C.; Zhao, X.; Li, L.; Kong, X. The related immunity responses of red swamp crayfish (*Procambarus clarkii*) following infection with *Aeromonas veronii*. *Aquac. Rep.* **2021**, *21*, 100849. [CrossRef]
51. Chen, Q.; Zhang, Z.; Tang, H.; Zhou, L.; Ao, S.; Zhou, Y.; Zhu, X.; Gao, X.; Jiang, Q.; Tu, C. *Aeromonas hydrophila* associated with red spot disease in *Macrobrachium nipponense* and host immune-related gene expression profiles. *J. Invertebr. Pathol.* **2021**, *182*, 107584. [CrossRef] [PubMed]
52. Sivakamavalli, J.; Selvaraj, C.; Singh, S.K.; Vaseeharan, B. In vitro and in silico studies on cell adhesion protein peroxinectin from Fenneropenaeus indicus and screening of heme blockers against activity. *J. Mol. Recognit.* **2016**, *29*, 186–198. [CrossRef] [PubMed]
53. Da Costa, C.U.P.; Wagner, H.; Miethke, T.C. Heat shock protein-mediated activation of innate immune cells. In *Heat Shock Proteins and Inflammation*; Springer: Basel, Switzerland, 2003; pp. 43–54.
54. Mu, C.; Ge, Q.; Li, J. Exposure to Prorocentrum minimum induces oxidative stress and apoptosis in the ridgetail white prawn, *Exopalaemon carinicauda*. *J. Ocean Univ. China* **2019**, *18*, 727–734. [CrossRef]
55. Park, K.; Kwak, I. Salinity and bisphenol A alter cellular homeostasis and immune defense by heat shock proteins in the intertidal crab *Macrophthalmus japonicus*. *Estuar. Coast. Shelf Sci.* **2019**, *229*, 106381. [CrossRef]
56. Cheng, C.; Ma, H.; Deng, Y.; Feng, J.; Jie, Y.; Guo, Z. Immune and physiological responses of mud crab (*Scylla paramamosain*) under air exposure. *Comp. Biochem. Physiol. Part C Toxicol. Pharmacol.* **2020**, *233*, 108767. [CrossRef] [PubMed]
57. Dorts, J.; Silvestre, F.; Tu, H.T.; Tyberghein, A.; Phuong, N.T.; Kestemont, P. Oxidative stress, protein carbonylation and heat shock proteins in the black tiger shrimp, Penaeus monodon, following exposure to endosulfan and deltamethrin. *Environ. Toxicol. Pharmacol.* **2009**, *28*, 302–310. [CrossRef] [PubMed]
58. Hong, Y.; Huang, Y.; Yan, G.; Pan, C.; Zhang, J. Antioxidative status, immunological responses, and heat shock protein expression in hepatopancreas of Chinese mitten crab, *Eriocheir sinensis* under the exposure of glyphosate. *Fish Shellfish Immunol.* **2019**, *86*, 840–845. [CrossRef] [PubMed]
59. De Groote, M.A.; Testerman, T.; Xu, Y.; Stauffer, G.; Fang, F.C. Homocysteine antagonism of nitric oxide-related cytostasis in *Salmonella typhimurium*. *Science* **1996**, *272*, 414–417. [CrossRef] [PubMed]
60. Jiang, G.; Yu, R.; Zhou, M. Studies on nitric oxide synthase activity in haemocytes of shrimps *Fenneropenaeus chinensis* and *Marsupenaeus japonicus* after white spot syndrome virus infection. *Nitric Oxide* **2006**, *14*, 219–227. [CrossRef] [PubMed]
61. Li, S.; Zhang, Z.; Li, C.; Zhou, L.; Liu, W.; Li, Y.; Zhang, Y.; Zheng, H.; Wen, X. Molecular cloning and expression profiles of nitric oxide synthase (NOS) in mud crab *Scylla paramamosain*. *Fish Shellfish Immunol.* **2012**, *32*, 503–512. [CrossRef] [PubMed]
62. Jianan, X.; Yutao, M.; Bin, L.; Hui, G.; Anli, W. Apoptosis of tiger shrimp (*Penaeus monodon*) haemocytes induced by *Escherichia coli* lipopolysaccharide. *Comp. Biochem. Physiol. A Mol. Integr. Physiol.* **2013**, *164*, 301–306.
63. Chen, X.; Feng, W.; Yan, F.; Li, W.; Xu, P.; Tang, Y. Alteration of antioxidant status, glucose metabolism, and hypoxia signal pathway in Eirocheir sinensis after acute hypoxic stress and reoxygenation. *Comp. Biochem. Physiol. Part C Toxicol. Pharmacol.* **2023**, *268*, 109604. [CrossRef] [PubMed]
64. Li, Y.; Wei, L.; Cao, J.; Qiu, L.; Jiang, X.; Li, P.; Song, Q.; Zhou, H.; Han, Q.; Diao, X. Oxidative stress, DNA damage and antioxidant enzyme activities in the pacific white shrimp (*Litopenaeus vannamei*) when exposed to hypoxia and reoxygenation. *Chemosphere* **2016**, *144*, 234–240. [CrossRef] [PubMed]
65. Jiao, L.; Dai, T.; Zhong, S.; Jin, M.; Sun, P.; Zhou, Q. Vibrio parahaemolyticus Infection Influenced Trace Element Homeostasis, Impaired Antioxidant Function, and Induced Inflammation Response in *Litopenaeus vannamei*. *Biol. Trace Element Res.* **2020**, *199*, 329–337. [CrossRef] [PubMed]
66. Duan, Y.; Zhang, Y.; Dong, H.; Zhang, J. Effect of desiccation on oxidative stress and antioxidant response of the black tiger shrimp Penaeus monodon. *Fish Shellfish Immunol.* **2016**, *58*, 10–17. [CrossRef] [PubMed]
67. Zhang, L.; Song, Z.; Zhong, S.; Gan, J.; Liang, H.; Yu, Y.; Wu, G.; He, L. Acute hypoxia and reoxygenation induces oxidative stress, glycometabolism, and oxygen transport change in red swamp crayfish (*Procambarus clarkii*): Application of transcriptome profiling in assessment of hypoxia. *Aquac. Rep.* **2022**, *23*, 101029. [CrossRef]
68. Zeng, Q.; Yang, Q.; Chai, Y.; Wei, W.; Luo, M.; Li, W. Polystyrene microplastics enhanced copper-induced acute immunotoxicity in red swamp crayfish (*Procambarus clarkii*). *Ecotoxicol. Environ. Saf.* **2023**, *249*. [CrossRef] [PubMed]
69. Ruan, G.; Li, S.; He, N.; Fang, L.; Wang, Q. Short-term adaptability to non-hyperthermal stress: Antioxidant, immune and gut microbial responses in the red swamp crayfish, *Procambarus clarkii*. *Aquaculture* **2022**, *560*. [CrossRef]
70. Hossain, M.; Huang, H.; Yuan, Y.; Wan, T.; Jiang, C.; Dai, Z.; Xiong, S.; Cao, M.; Tu, S. Silicone stressed response of crayfish (*Procambarus clarkii*) in antioxidant enzyme activity and related gene expression. *Environ. Pollut.* **2020**, *274*, 115836. [CrossRef] [PubMed]
71. Peng, Y.Q.; Wang, M.J.; Chen, H.G.; Chen, J.H.; Gao, H.; Huang, H.H. Immunological responses in haemolymph and histologic changes in the hepatopancreas of *Charybdis japonica* (A. Milne-Edwards, 1861) (Decapoda: Brachyura: Portunidae) exposed to bisphenol A. *J. Crustac. Biol.* **2018**, *38*, 489–496. [CrossRef]
72. Meng, X.-L.; Liu, P.; Li, J.; Gao, B.-Q.; Chen, P. Physiological responses of swimming crab *Portunus trituberculatus* under cold acclimation: Antioxidant defense and heat shock proteins. *Aquaculture* **2014**, *434*, 11–17. [CrossRef]

73. Wang, S.; Ji, C.; Li, F.; Wu, H. Toxicological responses of juvenile Chinese shrimp *Fenneropenaeus chinensis* and swimming crab *Portunus trituberculatus* exposed to cadmium. *Ecotoxicol. Environ. Saf.* **2022**, *234*, 113416. [CrossRef]
74. Lu, Y.; Qiu, Q.; Li, C.; Cheng, L.; Liu, J. Antioxidant responses of *Fenneropenaeus chinensis* to white spot syndrome virus challenge. *Aquac. Int.* **2019**, *28*, 139–151. [CrossRef]

antioxidants

MDPI

Article

Excessive Replacement of Fish Meal by Soy Protein Concentrate Resulted in Inhibition of Growth, Nutrient Metabolism, Antioxidant Capacity, Immune Capacity, and Intestinal Development in Juvenile Largemouth Bass (*Micropterus salmoides*)

Hualiang Liang [1], Mingchun Ren [1,*], Lu Zhang [2,*], Haifeng Mi [2], Heng Yu [1,3], Dongyu Huang [1], Jiaze Gu [3] and Tao Teng [2]

1 Key Laboratory of Integrated Rice-Fish Farming Ecology, Ministry of Agriculture and Rural Affairs, Freshwater Fisheries Research Center, Chinese Academy of Fishery Sciences, Wuxi 214081, China
2 Tongwei Agricultural Development Co., Ltd., Key Laboratory of Nutrition and Healthy Culture of Aquatic Livestock and Poultry, Ministry of Agriculture and Rural Affairs, Healthy Aquaculture Key Laboratory of Sichuan Province, Chengdu 610093, China
3 Wuxi Fisheries College, Nanjing Agricultural University, Wuxi 214081, China
* Correspondence: renmc@ffrc.cn (M.R.); zhangl21@tongwei.com (L.Z.)

Abstract: This study investigated the effects of replacing 0% (SPC0), 25% (SPC25), 50% (SPC50), 75% (SPC75), and 100% (SPC100) of fish meal (FM) with soy protein concentrate (SPC) on the growth, nutritional metabolism, antioxidant capacity, and inflammatory factors in juvenile largemouth bass (*Micropterus salmoides*) (17.03 ± 0.01 g). After 56 days of culturing, various growth parameters including FW, WGR, and SGR were not significantly different among SPC0, SPC25, and SPC50 groups; however, they were significantly higher than those in SPC75 and SPC100 groups. Conversely, significantly lower FCR were determined for the SPC0, SPC25, and SPC50 groups compared with that for the SPC100 group; specifically, no significant difference among SPC0, SPC25, and SPC50 groups was found. Moreover, compared with SPC75 and SPC100 groups, a significantly higher FI was observed in the SPC0 group, whereas a significantly lower SR was observed in SPC100 compared with that in SPC0 and SPC25 groups. Compared with the SPC0 group, significantly lower mRNA levels of *tor*, *rps6*, *4ebp1*, *ppar*γ, and *fas* were found in SPC75 and SPC100. Additionally, the mRNA levels of *cpt* were significantly higher in SPC0, SPC25, and SPC50 groups than in SPC75 and SPC100 groups. Moreover, the mRNA levels of *scd* and *acc* remained unchanged for all the groups. Replacement of FM with SPC did not significantly affect the mRNA levels of *gk*, *pk*, and *pepck*. Compared with the SPC0 group, significantly decreased activities of CAT were observed in the SPC50, SPC75, and SPC100 groups, and significantly decreased activities of GSH-Px were observed in the SPC75 and SPC100 groups. In addition, significantly lower activity of SOD was observed in SPC100 compared with the other groups. Moreover, compared with the other groups, the SPC75 and SPC100 groups had significantly decreased and increased contents of GSH and MDA, respectively, while significantly lower mRNA levels of *nrf2*, *cat*, *sod*, and *gsh-px* were found in SPC50, SPC75, and SPC100; however, significantly higher mRNA levels of *keap1* were observed in SPC75 and SPC100 groups. Additionally, significantly higher mRNA levels of *il-8* and *nf-κb* were found in the SPC50, SPC75, and SPC100 groups compared with the SPC0 group. Conversely, significantly lower mRNA levels of *il-10* and significantly higher mRNA levels of *tnf-α* were found in the SPC75 and SPC100 groups compared with the other groups. Compared with the SPC0 group, mucosal thickness and villus height were significantly decreased in the SPC75 and SPC100 groups. Collectively, SPC replacing 50% FM did not affect its growth of juvenile largemouth bass. However, SPC replacing 50% or more FM might inhibit antioxidant capacity and immune capacity to even threaten the SR, resulting in impaired intestinal development in replacing FM level of 75% or more.

Citation: Liang, H.; Ren, M.; Zhang, L.; Mi, H.; Yu, H.; Huang, D.; Gu, J.; Teng, T. Excessive Replacement of Fish Meal by Soy Protein Concentrate Resulted in Inhibition of Growth, Nutrient Metabolism, Antioxidant Capacity, Immune Capacity, and Intestinal Development in Juvenile Largemouth Bass (*Micropterus salmoides*). *Antioxidants* **2024**, *13*, 809. https://doi.org/10.3390/antiox13070809

Academic Editor: Alessandra Napolitano

Received: 29 May 2024
Revised: 27 June 2024
Accepted: 1 July 2024
Published: 4 July 2024

Keywords: replacement of fish meal; soy protein concentrate; nutrition metabolism; antioxidant capacity; immune capacity

1. Introduction

Sustainable aquaculture ensures food security by providing high-quality protein worldwide [1]. Compared with 2015, aquatic feed production is expected to increase by 33% by 2025 [2]. Therefore, reducing the use of raw materials in feed production is crucial for ensuring the sustainable development of the aquaculture industry [2,3]. The growth and health of fish are largely affected by the availability of adequate nutrients irrespective of the farming system used [4]. Alternatively, the quality of the main protein source in the feed affects the nutritional value of fish [5]. In aquaculture, fish meal (FM) is considered a high-quality protein source due to the balanced composition of amino acids, protein content, antinutritional factors, unsaturated fatty acids, and so on [6]. However, FM is a majorly unsustainable ingredient in aquatic feed due to the overfishing of the ocean and its rising cost resulting from the continuous development of the aquaculture industry [3]. Moreover, there are limited or no prospects for increasing FM production in the future [7]. Nevertheless, inconsistency in the supply and price of FM pose considerable risks; therefore, identification, development, and utilization of FM alternatives remain a high-priority strategy for risk management in sustainable aquaculture development [8].

Soybean protein concentrate (SPC) is obtained by extracting defatted soybean flakes using water, ethanol, or methanol, and it has digestible protein and energy with better palatability compared to soybean meal; therefore, it is considered a quality protein source to replace FM [9]. In addition, most of the antinutritional factors of SPC are inactivated or removed [10,11]. Some studies have found that SPC could successfully replace part of FM in the feed of aquatic animals, such as rice field eel (*Monopterus albus*) [12], golden crucian carp (*Cyprinus carpio* × *Carassius auratus*) [13], Coho Salmon (*Oncorhynchus kisutch*) [14], golden pompano (*Trachinotus ovatus*) [15], Atlantic salmon (*Salmo salar* L.) [16], Totoaba (*Totoaba macdonaldi*) [17], red drum (*Sciaenops ocellatus*) [18], and Florida pompano (*Trachinotus carolinus*) [19], suggesting the potential of SPC to replace FM in fish feed. However, excessive replacement of FM with SPC inhibited the growth, feed utilization, and feeding of the fish to even risk its survival rate [20–22]. The underlying mechanism could be plant-based protein-mediated inhibition of TOR, signaling transduction and downregulation of lipolysis-related factors, thereby inducing liver metabolic disorders and inhibition of metabolism [23,24]. Hence, it is important to determine the replacement proportion of FM with SPC in aquatic animal feed.

The healthy properties of soybeans are attributed to the soy isoflavones [25,26] and their metabolites which are capable of exerting anti-inflammatory effects [27]. Bitzer et al. [28] reported that SPC had a cellular protective role in vitro, alleviating the severity of inflammation and loss of intestinal barrier function in vivo. However, plant protein sources widely used in feed could induce intestinal inflammation in carnivorous fish [29,30]. The replacement of 20% FM with SPC did not affect the intestinal health of the fish; however, replacement above 40% negatively impacted the intestinal morphology-related indicators such as microvilli length, causing obvious symptoms of enteritis in pearl gentian groupers (*Epinephelus fuscoguttatus* ♀× *Epinephelus lanceolatus* ♂) [31]. Furthermore, the partial replacement of FM with SPC was beneficial for improving the serum antioxidant capacity in rice field eels [12]. Zhu et al. [13] demonstrated that when SPC replaced 40% FM, antioxidant enzyme activities and malondialdehyde (MDA) content were not significantly affected in golden crucian carp (*Cyprinus carpio* × *Carassius auratus*); however, a high proportion of SPC replacing FM could lead to decreased enzymes activities of glutathione peroxidase (GSH-Px) and catalase (CAT) in hybrid grouper (*Epinephelus fuscoguttatus* ♀× *Epinephelus lanceolatus* ♂) [32] and decreased enzymes activities of GSH-Px and total superoxide dismutase in starry flounder (*Platichthys stellatus*) [20].

Largemouth bass (*Micropterus salmoides*) is a native of North America. Due to its rapid growth rate, wide temperature tolerance, versatility in adapting to different conditions, and attractiveness as a food source, it has been widely cultivated in China [33] and other parts of the world [34]. As a carnivorous fish species, largemouth bass has a high demand for FM, constituting about 40–55% of dry matter [35,36]. SPC has immense potential in replacing FM and has been applied to other aquatic animals with good results [12–19]; however, research information on largemouth bass is insufficient. Hence, this experiment was designed to study the effects of the replacement of FM with SPC on the growth, nutritional metabolism, antioxidant status, and inflammatory factors of juvenile largemouth bass.

2. Materials and Methods

2.1. Diets

A total of five feed groups were designed as SPC0, SPC25, SPC50, SPC75, and SPC100 replacing 0%, 25%, 50%, 75%, and 100% of FM with SPC in the feed, respectively (Table 1). The main protein sources, including FM, SPC, soybean meal, and corn protein meal, were sieved through a screen size of 0.18 mm before meal preparation with sequential mixing of all the raw materials according to our previously reported method [37].

Table 1. Experimental formula (dry matter, %).

Ingredients	SPC0	SPC25	SPC50	SPC75	SPC100
Soy protein concentrate [1]	0	11.5	23.1	34.7	46.3
Fish meal [1]	45	33.75	22.5	11.25	0
Soybean meal [1]	12	12	12	12	12
corn gluten meal [1]	13	13	13	13	13
Cassava starch	5	5	5	5	5
Wheat flour	5	5	5	5	5
Rice bran	3.36	3.36	3.36	3.36	3.36
Microcrystalline cellulose	9.04	6.904	4.566	2.285	0
Fish oil	3.9	4.5	5.1	5.7	6.3
Vitamins premix [2]	1	1	1	1	1
Mineral premix [2]	1	1	1	1	1
Calcium dihydrogen phosphate	1.2	2.2	3.2	4.2	5.2
Choline chloride	0.5	0.5	0.5	0.5	0.5
Lysine	0	0.169	0.335	0.5	0.666
Methionine	0	0.117	0.234	0.35	0.467
Threonine	0	0	0.057	0.085	0.113
Valine	0	0	0.048	0.07	0.093
Proximate analysis (dry basis)					
Crude protein (%)	47.21	47.09	47.05	46.82	46.83
Crude lipids (%)	9.42	9.92	9.94	9.91	9.87
Gross energy (kJ/g)	17.13	16.98	17.01	17.08	16.99

Note: [1] The crude protein and crude lipids levels of the raw material are shown below, respectively. Fish meal, 67.63% and 9.46% g; soy protein concentrate, 63% and 4.1%; soybean meal, 53.26% and 4.25%; corn gluten meal, 59.24% and 3.3%. These materials were obtained from Wuxi Tongwei feedstuffs Co., Ltd. (Wuxi, China) [2] Mineral premix and vitamins premix were obtained from HANOVE Animal Health Products (Wuxi, China).

2.2. Culture Experiment

Largemouth bass was procured from the Freshwater Fisheries Research Center (FFRC) followed by the culture experiment at the FFRC base. The fish were acclimatized for two weeks before the breeding experiment. After 24 h of fasting, 300 fish (17.03 ± 0.01 g) were divided into 15 cages with 20 fish per cage by the randomness principle. They were fed twice a day (at 6:30 and 18:30) until they no longer surfaced to feed. The water quality conditions were water temperature of 24–29 °C with dissolved oxygen ≥6 mg/L (pH 7.4–8.0), respectively.

2.3. Sample Collection

At day 56 of the breeding experiment, fish were fasted for 24 h followed by counting and weighing to determine the growth parameters. Three fish were randomly selected in each cage and anesthetized (100 mg/L MS-222) to collect the liver and intestine samples, which were stored separately in cryopreservation tubes and immediately placed in liquid nitrogen for freezing. Thereafter, the samples were kept in the refrigerator at −80 °C for further use. In addition, the intestine tissues were fixed with 4% paraformaldehyde for hematoxylin and eosin (HE) staining analysis.

2.4. Experimental Determination Method

The experimental feed composition and whole fish components were measured as described by AOAC [38] and our previous study [39]. Briefly, the sample was dried in an oven at 105 °C to a constant weight for testing the moisture level, and the dried sample was ground into a powder for further analysis. The crude protein was quantified by the Kjeldahl nitrogen determination method on an automatic instrument (Haineng K1100, Jinan Haineng Instrument Co., Ltd., Jinan, China). The crude lipid in the sample was extracted by the Soxhlet extraction method in an automatic fat analyzer (Haineng SOX606, Jinan Haineng Instrument Co., Ltd., China). The ash content was analyzed at 550 °C for 5 h by burning in a Muffle furnace (XL-2A, Hangzhou Zhuochi Instrument Co., Ltd., Hangzhou, China). Additionally, energy in the feed was measured using an oxygen bomb calorimeter (IKA C6000, Stauffen, Germany). The activities of intestinal antioxidant enzymes and the levels of MDA and glutathione (GSH) were detected using a kit following our previous method [40]. In brief, the activities of superoxide dismutase (SOD), CAT, and GSH-Px were tested by hydroxylamine method, ammonium molybdenum acid method, and colorimetric method, respectively. The levels of GSH and MDA were tested by microplate method and thiobarbituric acid method, respectively. Assay kits purchased from Jian Cheng Bioengineering Institute (Nanjing, China).

The intestinal tissue samples were fixed in 4% paraformaldehyde for more than 48 h. After dehydration in the alcohol gradient, samples were embedded in wax, followed by sectioning and cooling at −20 °C in a refrigerator. The frozen sections were brought to room temperature, and the section of the intestine was performed through the following steps: fixation by 4% paraformaldehyde, dehydration by gradient alcohol and methyl salicylate clearing, paraffin embedding, slicing, staining, etc. Finally, pathological changes in intestine were analyzed with a Zeiss microscope (Axioplan 2, Oberkochen, Germany). The specific method can be found in our previous study [41].

Next, RNA was extracted from the liver and intestine samples by RNAiso Plus (Vazyme, Nanjing, China) reagent. The $A_{260/280}$ value of 1.8–2.0 served as a standard for further experiments by using the NanoDrop 2000 spectrophotometer. The primers were synthesized by Shengong Bioengineering Co., LTD (Shanghai, China). CFX96 Touch (Bio-Rad, Singapore) was used for quantitative real-time PCR detection. The *β-actin* gene was selected as the reference gene to calculate mRNA levels using the standard curve method [42] and the gene expression levels were further standardized. The primers for gene amplification are shown in Table 2.

Table 2. Experimental primer.

Genes	Forward (5′-3′)	Reverse (5′-3′)	Primer Source
tor [1]	TTTGGAACCAAACCCCGTCA	ATCAGCTCACGGCAGTATCG	XM_038723321.1
rps6 [2]	TCCAGAGACTCGTGACACCT	AGCTTGGCATACTCTGAGGC	XM_038713349.1
4ebp1 [3]	CCAGGATCATCTATGACCGAAAG	TGCAGCGATATTGTTGTTGTTC	XM_038703879.1
fas [4]	AGTTGAAGGCTGCTGATG	GCTGTGGATGATGTTGGT	XP_028423094.1
acc [5]	TTACATCGCAGCCAACAG	CTCTCCACCTTCCTCTACA	XP_022609673.1
scd [6]	CGATGCTGCTTCTTCACT	GACACGGTTCTGCCATTA	XM_038735580.1
cpt [7]	TTACCGTATGGCTATGACTG	GGCTCCGATAACACCTCT	XP_027141042.1

Table 2. *Cont.*

Genes	Forward (5′-3′)	Reverse (5′-3′)	Primer Source
pparγ [8]	GAGTTCTCAGTCAAGTTCAAC	AATGTAGCACCGTCTCCT	MK614721.1
gk [9]	CCCTTGTGGGCAGGAGAAAA	ACAACTGAGTCCTCCTTGCG	XP_023260296.1
pk [10]	CACGCAACACTGGCATCATC	TCGAAGCTCTCACATGCCTC	MT431526.1
pepck [11]	GGCAAAACCTGGAAGCAAGG	ATAATGGCGTCGATGGGGAC	MT431525.1
nrf2 [12]	CCACACGTGACTCTGATTTCTC	TCCTCCATGACCTTGAAGCAT	Transcriptome data
cat [13]	CTATGGCTCTCACACCTTC	TCCTCTACTGGCAGATTCT	MK614708.1
sod [14]	CCCCACAACAAGAATCATGC	TCTCAGCCTTCTCGTGGA	MK614709.1
gsh-px [15]	ATGGCTCTCATGACTGATCCAAA	GACCAACCAGGAACTTCTCAAA	MK614713.1
keap1 [16]	GCACCTAACCGTGGAACTCAA	CCAGTTTTAGCCAGTCATTGTTCC	[43]
nf-κb [17]	AGAAGACGACTCGGGGATGA	GCTTCTGCAGGTTCTGGTCT	[44]
il-8 [18]	GAGGGTACATGTCTGGGGGA	CCTTGAAGGTTTGTTCTTCATCGT	XM_038713529.1
tnf-α [19]	CTTCGTCTACAGCCAGGCATCG	TTTGGCACACCGACCTCACC	[36]
il-10 [20]	CGGCACAGAAATCCCAGAGC	CAGCAGGCTCACAAAATAAACATCT	[36]
β-actin	ATGCAGAAGGAGATCACAGCCT	AGTATTTACGCTCAGGTGGGG	AF253319.1

Note: [1] *tor*, target proteins rapamycin; [2] *rps6*, ribosomal protein S6 kinase; [3] *4ebp1*, eukaryotic initiation factor 4E-binding protein 1; [4] *fas*, fatty acid synthetase; [5] *acc*, acetyl-CoA carboxylase; [6] *scd*, stearoyl-CoA desaturase; [7] *cpt*, carnitine palmitoyl transferase; [8] *pparγ*, peroxisome proliferator-activated receptor-γ; [9] *gk*, glucokinase; [10] *pk*, pyruvate kinase; [11] *pepck*, phosphoenolpyruvate carboxylase; [12] *nrf2*, nuclear factor E2 related factor 2; [13] *cat*, catalase; [14] sod, superoxide dismutase; [15] *gsh-px*, glutathione peroxidase; [16] *keap1*, Kelch-like ECH associated protein 1; [17] *nf-κb*, nuclear factor kappa B; [18] *il-8*, interleukin-8; [19] *tnf-α* tumor necrosis factor α; [20] *il-10*, interleukin-10.

2.5. Data Analysis

SPSS (20.0) was used for one-way ANOVA, and the method of Tukey was used to analyze the significant difference among all groups ($p < 0.05$). Results were expressed as mean ± standard error, with different superscript letters representing significant differences ($p < 0.05$).

3. Results

3.1. Growth Performance

Table 3 shows the results of growth performance. Various growth parameters including FW, WGR, and SGR were not significantly different among SPC0, SPC25, and SPC50 groups ($p > 0.05$); however, they were significantly higher than those in SPC75 and SPC100 groups ($p < 0.05$). Conversely, significantly lower FCR was determined for the SPC0, SPC25, and SPC50 groups compared with that for the SPC100 group ($p < 0.05$); specifically, no significant difference among SPC0, SPC25, and SPC50 groups was found ($p > 0.05$). Moreover, compared with SPC75 and SPC100 groups, a significantly higher FI and SR was observed in the SPC0 group ($p < 0.05$).

Table 3. Growth performance.

Groups	IW (g) [1]	FW (g) [2]	WGR (%) [3]	SGR (%/d) [4]	FCR [5]	FI (g fish^{-1}d^{-1}) [6]	SR (%) [7]
SPC0	17.07 ± 0.02	58.38 ± 0.13 ^{c}	242.08 ± 0.91 ^{c}	2.20 ± 0.00 ^{c}	1.40 ± 0.03 ^{a}	0.52 ± 0.00 ^{c}	96.67 ± 1.67 ^{c}
SPC25	17.03 ± 0.02	57.96 ± 0.75 ^{c}	240.25 ± 4.17 ^{c}	2.19 ± 0.02 ^{c}	1.41 ± 0.04 ^{a}	0.51 ± 0.01 ^{bc}	95.00 ± 2.89 ^{bc}
SPC50	17.02 ± 0.02	57.18 ± 1.95 ^{c}	236.06 ± 11.71 ^{c}	2.16 ± 0.06 ^{c}	1.54 ± 0.04 ^{a}	0.50 ± 0.01 ^{bc}	88.33 ± 1.67 ^{abc}
SPC75	17.03 ± 0.02	51.07 ± 0.63 ^{b}	199.84 ± 3.66 ^{b}	1.96 ± 0.02 ^{b}	1.76 ± 0.12 ^{ab}	0.48 ± 0.00 ^{ab}	85.00 ± 2.89 ^{ab}
SPC100	17.02 ± 0.02	40.78 ± 1.31 ^{a}	139.65 ± 7.76 ^{a}	1.56 ± 0.06 ^{a}	1.99 ± 0.13 ^{b}	0.45 ± 0.01 ^{a}	83.33 ± 1.67 ^{a}

Note: All data are expressed as mean ± standard error. Means in the same column with different superscripts are significantly different ($p < 0.05$). [1] IW, initial average weight. [2] FW, final average weight. [3] Weight gain rate (WGR, %) = 100 × (final body weight (g) − initial body weight (g))/initial body weight (g). [4] Specific growth rate (SGR, %/d) = 100 × ((Ln (final body weight (g)) − Ln (initial body weight (g)))/days). [5] Feed conversion ratio (FCR) = dry feed fed (g)/wet weight gain (g). [6] Feed intake rate (FI, g fish^{-1}d^{-1}) = dry feed fed (g)/((total initial weight (g) + total final weight (g))/2/days). [7] Survival rate (SR, %) = 100 × (survival fish number/total fish number).

3.2. Whole Fish Composition

The results of whole fish composition revealed no significant differences in the whole fish composition in the groups ($p > 0.05$; Table 4).

Table 4. Whole fish composition.

Groups	Moisture (%)	Ash (%)	Lipids (%)	Protein (%)
SPC0	71.72 ± 0.33	3.49 ± 0.12	6.94 ± 0.23	16.59 ± 0.35
SPC25	71.08 ± 0.32	4.18 ± 0.31	6.45 ± 0.32	16.59 ± 0.37
SPC50	71.64 ± 0.21	4.42 ± 0.36	6.47 ± 0.78	16.92 ± 2.22
SPC75	71.81 ± 0.29	4.14 ± 0.42	5.69 ± 0.22	16.43 ± 2.11
SPC100	72.42 ± 0.29	4.18 ± 0.30	5.05 ± 0.21	16.27 ± 1.62

Note: All data are expressed as mean ± standard error.

3.3. The mRNA Expression of Protein Metabolism-Related Genes in the Liver

Results revealed no significant differences between SPC0, SPC25, and SPC50 groups in the mRNA levels of *tor*, *rps6*, and *4ebp1* ($p > 0.05$) (Figure 1a–c); however, significantly lower mRNA levels of those were found in SPC75 and SPC100 compared with the other groups ($p < 0.05$) (Figure 1a–c).

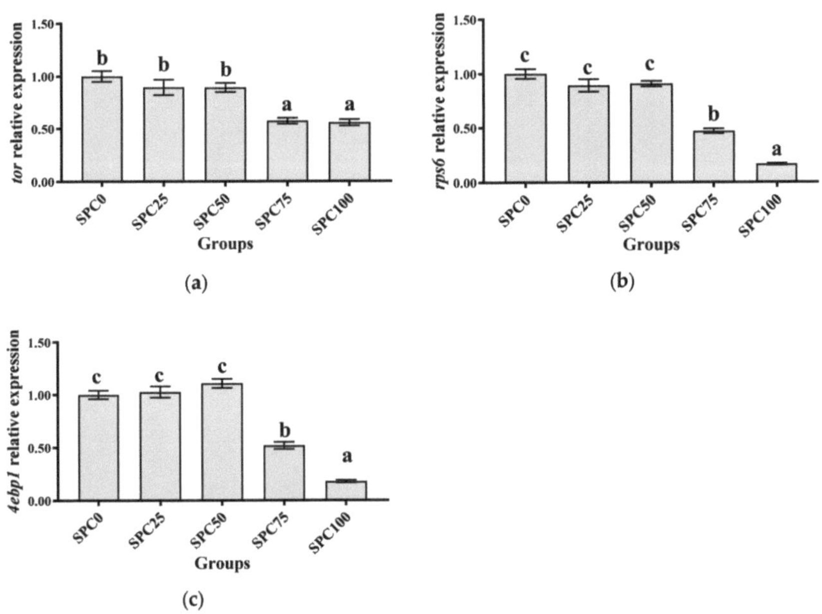

Figure 1. The mRNA expression of protein metabolism-related genes in the liver: (**a**) *tor*; (**b**) *rps6*; (**c**) *4ebp1*. The different letters of means are significantly different.

3.4. The mRNA Expression of Lipid and Glucose Metabolism-Related Genes in the Liver

The SPC0, SPC25, and SPC50 groups revealed no significant differences in the mRNA levels of *ppar*γ and *fas* ($p > 0.05$), which were significantly lower than those in the other groups ($p < 0.05$) (Figure 2a,b). Additionally, the mRNA levels of *cpt* were significantly higher in SPC0, SPC25, and SPC50 groups than in SPC75 and SPC100 groups ($p < 0.05$) (Figure 2c). Moreover, the mRNA levels of *scd* and *acc* remained unchanged for all the groups ($p > 0.05$) (Figure 2d,e). Replacement of FM with SPC did not significantly affect the mRNA levels of *gk*, *pk*, and *pepck* ($p > 0.05$) (Figure 2f–h).

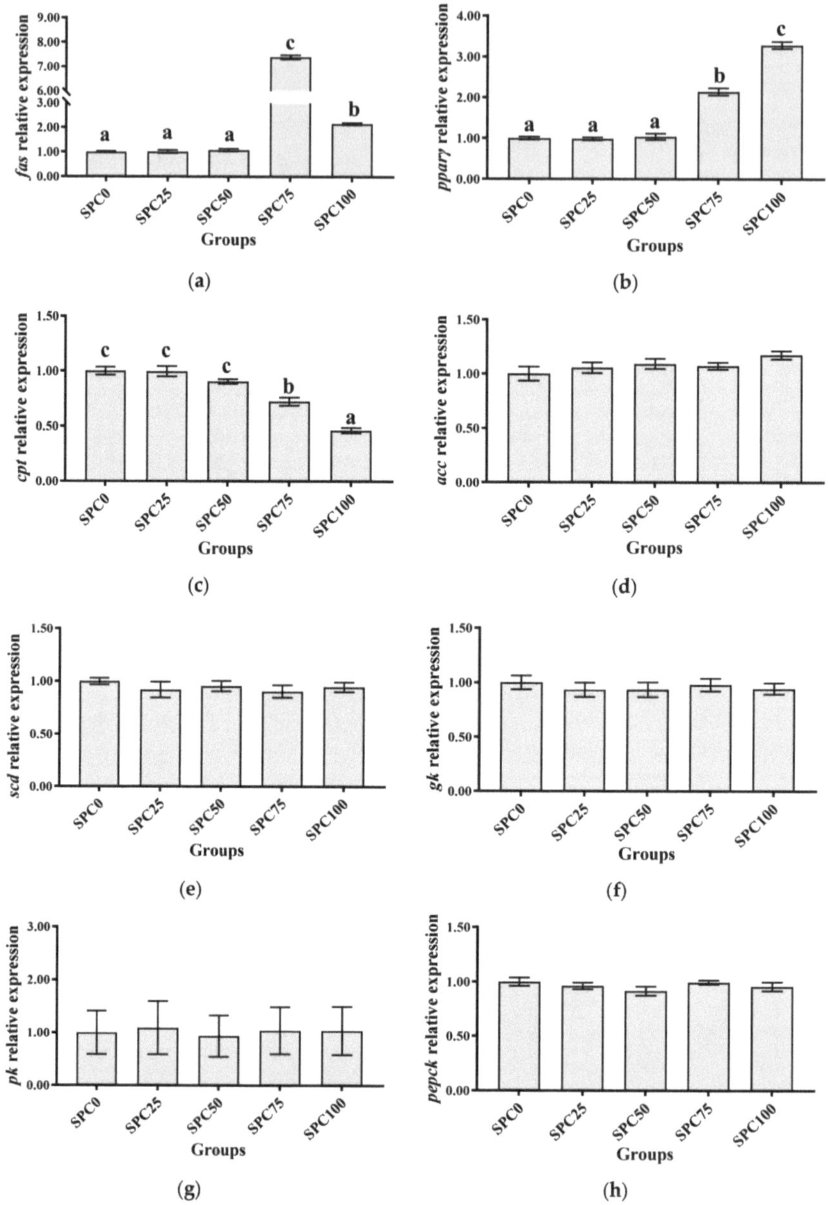

Figure 2. The mRNA expression of lipid and glucose metabolism-related genes in the liver: (**a**) *fas*; (**b**) *pparγ*; (**c**) *cpt*; (**d**) *acc*; (**e**) *scd*; (**f**) *gk*; (**g**) *pk*; (**h**) *pepck*. The different letters of means are significantly different.

3.5. Intestinal Antioxidant Parameters

Compared with the SPC0 group, significantly decreased activities of CAT were observed in the SPC50, SPC75, and SPC100 groups ($p < 0.05$) (Figure 3a), and significantly decreased activities of GSH-Px were observed in the SPC75 and SPC100 groups ($p < 0.05$) (Figure 3b). In addition, significantly lower activity of SOD was observed in SPC100 compared with the other groups ($p < 0.05$) (Figure 3c). Moreover, compared with the other groups, the SPC75 and SPC100 groups had significantly decreased and increased contents of GSH and MDA, respectively ($p < 0.05$) (Figure 3d,e).

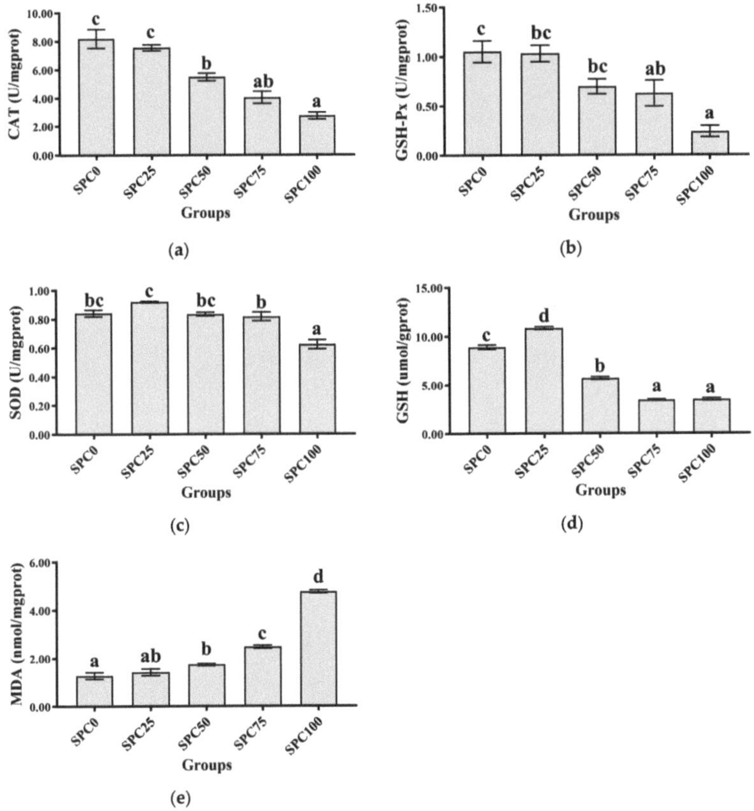

Figure 3. Intestinal antioxidant parameters: (**a**) CAT; (**b**) GSH-Px; (**c**) SOD; (**d**) GSH; (**e**) MDA. The different letters of means are significantly different.

3.6. The mRNA Expression of Antioxidant Genes in the Intestine

Compared with the SPC0 group, significantly lower mRNA levels of *nrf2*, *cat*, *sod*, and *gsh-px* were found in SPC50, SPC75, and SPC100 ($p < 0.05$) (Figure 4a,c–e); however, significantly higher mRNA levels of *keap1* were observed in SPC75 and SPC100 groups ($p < 0.05$) (Figure 4b).

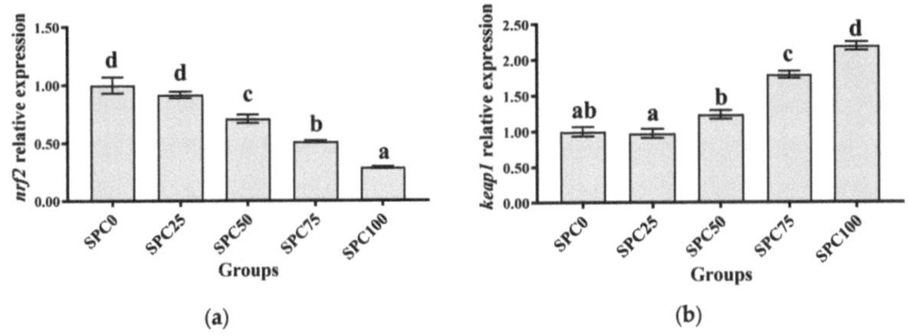

Figure 4. *Cont.*

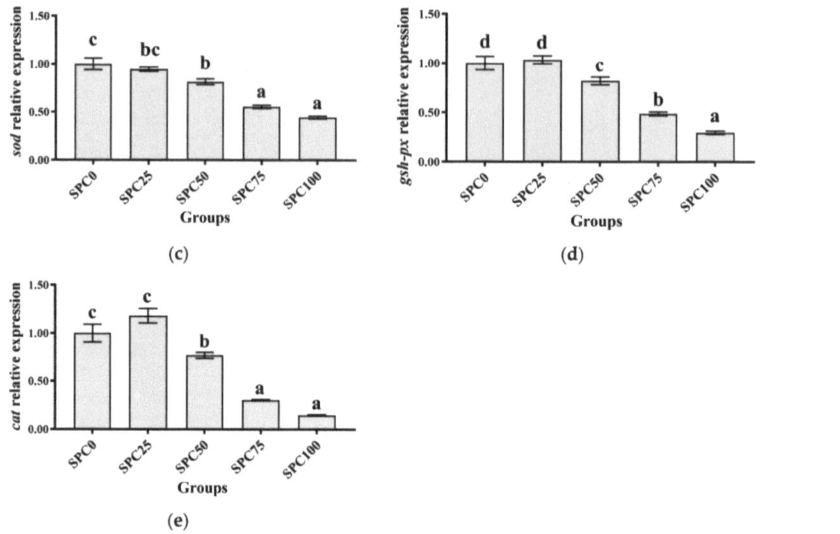

Figure 4. The mRNA expression of antioxidant genes in the intestine: (**a**) *nrf2*; (**b**) *keap1*; (**c**) *sod*; (**d**) *gsh-px*; (**e**) *cat*. The different letters of means are significantly different.

3.7. The mRNA Expression of Inflammatory Response-Related Genes in the Intestine

Compared with the SPC0 group, significantly higher mRNA levels of *il-8* and *nf-κb* were found in the SPC50, SPC75, and SPC100 groups ($p < 0.05$) (Figure 5a–c). Conversely, significantly lower mRNA levels of *il-10* and significantly higher mRNA levels of *tnf-α* were found in the SPC75 and SPC100 groups compared with the other groups ($p < 0.05$) (Figure 5d).

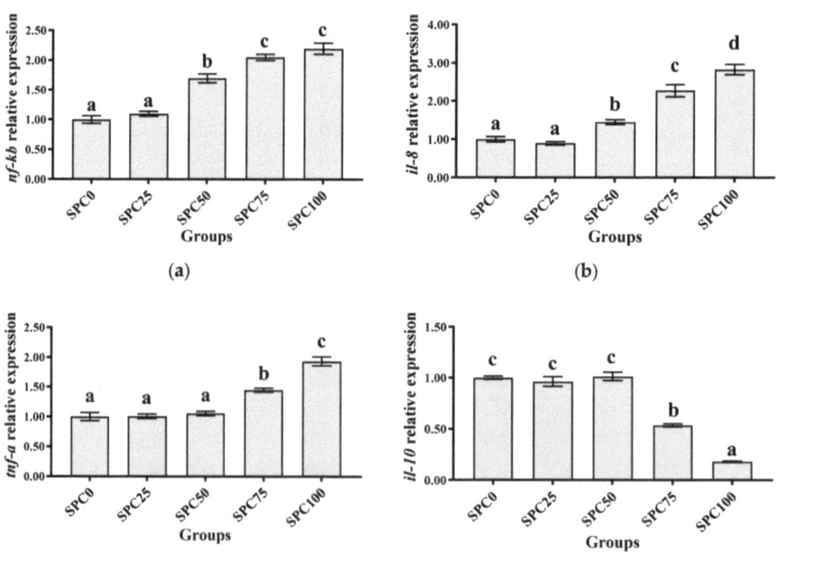

Figure 5. The mRNA expression of inflammatory response-related genes in the intestine: (**a**) *nf-κb*; (**b**) *il-8*; (**c**) *tnf-α*; (**d**) *il-10*. The different letters of means are significantly different.

3.8. Intestinal Morphology

Compared with the SPC0 group, mucosal thickness and villus height were significantly decreased in the SPC75 and SPC100 groups ($p < 0.05$; Figure 6).

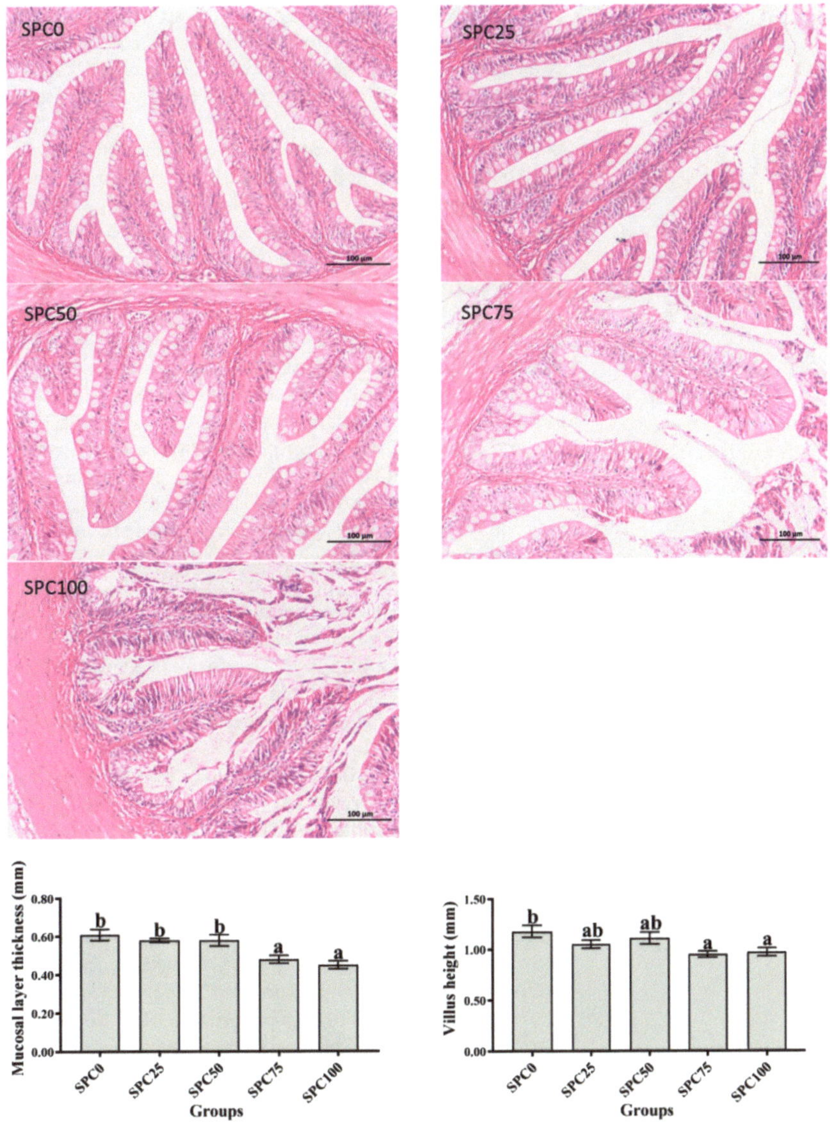

Figure 6. Intestinal morphology. The different letters of means are significantly different.

4. Discussion

In the study, varying concentrations of SPC (0%, 25%, and 50%) did not affect growth-performance-related parameters, including FW, WGR, SGR, FCR, FI, and SR. This is consistent with the previous findings of a study on largemouth bass by Cui et al. [45]. Metochis et al. [46] found that replacing 35% of FM with SPC improved the growth performance of Atlantic salmon significantly, indicating the potential of SPC in replacing FM in fish feed. However, when SPC replaced 60% FM in the feed, the growth was significantly decreased, while the FCR was significantly increased [47]. Our study has shown similar results where significant inhibition of the growth and feed utilization of juvenile largemouth bass was observed when SPC replaced 75% or higher proportion of FM. Therefore, it is evident that only a specific proportion of FM should be replaced with SPC as excessive SPC negatively impacts fish growth, feed utilization, and SR. In addition, no significant difference was observed in the whole body composition between the groups, which was consistent with the previous studies of largemouth bass [45].

Protein source replacement of FM can affect the protein synthesis in largemouth bass [41,48]. Hay and Sonenberg [49] showed that the TOR signaling pathway affects protein synthesis and controls cell growth. Our results demonstrated that SPC75 and SPC100 groups inhibited the mRNA levels of *tor*, *rps6*, and *4ebp1* in the liver relative to other groups, suggesting that at 75% or higher proportion of SPC replacing FM, the expression of the TOR signaling pathway-related genes would be inhibited, thereby reducing the protein synthesis. Similar results were demonstrated in previous studies when high levels of plant protein supplementation in feed reduced mRNA levels of core genes of the TOR signaling pathway [50]. The inhibition of the TOR signaling pathway and protein synthesis could also describe the decreased growth of largemouth bass after feeding SPC75 and SPC100 diets. In addition, FM replacement with other protein sources will affect the lipid metabolism in fish [51]. Our study showed that the mRNA levels of *fas* and *ppar*γ (related to lipid synthesis) were significantly upregulated in the SPC75 and SPC100 groups, while those of *cpt* (related to lipid decomposition) were significantly downregulated. Alternatively, the replacement of FM with 75% or more SPC might limit lipolysis in largemouth bass by upregulating lipid synthesis and downregulating lipolysis-related genes, thereby causing fat accumulation. In Japanese perch (*Lateolabrax japonicas*), feeding a whole plant protein diet could cause fatty liver [52]. In summary, this study revealed that a 75% or higher proportion of SPC replacing FM significantly downregulated the expression of genes related to protein anabolism and lipolysis, thereby affecting the nutritional metabolic capacity of fish and inducing liver metabolic disorders. This might be the possible reason for the significantly decreased growth performance and feed utilization of largemouth bass fed with SPC75 and SPC100 diets. Previous studies on red sea bream (*Pagrus major*) have also reported that FM-free feed with SPC resulted in poor nutrient utilization and growth inhibition in fish [53].

Furthermore, GSH-Px, CAT, and SOD are the main antioxidant enzymes in fish [54]. GSH balances redox reactions by eliminating excess ROS, thereby protecting cells from oxidative stress [55]. Our results showed that significantly decreased activities of CAT were observed in the SPC50, SPC75, and SPC100 groups, and significantly decreased activities of GSH-Px and contents of GSH were observed in the SPC75 and SPC100 groups. In addition, significantly lower activity of SOD was observed in SPC100 compared with the other groups. In addition, MDA, a lipid peroxidation product, is an important index to measure cell damage and biotoxicity [56]. Our results showed significantly increased levels of MDA in the groups of SPC75 and SPC100, which was consistent with the previous studies. SPC replacing 60% or higher proportion of FM decreased SOD activity in the serum of starry flounder, and GSH-Px activity was significantly decreased when SPC replaced 80% and 100% of FM, which also presented lower MDA level in serum [20]. Moreover, the activities of antioxidant enzymes and their related mRNA levels are correlated [57]. The Nrf2 system could regulate the expression of antioxidant genes in fish [58]. We observed that significantly lower mRNA levels of *nrf2*, *cat*, *sod*, and *gsh-px* were found in SPC50, SPC75, and SPC100 and significantly higher mRNA levels of *keap1* were observed in SPC75 and SPC100 groups. These results showed that SPC replacing 50% or more of FM could significantly decrease the intestinal antioxidant capacity of largemouth bass. In a similar study, when SPC replaced 60% of FM, the antioxidant capacity was significantly restrained in golden crucian carp [13]. In addition, our results showed that replacing 50% of FM with SPC did not affect SOD and GSH-Px activities in the intestine of largemouth bass; however, at a replacement ratio of 25%, GSH levels were significantly increased, which might be related to the presence of soybean isoflavones in SPC. Yang et al. [59] found that appropriate supplementation level of soybean isoflavones to the feed could increase the levels of GSH, SOD, and CAT in the muscle of grass carp while downregulating and upregulating the mRNA levels of *keap1* and *nrf2*, respectively. Zhou et al. [60] also found that feed supplemented with an appropriate concentration of soy isoflavones could increase the SOD and CAT activities in golden pompano.

Conversely, the decreased FM content in feed can reduce the antioxidant capacity of fish, leading to intestinal inflammation [52]. Our results demonstrated that significantly higher mRNA levels of *il-8* and *nf-κb* were found in the SPC50, SPC75, and SPC100 groups and significantly lower mRNA levels of *il-10* and significantly higher mRNA levels of *tnf-α* were found in the SPC75 and SPC100 groups. These results indicated that high levels of replacement FM through SPC might cause inflammation. Similarly, in pearl gentian groupers, replacement of 20% FM with SPC did not affect *il-10* and *tnf-α* significantly; however, significantly lower mRNA levels of anti-inflammatory factors and higher levels of proinflammatory factors were observed at 40% SPC [31]. Zhang et al. [52] found that Japanese seabass fed on whole plant protein feed caused inflammation. Chen et al. [61] also reported that 75% FM replaced with SPC in the diet of hybrid grouper caused inflammation in the fish intestines. Collectively, these findings suggested that a large proportion of SPC replacing FM might lead to an intestinal inflammatory response in largemouth bass, causing intestinal damage. Moreover, the morphology and structure of the intestine are critical for maintaining normal function and nutrient absorption [62–64]. Our results did not show any changes in the intestinal mucosal thickness and villus height in the SPC0, SPC25, and SPC50 groups; however, significant decreases in these parameters were found in the SPC75 and SPC100 groups relative to the SPC0 group. This might have resulted from the lack of active substances such as small peptides [65], which are abundantly present in FM and necessary for intestinal development and health; therefore, a significant reduction in FM decreased villous height in the intestines of largemouth bass. However, our results showed that reduced intestinal mucosal thickness and villi height in the SPC75 and SPC100 groups might cause malabsorption of nutrients in fish, thereby leading to slow growth of largemouth bass.

5. Conclusions

SPC replacing 50% FM in the feed did not affect growth. Furthermore, SPC replacing 75% or more FM could inhibit protein synthesis by lowing the gene expressions of *tor*, *rps6*, and *4ebp1*, and might limit lipolysis in largemouth bass by upregulating lipid synthesis and downregulating lipolysis-related genes including *pparγ*, *fas*, and *cpt*, while there was no effect on the expression of genes related to glucose metabolism including *gk*, *pk*, and *pepck*. However, SPC replacing 50% or more FM might inhibit antioxidant capacity and immune capacity by regulating antioxidant enzyme activity and gene expressions. Furthermore, SPC replacing 75% or more FM could reduce the thickness of intestinal mucosa and villus height.

Author Contributions: Formal analysis, H.L. and H.Y.; writing—original draft preparation, H.L.; writing—review and editing, M.R. and L.Z.; project administration, M.R.; methodology, H.M. and T.T.; investigation, D.H. and J.G. All authors have read and agreed to the published version of the manuscript.

Funding: This study was financially supported by the National Key R & D Program of China (2023YFD2400601), the earmarked fund for CARS (CARS-46), the National Natural Science Foundation of China (32102806), and Central Public-interest Scientific Institution Basal Research Fund, Freshwater Fisheries Research Center, CAFS (NO. 2024JBFR01).

Institutional Review Board Statement: The study was conducted according to Management Rule of Laboratory Animals (Chinese Order No. 676 of the State Council, revised 1 March 2017). The study was approved by the Laboratory Animal Ethics Committee of the Freshwater Fisheries Research Center (LAECFFRC-2023-05-13).

Informed Consent Statement: Not applicable.

Data Availability Statement: Data are contained within the article.

Conflicts of Interest: Lu Zhang, Haifeng Mi, and Tao Teng are employed by Tongwei Agricultural Development Co., Ltd. The remaining authors declare that the research was conducted in the absence of any commercial or financial relationships that could be construed as potential conflicts of interest. Lu Zhang made important contributions to the revision of the draft manuscript. Haifeng Mi and Tao Teng made important contributions to experimental technique.

References

1. Boyd, C.E.; McNevin, A.A.; Davis, R.P. The contribution of fisheries and aquaculture to the global protein supply. *Food Secur.* **2022**, *14*, 805–827. [CrossRef] [PubMed]
2. Salin, K.R.; Arun, V.V.; Mohanakumaran Nair, C.; Tidwell, J.H. Sustainable Aquafeed. In *Sustainable Aquaculture. Applied Environmental Science and Engineering for a Sustainable Future*; Hai, F., Visvanathan, C., Boopathy, R., Eds.; Springer: Cham, Switzerland, 2018. [CrossRef]
3. Gasco, L.; Gai, F.; Maricchiolo, G.; Genovese, L.; Ragonese, S.; Bottari, T.; Caruso, G. Fishmeal Alternative Protein Sources for Aquaculture Feeds. In *Feeds for the Aquaculture Sector*; Springer Briefs in Molecular Science; Springer: Cham, Switzerland, 2018. [CrossRef]
4. Tacon, A.G.J.; Metian, M. Global overview on the use of fish meal and fish oil in industrially compounded aquafeeds: Trends and future prospects. *Aquaculture* **2008**, *285*, 146–158. [CrossRef]
5. Glencross, B.D.; Booth, M.; Allan, G.L. A feed is only as good as its ingredients—A review of ingredient evaluation strategies for aquaculture feeds. *Aquac. Nutr.* **2007**, *13*, 17–34. [CrossRef]
6. Olsen, R.L.; Hasan, M.R. A limited supply of fishmeal: Impact on future increases in global aquaculture production. *Trends Food Sci. Technol.* **2012**, *27*, 120–128. [CrossRef]
7. FAO. *The State of World Fisheries and Aquaculture-Sustainability in Action in the State of World Fisheries and Aquaculture (SOFIA)*; FAO: Rome, Italy, 2020.
8. Hardy, R.W. Utilization of plant proteins in fish diets: Effects of global demand and supplies of fishmeal. *Aquac. Res.* **2010**, *41*, 770–776. [CrossRef]
9. Bauer, W.; Prentice-Hernandez, C.; Tesser, M.B.; Wasielesky, W., Jr.; Poersch, L.H.S. Substitution of fishmeal with microbial floc meal and soy protein concentrate in diets for the pacific white shrimp *Litopenaeus vannamei*. *Aquaculture* **2012**, *342–343*, 112–116. [CrossRef]
10. Mambrini, M.; Roem, A.J.; Carvèdi, J.P.; Lallès, J.P.; Kaushik, S.J. Effects of replacing fish meal with soy protein concentrate and of DL-methionine supplementation in high-energy, extruded diets on the growth and nutrient utilization of rainbow trout, *Oncorhynchus mykiss*. *J. Anim. Sci.* **1999**, *77*, 2990–2999. [CrossRef]
11. Lusas, E.W.; Riaz, M.N. Soy protein products: Processing and use. *J. Nutr.* **1995**, *125* (Suppl. 3), 573S–580S. [PubMed]
12. Zhang, J.Z.; Zhong, L.; Peng, M.; Chu, W.Y.; Liu, Z.P.; Dai, Z.Y.; Hu, Y. Replacement of fish meal with soy protein concentrate in diet of juvenile rice field eel *Monopterus albus*. *Aquac. Rep.* **2019**, *15*, 100235. [CrossRef]
13. Zhu, R.; Li, L.; Li, M.; Yu, Z.; Wang, H.H.; Wu, L.F. The effects of substituting fish meal with soy protein concentrate on growth performance, antioxidant capacity and intestinal histology in juvenile golden crucian carp, *Cyprinus carpio* × *Carassius auratus*. *Aquac. Rep.* **2020**, *18*, 100435. [CrossRef]
14. Yoo, G.; Choi, W.; Bae, J.; Yu, H.; Lee, S.; Bai, S.C. Effects on Growth and Body Composition to Soy Protein Concentrate as a Fishmeal Replacement in Coho Salmon *Oncorhynchus kisutch*. *Korean J. Fish. Aquat. Sci.* **2021**, *54*, 118–123. [CrossRef]
15. Wu, Y.; Han, H.; Qin, J.; Wang, Y. Replacement of fishmeal by soy protein concentrate with taurine supplementation in diets for golden pompano (*Trachinotus ovatus*). *Aquac. Nutr.* **2015**, *21*, 214–222. [CrossRef]
16. Hartviksen, M.; Bakke, A.M.; Vecino, J.G.; Ringø, E.; Krogdahl, Å. Evaluation of the effect of commercially available plant and animal protein sources in diets for Atlantic salmon (*Salmo salar* L.): Digestive and metabolic investigations. *Fish Physiol. Biochem.* **2014**, *40*, 1621–1637. [CrossRef]
17. Trejo-Escamilla, I.; Galaviz, M.A.; Flores-Ibarra, M.; González, C.A.Á.; López, L.M. Replacement of fishmeal by soya protein concentrate in the diets of *Totoaba macdonaldi* (Gilbert, 1890) juveniles: Effect on the growth performance, in vitro digestibility, digestive enzymes and the haematological and biochemistry parameters. *Aquac. Res.* **2017**, *48*, 4038–4057. [CrossRef]
18. Rossi, W.; Moxely, D.; Buentello, A.; Pohlenz, C.; Gatlin, D.M. Replacement of fishmeal with novel plant feedstuffs in the diet of red drum *Sciaenops ocellatus*: An assessment of nutritional value. *Aquac. Nutr.* **2013**, *19*, 72–81. [CrossRef]
19. Quintero, H.E.; Davis, D.A.; Rhodes, M.A. Soy protein concentrate as an alternative ingredient in Florida pompano (*Trachinotus carolinus*) diets. *J. Appl. Aquac.* **2012**, *24*, 247–261. [CrossRef]
20. Li, P.Y.; Wang, J.Y.; Song, Z.D.; Zhang, L.M.; Zhang, H.; Li, X.X.; Pan, Q. Evaluation of soy protein concentrate as a substitute for fishmeal in diets for juvenile starry flounder (*Platichthys stellatus*). *Aquaculture* **2015**, *448*, 578–585. [CrossRef]
21. Hien, T.T.T.; Phu, T.M.; Tu, T.L.C.; Tien, N.V.; Duc, P.M.; Bengtson, D.A. Effects of replacing fish meal with soya protein concentrate on growth, feed efficiency and digestibility in diets for snakehead, *Channa striata*. *Aquac. Res.* **2017**, *48*, 3174–3181. [CrossRef]
22. Freitas, L.E.L.; Nunes, A.J.P.; do Carmo Sá, M.V. Growth and feeding responses of the mutton snapper, *Lutjanus analis* (Cuvier 1828), fed on diets with soy protein concentrate in replacement of Anchovy fish meal. *Aquac. Res.* **2011**, *42*, 866–877. [CrossRef]
23. Deng, J.M.; Mai, K.S.; Ai, Q.H.; Zhang, W.B.; Wang, X.J.; Xu, W.; Liufu, Z.G. Effects of replacing fish meal with soy protein concentrate on feed intake and growth of juvenile Japanese flounder, *Paralichthys olivaceus*. *Aquaculture* **2006**, *258*, 503–513. [CrossRef]

24. Liu, X.; Chi, S.Y.; Li, S.; Cheng, X.L.; Gao, W.H.; Xu, Q.Q.; Zhang, W.B.; Zhou, X.Q. Substitution of fish meal with enzyme-treated soybean in diets for juvenile largemouth bass (*Micropterus salmoides*). *Aquac. Nutr.* **2021**, *27*, 1569–1577. [CrossRef]
25. Mace, T.A.; Ware, M.B.; King, S.A.; Loftus, S.; Farren, M.R.; McMichael, E.; Scoville, S.; Geraghty, C.; Young, G.; Carson, W.E., 3rd; et al. Soy isoflavones and their metabolites modulate cytokine-induced natural killer cell function. *Sci. Rep.* **2019**, *9*, 5068. [CrossRef]
26. Sakai, T.; Kogiso, M. Soy isoflavones and immunity. *J. Med. Investig.* **2008**, *55*, 167–173. [CrossRef] [PubMed]
27. Nagata, C. Soy intake and chronic disease risk: Findings from prospective cohort studies in Japan. *Eur. J. Clin. Nutr.* **2021**, *75*, 890–901. [CrossRef]
28. Bitzer, Z.T.; Wopperer, A.L.; Chrisfield, B.J.; Tao, L.; Cooper, T.K.; Vanamala, J.; Elias, R.J.; Hayes, J.E.; Lambert, J.D. Soy protein concentrate mitigates markers of colonic inflammation and loss of gut barrier function in vitro and in vivo. *J. Nutr. Biochem.* **2017**, *40*, 201–208. [CrossRef]
29. Urán, P.A.; Gonçalves, A.A.; Taverne-Thiele, J.J.; Schrama, J.W.; Verreth, J.A.J.; Rombout, J.H.W.M. Soybean meal induces intestinal inflammation in common carp (*Cyprinus carpio* L.). *Fish Shellfish Immunol.* **2008**, *25*, 751–760. [CrossRef] [PubMed]
30. Yin, B.; Liu, H.Y.; Tan, B.P.; Dong, X.H.; Chi, S.Y.; Yang, Q.H.; Zhang, S.; Chen, L.Q. Cottonseed protein concentrate (CPC) suppresses immune function in different intestinal segments of hybrid grouper ♀*Epinephelus fuscoguttatus* × ♂*Epinephelus lanceolatu* via TLR-2/MyD88 signaling pathways. *Fish Shellfish Immunol.* **2018**, *81*, 318–328. [CrossRef]
31. Zhang, W.; Tan, B.P.; Deng, J.M.; Yang, Q.H.; Chi, S.Y.; Pang, A.B.; Xin, Y.; Liu, Y.; Zhang, H.T. Soybean protein concentrate causes enteritis in juvenile pearl gentian groupers (*Epinephelus fuscoguttatus* ♀× *Epinephelus lanceolatus* ♂). *Anim. Nutr.* **2023**, *12*, 171–185. [CrossRef]
32. Wang, J.X.; Liang, D.Z.; Yang, Q.H.; Tan, B.P.; Dong, X.H.; Chi, S.Y.; Liu, H.Y.; Zhang, S. The effect of partial replacement of fish meal by soy protein concentrate on growth performance, immune responses, gut morphology and intestinal inflammation for juvenile hybrid grouper (*Epinephelus fuscoguttatus* ♀× *Epinephelus lanceolatus* ♂). *Fish Shellfish Immunol.* **2020**, *98*, 619–631. [CrossRef]
33. Bai, J.J.; Lutz-Carrillo, D.J.; Quan, Y.C.; Liang, S.X. Taxonomic status and genetic diversity of cultured largemouth bass *Micropterus salmoides* in China. *Aquaculture* **2008**, *278*, 27–30. [CrossRef]
34. Maezono, Y.; Miyashita, T. Community-level impacts induced by introduced largemouth bass and bluegill in farm ponds in Japan. *Biol. Conserv.* **2002**, *109*, 111–121. [CrossRef]
35. Huang, D.; Wu, Y.B.; Lin, Y.Y.; Chen, J.M.; Karrow, N.; Ren, X.; Wang, Y. Dietary protein and lipid requirements for juvenile largemouth bass, *Micropterus salmoides*. *J. World Aquac. Soc.* **2017**, *48*, 782–790. [CrossRef]
36. Gu, J.Z.; Liang, H.L.; Ge, X.P.; Xia, D.; Pan, L.K.; Mi, H.F.; Ren, M.C. A study of the potential effect of yellow mealworm (*Tenebrio molitor*) substitution for fish meal on growth, immune and antioxidant capacity in juvenile largemouth bass (*Micropterus salmoides*). *Fish Shellfish Immunol.* **2022**, *120*, 214–221. [CrossRef] [PubMed]
37. Liang, H.L.; Xu, G.C.; Xu, P.; Zhu, J.; Li, S.L.; Ren, M.C. Dietary histidine supplementation maintained amino acid homeostasis and reduced hepatic fat accumulation of juvenile largemouth bass, *Mieropterus salmoides*. *Aquac. Nutr.* **2022**, *2022*, 4034922. [CrossRef]
38. AOAC (Association of Official Analytical Chemists). *Official Methods of Analysis of the Association of Official Analytical Chemists*, 15th ed.; Association of Official Analytical Chemists: Washington, DC, USA, 2003.
39. Ren, M.C.; Habte-Tsion, H.M.; Liu, B.; Miao, L.H.; Ge, X.P.; Xie, J.; Liang, H.L.; Zhou, Q.L.; Pan, L.K. Dietary leucine level affects growth performance, whole body composition, plasma parameters and relative expression of TOR and TNF-α in juvenile blunt snout bream, *Megalobrama amblycephala*. *Aquaculture* **2015**, *448*, 162–168. [CrossRef]
40. Liang, H.L.; Xu, P.; Xu, G.C.; Zhang, L.; Huang, D.Y.; Ren, M.C.; Zhang, L. Histidine Deficiency Inhibits Intestinal Antioxidant Capacity and Induces Intestinal Endoplasmic-Reticulum Stress, Inflammatory Response, Apoptosis, and Necroptosis in Largemouth Bass (*Micropterus salmoides*). *Antioxidants* **2022**, *11*, 2399. [CrossRef] [PubMed]
41. Li, M.Y.; Liang, H.L.; Xie, J.; Chao, W.; Zou, F.Q.; Ge, X.P.; Ren, M.C. Diet supplemented with a novel *Clostridium autoethanogenum* protein have a positive effect on the growth performance, antioxidant status and immunity in juvenile Jian carp (*Cyprinus carpio* var. Jian). *Aquac. Rep.* **2021**, *19*, 100572. [CrossRef]
42. Stone, D.A.; Gaylord, T.G.; Johansen, K.A.; Overturf, K.; Sealey, W.M.; Hardy, R.W. Evaluation of the effects of repeated fecal collection by manual stripping on the plasma cortisol levels, TNF-α gene expression, and digestibility and availability of nutrients from hydrolyzed poultry and egg meal by rainbow trout, *Oncorhynchus mykiss* (Walbaum). *Aquaculture* **2008**, *275*, 250–259. [CrossRef]
43. Zhao, F.; Xu, P.; Xu, G.; Huang, D.; Zhang, L.; Ren, M.; Liang, H. Dietary valine affects growth performance, intestinal immune and antioxidant capacity in juvenile largemouth bass (*Micropterus salmoides*). *Anim. Feed Sci. Technol.* **2023**, *295*, 115541. [CrossRef]
44. Xv, Z.C.; He, G.L.; Wang, X.L.; Hao, S.; Chen, Y.J.; Lin, S.M. Mulberry leaf powder ameliorate high starch-induced hepatic oxidative stress and inflammation in fish model. *Anim. Feed Sci. Technol.* **2021**, *278*, 115012. [CrossRef]
45. Cui, Z.H.; Zhang, J.Y.; Ren, X.; Wang, Y. Replacing dietary fish meal improves ecosystem services of largemouth bass (*Micropterus salmoides*) farming. *Aquaculture* **2022**, *550*, 737830. [CrossRef]
46. Metochis, C.P.; Spanos, I.; Auchinachie, N.; Crampton, V.O.; Bell, J.G.; Adams, A.; Thompson, K.D. The effects of increasing dietary levels of soy protein concentrate (SPC) on the immune responses and disease resistance (furunculosis) of vaccinated and non-vaccinated Atlantic salmon (*Salmo salar* L.) parr. *Fish Shellfish Immunol.* **2016**, *59*, 83–94. [CrossRef] [PubMed]
47. Mohd Faudzi, N.; Yong, A.S.K.; Shapawi, R.; Senoo, S.; Biswas, A.; Takii, K. Soy protein concentrate as an alternative in replacement of fish meal in the feeds of hybrid grouper, brown-marbled grouper (*Epinephelus fuscoguttatus*) × giant grouper (*E. lanceolatus*) juvenile. *Aquac. Res.* **2018**, *49*, 431–441. [CrossRef]

48. Irm, M.; Taj, S.; Jin, M.; Luo, J.X.; Andriamialinirina, H.J.T.; Zhou, Q.C. Effects of replacement of fish meal by poultry by-product meal on growth performance and gene expression involved in protein metabolism for juvenile black sea bream (*Acanthoparus schlegelii*). *Aquaculture* **2020**, *528*, 735544. [CrossRef]
49. Hay, N.; Sonenberg, N. Upstream and downstream of mTOR. *Genes Dev.* **2004**, *18*, 1926–1945. [CrossRef] [PubMed]
50. Zhou, Q.L.; Habte-Tsion, H.M.; Ge, X.P.; Xie, J.; Ren, M.C.; Liu, B.; Miao, L.H.; Pan, L.K. Graded replacing fishmeal with canola meal in diets affects growth and target of rapamycin pathway gene expression of juvenile blunt snout bream, *Megalobrama amblycephala*. *Aquac. Nutr.* **2018**, *24*, 300–309. [CrossRef]
51. Peng, K.; Mo, W.Y.; Xiao, H.F.; Wang, G.X.; Huang, Y.H. Effects of black soldier fly pulp on growth performance, histomorphology and lipid metabolism gene expression of *Micropterus salmoides*. *Aquac. Rep.* **2021**, *20*, 100737. [CrossRef]
52. Zhang, Y.; Chen, P.; Liang, X.F.; Han, J.; Wu, X.F.; Yang, Y.H.; Xue, M. Metabolic disorder induces fatty liver in Japanese seabass, *Lateolabrax japonicas* fed a full plant protein diet and regulated by cAMP-JNK/NF-κB-caspase signal pathway. *Fish Shellfish Immunol.* **2019**, *90*, 223–234. [CrossRef]
53. Tola, S.; Fukada, H.; Masumoto, T. Effects of feeding a fish meal-free soy protein concentrate-based diet on the growth performance and nutrient utilization of red sea bream (*Pagrus major*). *Aquac. Res.* **2019**, *50*, 1087–1095. [CrossRef]
54. Kohen, R.; Nyska, A. Invited review: Oxidation of biological systems: Oxidative stress phenomena, antioxidants, redox reactions, and methods for their quantification. *Toxicol. Pathol.* **2002**, *30*, 620–650. [CrossRef]
55. Chakravarthi, S.; Jessop, C.E.; Bulleid, N.J. The role of glutathione in disulphide bond formation and endoplasmic-reticulum-generated oxidative stress. *EMBO Rep.* **2006**, *7*, 271–275. [CrossRef]
56. Parvez, S.; Raisuddin, S. Protein carbonyls: Novel biomarkers of exposure to oxidative stress-inducing pesticides in freshwater fish *Channa punctata* (Bloch). *Environ. Toxicol. Pharmacol.* **2005**, *20*, 112–117. [CrossRef] [PubMed]
57. Fontagné-Dicharry, S.; Lataillade, E.; Surget, A.; Larroquet, L.; Cluzeaud, M.; Kaushik, S. Antioxidant defense system is altered by dietary oxidized lipid in first-feeding rainbow trout (*Oncorhynchus mykiss*). *Aquaculture* **2014**, *424*, 220–227. [CrossRef]
58. Ma, Q. Role of nrf2 in oxidative stress and toxicity. *Annu. Rev. Pharmacol. Toxicol.* **2013**, *53*, 401–426. [CrossRef]
59. Yang, B.; Jiang, W.D.; Wu, P.; Liu, Y.; Zeng, Y.Y.; Jiang, J.; Kuang, S.Y.; Tang, L.; Tang, W.N.; Wang, S.W. Soybean isoflavones improve the health benefits, flavour quality indicators and physical properties of grass carp (*Ctenopharygodon idella*). *PLoS ONE* **2019**, *14*, e0209570. [CrossRef] [PubMed]
60. Zhou, C.P.; Lin, H.Z.; Ge, X.P.; Niu, J.; Wang, J.; Wang, Y.; Chen, L.X.; Huang, Z.; Yu, W.; Tan, X.H. The Effects of dietary soybean isoflavones on growth, innate immune responses, hepatic antioxidant abilities and disease resistance of juvenile golden pompano *Trachinotus ovatus*. *Fish Shellfish Immunol.* **2015**, *47*, 1043–1053. [CrossRef] [PubMed]
61. Chen, Y.; Liu, W.K.; Ma, J.; Wang, Y.R.; Huang, H. Comprehensive physiological and transcriptomic analysis revealing the responses of hybrid grouper (*Epinephelus fuscoguttatus* ♀× *E. lanceolatus* ♂) to the replacement of fish meal with soy protein concentrate. *Fish Physiol. Biochem.* **2020**, *46*, 2037–2053. [CrossRef]
62. Gao, Y.; Han, F.; Huang, X.; Rong, Y.; Wang, Y. Changes in gut microbial populations, intestinal morphology, expression of tight junction proteins, and cytokine production between two pig breeds after challenge with *Escherichia coli* K88: A comparative study. *J. Anim. Sci.* **2013**, *91*, 5614–5625. [CrossRef]
63. Fang, H.H.; Xie, J.J.; Liao, S.Y.; Guo, T.Y.; Xie, S.W.; Liu, Y.G.; Tian, S.Y.; Niu, J. Effects of dietary inclusion of shrimp paste on growth performance, digestive enzymes activities, antioxidant and immunological status and intestinal morphology of hybrid snakehead (*Channa maculata* ♀× *Channa argus* ♂). *Front. Physiol.* **2019**, *10*, 472899. [CrossRef]
64. Al-Fataftah, A.R.; Abdelqader, A. Effects of dietary Bacillus subtilis on heat-stressed broilers performance, intestinal morphology and microflora composition. *Anim. Feed Sci. Technol.* **2014**, *198*, 279–285. [CrossRef]
65. Yang, P.X.; Li, X.Q.; Song, B.W.; He, M.; Wu, C.Y.; Leng, X.J. The potential of *Clostridium autoethanogenum*, a new single cell protein, in substituting fish meal in the diet of largemouth bass (*Micropterus salmoides*): Growth, feed utilization and intestinal histology. *Aquac. Fish.* **2023**, *8*, 67–75. [CrossRef]

antioxidants

MDPI

Article

Optimizing Anesthetic Practices for Mud Crab: A Comparative Study of Clove Oil, MS-222, Ethanol, and Magnesium Chloride

Lulu Zhu [1,2,3,4], Shanshan Qi [1,2,3,4], Ce Shi [1,2,3,4,*], Shujian Chen [1,2,3,4], Yangfang Ye [2,3,4], Chunlin Wang [2,3,4], Changkao Mu [2,3,4], Ronghua Li [2,3,4], Qingyang Wu [2,3,4], Xiaopeng Wang [2,3,4] and Yueyue Zhou [2,3,4]

1 Marine Economic Research Center, Donghai Academy, Ningbo University, Ningbo 315000, China; zhululu0426@163.com (L.Z.); qishanshan0616@163.com (S.Q.); chenshujian@nbu.edu.cn (S.C.)
2 Key Laboratory of Aquacultural Biotechnology, Ningbo University, Chinese Ministry of Education, Ningbo 315000, China; yeyangfang@nbu.edu.cn (Y.Y.); wangchunlin@nbu.edu.cn (C.W.); muchangkao@nbu.edu.cn (C.M.); lironghua@nbu.edu.cn (R.L.); wuqingyang@nbu.edu.cn (Q.W.); wangxiaopeng@nbu.edu.cn (X.W.); zhouyueyue@nbu.edu.cn (Y.Z.)
3 Key Laboratory of Green Mariculture (Co-Construction by Ministry and Province), Ministry of Agriculture and Rural, Ningbo 315000, China
4 Collaborative Innovation Center for Zhejiang Marine High-Efficiency and Healthy Aquaculture, Ningbo 315000, China
* Correspondence: shice@nbu.edu.cn

Abstract: Anesthesia serves as an effective method to mitigate the stress response in aquatic animals during aquaculture and product transportation. In this study, we assessed the anesthetic efficacy of clove oil, tricaine methane-sulfonate (MS-222), ethanol, and magnesium chloride by anesthesia duration, recovery time, 24-hour survival rate, and the behavior of mud crabs (*Scylla paramamosain*). Additionally, the optimal anesthetic concentration for varying body weights of mud crabs was also investigated. The results revealed that clove oil emerged as the optimal anesthetic for mud crabs, with a 24-hour survival rate surpassing those observed in MS-222 and magnesium chloride treatments. Ethanol caused amputation and hyperactivity in mud crabs. Regression analyses between the optimal anesthetic concentration of clove oil and the weight categories of 0.03–27.50 g and 27.50–399.73 g for mud crabs yielded the following equations: $y = 0.0036\ x^3 - 0.1629\ x^2 + 1.7314\ x + 4.085$ ($R^2 = 0.7115$) and $y = 0.0437\ x + 2.9461$ ($R^2 = 0.9549$). Clove oil exhibited no significant impact on serum cortisol, glucose, lactate content, aspartate aminotransferase (AST), alanine aminotransferase (ALT) activities, or superoxide dismutase (SOD), catalase (CAT), glutathione peroxidase (GSH-Px), and malondialdehyde (MDA) levels in mud crabs across different treatment groups. Anesthesia induced by clove oil in mud crabs resulted in an increase in inhibitory neurotransmitters such as glycine. However, the recovery from anesthesia was associated with elevated levels of the excitatory neurotransmitters L-aspartic acid and glutamate. In conclusion, clove oil proves to be a safe and optimal anesthetic agent for mud crabs, exerting no physiological stress on the species.

Keywords: *Scylla paramamosain*; anesthetic; physiological response; antioxidant capacity; neurotransmitters; metabolomics

Citation: Zhu, L.; Qi, S.; Shi, C.; Chen, S.; Ye, Y.; Wang, C.; Mu, C.; Li, R.; Wu, Q.; Wang, X.; et al. Optimizing Anesthetic Practices for Mud Crab: A Comparative Study of Clove Oil, MS-222, Ethanol, and Magnesium Chloride. *Antioxidants* **2023**, *12*, 2124. https://doi.org/10.3390/antiox12122124

Academic Editor: Alessandra Napolitano

Received: 23 November 2023
Revised: 12 December 2023
Accepted: 14 December 2023
Published: 16 December 2023

1. Introduction

In aquaculture and the transportation of aquatic products, interventions such as measuring, vaccinating, handling, and classifying are essential for optimizing harvest outcomes. However, these processes can induce physical damage and physiological stress in aquatic animals, which increases susceptibility to pathogen infections and eventual mortality [1,2]. Consequently, mitigating damage and stress is crucial for enhancing farmed animals' survival rate and welfare [3]. Anesthetic use is a practical approach to reducing animal injury and stress resulting from human operations.

Anesthetics are broadly categorized into synthetic and natural types [4]. Ethanol, MS-222, and magnesium chloride are commonly used synthetic anesthetics in aquaculture research. Clove oil, a primary natural anesthetic in aquaculture, is renowned for its effectiveness, low risk of poisoning, antibacterial properties, and antioxidant effects [5]. Clove oil and its derivatives as anesthetic agents for decapod crustaceans have been investigated in other crustaceans for many years [6]. MS-222 is one of the most widely used anesthetics for poikilotherms worldwide [7]. Some studies have reported the anesthetic effect of MS-222 on crustaceans such as blue crabs (*Callinectes sapidus*), red swamp crayfish (*Procambarus clarkii*), and gammarids (*Gammarus pulex*) [8,9]. Ethanol, as an anesthetic in the neurotransmitter signaling pathway, could bind to specific receptors and promote the opening of ion channels, which leads to changes in the content of neurotransmitters, and it achieved anesthesia in crustaceans such as pacific white shrimp (*Litopenaeus vannamei*) and river prawn (*Macrobrachium tenellum*) [10–12]. Magnesium chloride was used in adult crayfish (*Astacus astacus, and Astacus leptodactilus*) and juvenile lobsters of European lobsters (*Homarus gammarus*) to investigate neural responses during stunning and killing [13]. The ideal properties of anesthetics are chemically stable, safe, effective, and without side effects [14]. Various factors influence their effectiveness and safety. These factors include abiotic elements such as temperature and salinity and biological factors like species, growth stage, and weight. For instance, transportation at high temperatures using magnesium chloride led to increased mortality in sea urchins (*Paracentrotus lividus*) [15]. Clove oil was only partially effective at the concentration of 20 mL L^{-1}, and MS-222 had no anesthetic effects on the Chinese mitten crab (*Eriocheir sinensis*) [16]. Thus, exploring the anesthetic effects of different anesthesia treatments with specific species under various conditions is necessary.

Assessing the efficacy of anesthesia involves observing behaviors such as active movement and aggression [17]. Clove oil, for example, induced struggling in cuttlefish but limited their control over arm movement, while magnesium chloride caused erratic behavior [18]. Behavioral analysis software has also been employed to quantitatively analyze movement characteristics in abalone, providing insights into the effects of environmental factors [19]. Moreover, metabolomics, particularly NMR-based metabolomics, is a powerful tool to identify metabolite changes influenced by environmental factors [20]. It has been used to analyze many small molecules involved in physiological and pathological processes, such as the metabolic response of swimming crab (*Portunus trituberculatus*) under dibutyl phthalate pollution [21]. In this study, NMR-based metabolomics analysis was applied to investigate metabolite changes in the nervous system of mud crabs under the influence of clove oil anesthesia.

The mud crab, a significant economic marine species, is widely distributed in the Indo-West-Pacific region, with *S. paramamosain* being the predominant aquaculture species in China [22]. Despite advancements in mud crab breeding technology, reliance on wild seedlings remains due to specific challenges in artificial breeding technology. Modern genetic breeding necessitates the collection and analysis of parental phenotype data. With its non-invasive and non-direct contact advantages, computer vision technology has been widely adopted for this purpose. However, the unpredictable nature of mud crab activity poses challenges, making anesthesia a valuable tool for improving data accuracy and efficiency while minimizing animal stress.

While anesthetics are routinely employed in scientific research on crustaceans and the methods to induce analgesia and anesthesia were recently reviewed [3], the field of mud crab anesthesia remains underexplored. This study sought to address this gap by evaluating the efficacy of clove oil, MS-222, ethanol, and magnesium chloride. The assessment, based on parameters such as anesthesia duration, recovery time, 24-hour survival rate, and observed behavior, provides valuable insights into the optimal use of anesthetics in mud crab aquaculture and holds significant implications for their cultivation and circulation. These findings contribute to the broader understanding of mud crab welfare and may inform practices that enhance the overall success of mud crab farming.

2. Materials and Methods

2.1. Experimental Animal and Rearing Conditions

The mud crabs used in the present study were obtained from Jiashun Aquatic Cooperative (Taizhou City, China). The crabs were transferred to the pilot base at Ningbo University Meishan Campus. Before the experiment, crabs were acclimatized for seven days in the canvas pools with 0.6 m^3 of water. During the acclimation, crabs were fed a commercial diet at 20:00, and 10% of the water was changed daily (8:00 a.m.). In this process, the salinity ranged from 23 to 25 ppt, dissolved oxygen (DO) was >8 mg L^{-1}, water temperature was 22–24 °C, and the total ammonia nitrogen (TAN) and NO_2-N was <0.5 mg L^{-1}. After acclimation, the 520 healthy and intact crabs were randomly selected and starved for 24 hours before the experiments.

2.2. Anesthetic Agents

Clove oil (98.5% Eugenol), ethanol (100%), magnesium chloride ($MgCl_2$), and MS-222 were purchased from Sinopharm Chemical Reagent Company Limited (Shanghai City, China). The clove oil was initially diluted in ethanol with a 1:9 ratio to prepare a stock solution [23]. In addition, magnesium chloride was prepared with distilled water to avoid the formation of precipitation [24]. MS-222 and ethanol were added directly to seawater to the desired test concentration.

2.3. Criteria and Definitions

The criteria for determining the efficacy of different anesthetics were based on the behavioral responses [25]. Detailed information about the stages of anesthesia and recovery-associated behaviors in mud crabs is listed in Table 1. The mud crabs were exposed to different concentrations (details in Section 2.4) of clove oil, MS-222, ethanol, and magnesium chloride. The details of anesthesia time, recovery time, and behavioral animal trajectory tracking system (EthoVision XT, https://www.noldus.com/ethovision-xt, accessed on 10 December 2023) analysis indicators involved in the experiment are shown in Table S1.

Table 1. Standard of anesthesia and recovery-associated behaviors.

Behaviors/Response	
Stages of anesthesia	
I	Crabs partially lose their balance and righting reflexes, and their response to stimuli persists
II	Crabs lose balance and righting reflexes, respond weakly to stimuli, do not retract their limbs, and lose their defensive behavior
III	Unconscious, immobile, unresponsive to stimuli
IV	Excessive anesthesia leads to crab death
Stages of recovery	
I	Partial regain of equilibrium, response to external stimuli
II	Crabs have defensive behavior respond quickly to stimulation, and resume feeding behavior

2.4. Experimental Design and Sampling

2.4.1. Discovery of Optimal Anesthetics

As established in previous studies, the criteria for optimal anesthesia necessitate the aquatic animals to be anesthetized within 3 min and recover within 5 min [26]. Therefore, 160 mud crabs (2.41 ± 0.65 g) were randomly selected after 24 hours of starvation. Separately, 8 mud crabs were exposed to different concentrations of clove oil (1.2, 2.4, 4.8, 7.2, 9.6 mg L^{-1}), MS-222 (1, 2, 3, 4, 5 g L^{-1}), ethanol (50, 100, 200, 300, 400 mg L^{-1}), and magnesium chloride (200, 300, 400, 500, 600 g L^{-1}). Each treatment involved 8 mud crabs, and individual observations were conducted throughout the experimental process. Specifically, the crabs were immersed in the respective anesthetic solutions, and after the anesthesia induction test, they were promptly transferred to a recovery device containing seawater without anesthetic. Anesthesia and recovery times were recorded during this process. After recovery, the mud crabs were placed in a temporary sample device, and mortality was monitored for 24 hours. Subsequently, 80 mud crabs (17.31 ± 3.98 g) were exposed to clove oil (2, 4, 6, 8, 10 mg L^{-1}) and ethanol (50, 100, 200, 300, 400 mg L^{-1}). To further screen the optimal anesthetic, each treatment was composed of 8 mud crabs with digital video cameras to compare the behavioral response.

All crabs, both males and females, were randomly selected and subjected to a 24-hour period of starvation before the commencement of the experiment. Subsequently, the crabs were captured and immersed in 1000 mL of water containing an anesthetic solution. Upon reaching anesthetic stage II, the crabs were promptly removed and transferred to 1000 mL of seawater without any anesthetic. The aquaria used for both anesthesia and recovery had dimensions of 20 cm × 20 cm × 20 cm. It is worth noting that the size and volume of the aquaria remained consistent across all experiments.

2.4.2. Effects of Optimal Anesthetic on Physiology and Metabolism of Mud Crabs and Efficacy

To assess the impact of clove oil concentration on mud crabs of varying sizes, 280 mud crabs (0.03 ± 0.01 g) were exposed to different levels of clove oil (0.2, 0.4, 0.8, 1.5, 3.0 mg L^{-1}), and mud crabs with higher body weights (7.04 ± 1.54 g, 15.10 ± 2.25 g, 27.50 ± 3.26 g, 104.83 ± 12.56 g, 159.23 ± 37.66 g, 399.73 ± 56.36 g) were exposed to clove oil concentrations of 2, 4, 6, 8, and 10 mg L^{-1}, respectively. Eight mud crabs were observed in each treatment, and anesthesia and recovery times were recorded. After recovery, the crabs were temporarily relocated to a prepared sample device, with mortality monitored for 24 hours. After confirming clove oil's anesthetic efficacy, its effects on the serum physiology, biochemistry, and antioxidant capacity of mud crab hepatopancreas were examined. Mud crabs (7.04 ± 1.54 g) were exposed to the optimal clove oil concentration (8 mg L^{-1}). In this study, the hemolymph and hepatopancreas samples were collected from crabs before anesthesia (CG), after 3 min of anesthesia with clove oil (AG), upon initial return to the normal active stage (RG), and at 3 h (R3G) and 6 h (R6G), and 6 crabs were sampled at each time point.

Hemolymph was extracted from the base of the swimmeret of mud crabs, transferred to tubes containing an anti-coagulant (heparin), and centrifuged at 4 °C to obtain serum. The serum was stored at −80 °C for the analysis of cortisol, glucose, lactate concentration, aspartate aminotransferase (AST), and alanine aminotransferase (ALT). The hepatopancreas was quickly dissected, frozen in liquid nitrogen, and stored at −80 °C for hepatic superoxide dismutase (SOD), catalase (CAT), and glutathione peroxidase (GSH-Px) activity and malondialdehyde (MDA) content analysis.

For ^{1}H NMR-based metabolomic analysis, the effects of clove oil on the metabolic phenotype of the thoracic ganglion tissue of mud crabs were investigated. Mud crabs (399.73 ± 56.36 g) were exposed to optimal concentrations of clove oil (20 mg L^{-1}). In this study, the thoracic ganglia samples were collected from crabs before anesthesia (CG), after 3 min of anesthesia with clove oil (AG), and upon returning to complete recovery (RG), and 6 crabs were sampled at each time point. The mud crabs were placed on ice, and thoracic ganglia were removed, immediately snap-frozen in liquid nitrogen, and stored at −80 °C for ^{1}H NMR analysis.

2.5. *Analysis of Serum Biochemical and Hepatic Antioxidant Capacity*

Six hemolymph samples in each replicate from each treatment were selected for physiological and biochemical analysis. Hemolymph cortisol levels were determined by an ELISA kit (Enzyme-linked Biotechnology, Shanghai, China). The assay was carried out on microplates based on the principle of competitive binding. Firstly, samples, standards, and horseradish peroxidase-labeled detection antibodies were added to the microplate in sequence. Then, the microplate was incubated at 37 °C for 60 min and thoroughly washed with detergent. The substrate for color development, tetramethylbenzidine (TMB), changed from blue to yellow, and the yellow depth of the reaction's final solution was proportional to the measured cortisol content. The absorbance at 450 nm was read within 15 min using an Absorbance Microplate Reader (Spectra Max 190, Molecular Device, Sunnyvale City, CA, USA) [27]. The glucose (A154-1-1), lactate (A019-2-1) levels and AST (C00-1-1) and ALT (C010-2-1) activities in the hemolymph were determined using commercially available kits (Nanjing Jiancheng Bioengineering Institute, Nanjing, China).

Six hepatopancreas samples in each replicate from each treatment were selected for antioxidant capacity analysis. Before antioxidant capacity analysis, samples were homogenized in ice-cold normal saline and centrifuged at 825× *g* at 4 °C for 10 min. The SOD (A001-3-2), CAT (A007-1-1), and GSH-Px (A006-2-1) activity and the MDA (A003-1-2) content were measured using commercial detection kits (Nanjing Jiancheng Bioengineering Institute, Nanjing, China). The operation steps were according to the corresponding commercial kits' instructions.

2.6. NMR-Based Metabolomic Analysis

Each of the nervous systems (about 0.1 g) from 18 crab samples were added into a 2.5 mL Eppendorf tube with crushing beads and 1 mL aqueous methanol (methanol/water = 2:1). The supernatants of each sample were collected after homogenization and centrifugation. After repeating the above extraction, the resultant two supernatants were combined and freeze-dried in a vacuum pump after removing the methanol. The extracts of each sample were then dissolved separately in 600 μL phosphate buffer. Following centrifugation, 550 μL of the supernatant from each extract was moved into a 5 mm NMR tube for NMR analysis. NMR spectroscopic analysis was described previously [28]. For each of ^{1}H NMR spectra, the region of δ 9.3–0.7 without residual water signal of δ 5.2–4.7 and methanol signal of δ 3.38–3.34 was binned with an equal width of 0.004 ppm (2.4 Hz). To compensate for overall concentration differences, all bins of NMR data were averaged and normalized to the wet weight of the corresponding sample. Before the multivariable data analysis, these binned data were averaged and normalized to the wet weight of the corresponding sample to compensate for overall concentration differences.

The normalized NMR data were imported into SIMCA-P$^+$ (version 12.0, Umetrics, Umeå, Sweden) for multivariate data analysis. The normalized NMR data were used for principal component analysis (PCA). NMR data were automatically normalized for orthogonal partial to latent structure discriminant analysis (OPLS-DA) with NMR data as X and group information as Y matrixes. The reliability of the OPLS-DA model was verified by 9-fold cross-validation [29]. The higher the Q^2 value of the model quality parameter, the more reliable the quality of the model. The reliability of the OPLS-DA model required further evaluation, with cross-validation of the analysis of variance approach, and only OPLS-DA models with $p < 0.05$ were reliable. To obtain metabolites that contribute significantly to the discrimination between the two groups, the loadings of the NMR data were back-scaled transformed and automatically normalized [30]. The correlation coefficient graph of the OPLS-DA model was drawn using MATLAB software (V 7.10). The color of each point on the correlation coefficient plot indicates the NMR data weight size, and the weight represents Pearson's product-moment correlation coefficient (r) value between the X and Y variables. The absolute value of the correlation coefficient ($|R|$) was used in this study, and each metabolite with $|r| > 0.522$ was considered statistically significant ($p < 0.05$) [31,32].

2.7. Statistical Analysis

All data were expressed as mean ± standard deviation. The data do not conform to the analysis of variance. The results were tested via the Kruskal–Wallis test using SPSS20.0 software. Differences were considered statistically significant at $p < 0.05$. Prism 8 software (version 8.01, GraphPad Software Inc., La Jolla, CA, USA) was used for data visualization.

3. Results

3.1. Anesthesia and Recovery Times for Four Anesthetics

The anesthetic effects of clove oil, MS-222, ethanol, and magnesium chloride on the mud crabs are presented in Figure 1. After exposure to clove oil, the anesthesia time for mud crabs decreased with the increasing anesthetic concentration (Figure 1A, $p < 0.05$). There was no significant difference in recovery time between anesthetic concentrations from 1.2 to 7.2 mg L^{-1} (Figure 1A, $p > 0.05$), and the recovery time decreased significantly in the 9.6 mg L^{-1} group (Figure 1A, $p < 0.05$). Moreover, the 24-hour survival rate of crabs under clove oil treatment was 100% (Figure 1B). Similarly, with the MS-222, ethanol, and magnesium chloride treatment, the anesthesia time for mud crabs decreased with the increase in anesthetic concentration (Figure 1C,E,G, $p < 0.05$). The anesthesia time for 5 g L^{-1} MS-222 was significantly shorter than that for 1 g L^{-1} and 2 g L^{-1} MS-222 (Figure 1C, $p < 0.05$). The anesthesia time for 400 mg L^{-1} ethanol was significantly shorter than that for 50 mg L^{-1} and 100 mg L^{-1} ethanol (Figure 1E, $p < 0.05$). The anesthesia time for 600 g L^{-1} magnesium chloride was significantly shorter than that for 200 g L^{-1} and 300 g L^{-1} magnesium chloride (Figure 1G, $p < 0.05$). The recovery time was not affected by the concentration of MS-222 and magnesium chloride (Figure 1C,G, $p > 0.05$). The recovery time for 400 mg L^{-1} ethanol was significantly shorter than that for 50 mg L^{-1} and 200 mg L^{-1} ethanol (Figure 1E, $p < 0.05$). The mortality was found when crabs were treated with MS-222 and magnesium chloride (Figure 1D,H); however, the 24-hour survival rate of mud crabs exposed to ethanol solutions was 100% (Figure 1F).

3.2. Effects of Clove Oil and Ethanol Concentration on the Behavior of Mud Crab

An increase in the concentration of clove oil and ethanol resulted in a significant decrease in the moving distance of mud crabs during the anesthesia stage (Figure 2A,B, $p < 0.05$). Similarly, a decrease in the moving speed of mud crabs was observed during the recovery stage (Figure 2C,D, $p < 0.05$). Notably, the moving distance and moving speed of mud crabs anesthetized by clove oil was significantly higher than that of ethanol (Figure 2). At the same time, it was found that the manic and active behavior frequency of mud crabs was more intense with the ethanol solutions (Figure 3), and the phenomenon of amputation occurring in mud crabs was observed.

3.3. Effect of Clove Oil Concentration on Mud Crab of Different Body Weight

The effect of clove oil concentrations on mud crabs of varying sizes was evaluated. The results showed that the anesthesia time for crabs decreased with the increase in anesthetic concentrations (Figure 4A–G, $p < 0.05$) with 100% 24-hour survival rates in all groups. The recovery time was not affected by body weight and the concentrations of clove oil (Figure 4A–G, $p > 0.05$). They were found that the optimal anesthetic concentrations for 0.03 ± 0.01 g, 7.04 ± 1.54 g, 15.10 ± 2.25 g, 15.10 ± 2.25 g, 104.83 ± 12.56 g, 159.23 ± 37.66 g, and 399.73 ± 56.36 g mud crabs were 3 mg L^{-1}, 8 mg L^{-1}, 6 mg L^{-1}, 4 mg L^{-1}, 6 mg L^{-1},12 mg L^{-1}, and 20 mg L^{-1}, respectively. Through regression analysis, the correlation between the optimal anesthetic concentration of clove oil depends on the weight of the crab, for crabs weighted from 0.03 to 27.50 g, the cubic function relationship was $y = 0.0036 x^3 - 0.1629 x^2 + 1.7314 x + 4.085$ ($R^2 = 0.7115$); for those weighted from 27.50 to 399.73 g, the primary function relationship was $y = 0.0437 x + 2.9461$ ($R^2 = 0.9549$) (Figure 5).

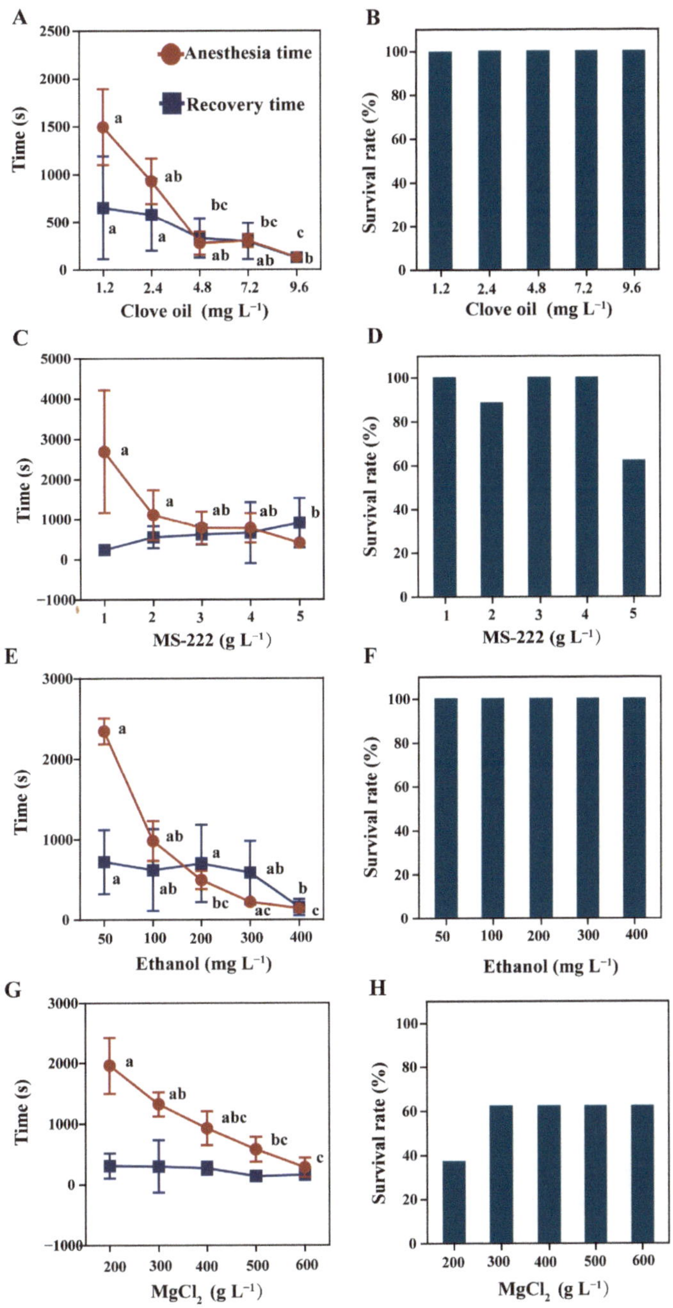

Figure 1. Anesthesia and recovery time of mud crab with clove oil (**A**), MS-222 (**C**), ethanol (**E**), and magnesium chloride ($MgCl_2$) (**G**). 24-hour survival rate after anesthesia with clove oil (**B**), MS-222 (**D**), ethanol (**F**), and magnesium chloride (**H**). Time is referred to as seconds. Data are presented as mean ± SD. There were statistically significant differences in anesthesia timelines and recovery timelines plots with different lowercase letters ($p < 0.05$, $n = 8$). The same applies below.

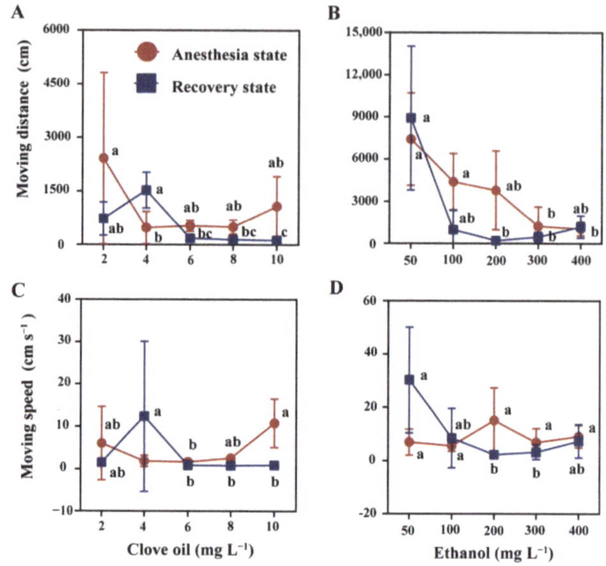

Figure 2. The behavior of the mud crab during anesthesia and recovery. Moving distance and moving speed of mud crab during anesthesia and recovery state with clove oil (**A**,**C**) and ethanol (**B**,**D**). There were statistically significant differences in anesthesia timelines and recovery timelines plots with different lowercase letters ($p < 0.05$, $n = 8$).

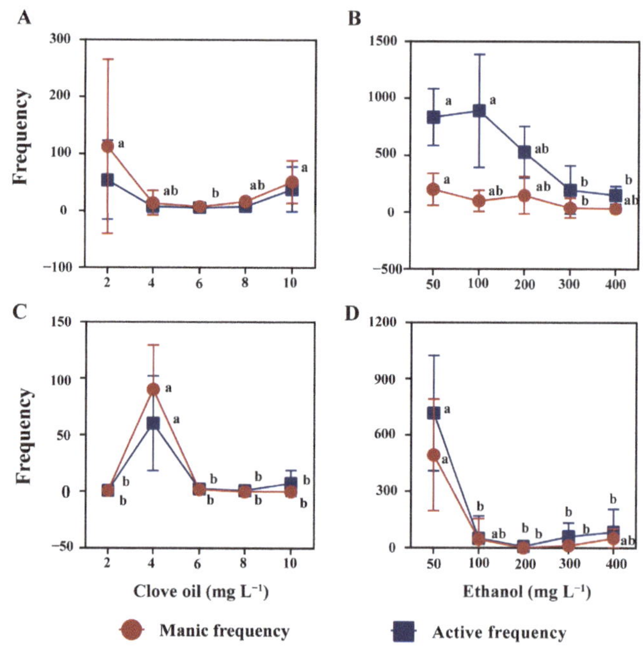

Figure 3. Manic and active behavior frequency of mud crab during anesthesia and recovery. Clove oil and ethanol anesthesia state (**A**,**B**). Clove oil and ethanol recovery state (**C**,**D**). There were statistically significant differences in anesthesia timelines and recovery timelines plots with different lowercase letters ($p < 0.05$, $n = 8$).

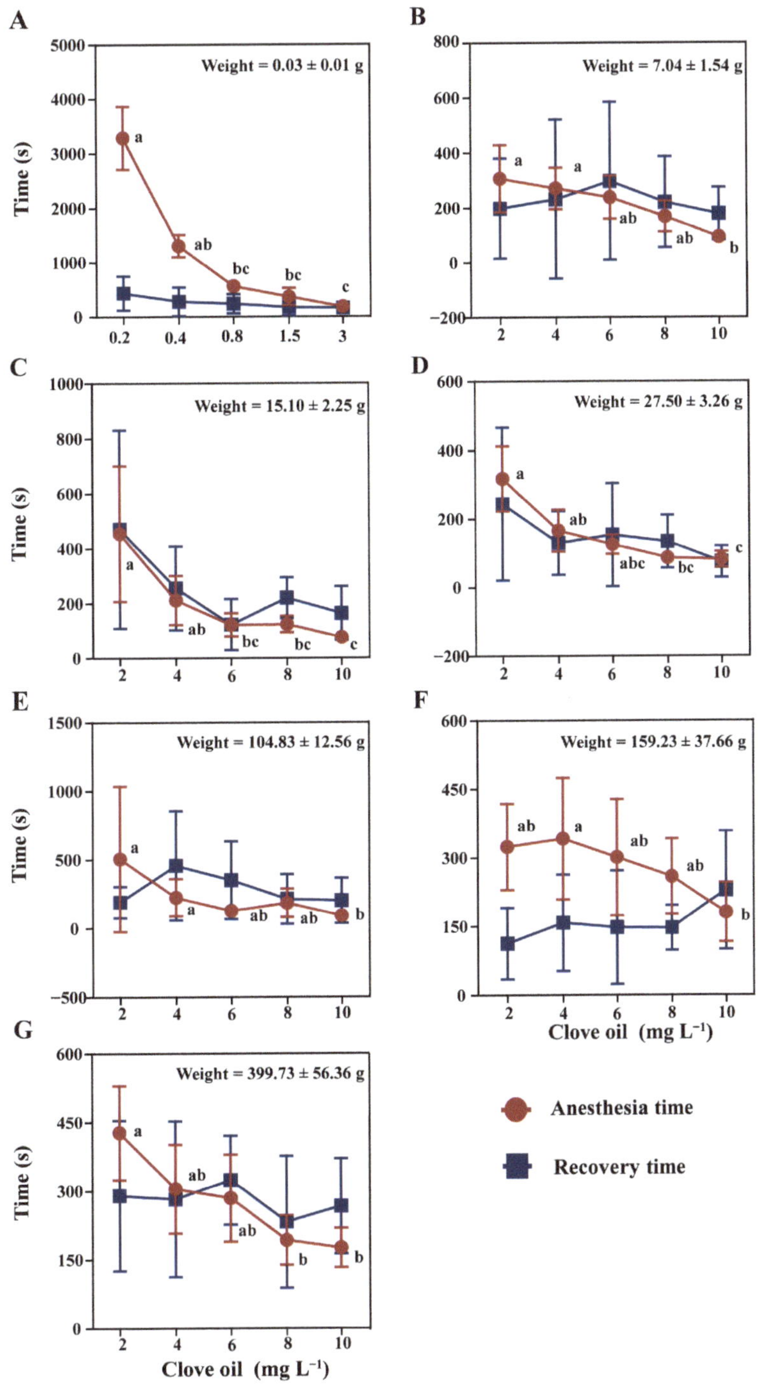

Figure 4. Anesthesia and recovery time of clove oil for mud crabs of different weights (**A–G**). There were statistically significant differences in anesthesia timelines and recovery timelines plots with different lowercase letters ($p < 0.05$, $n = 8$).

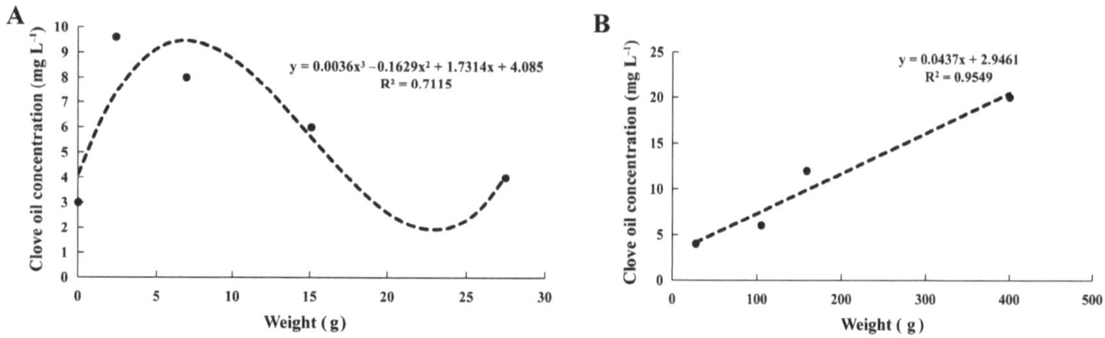

Figure 5. Cubic function relationship between 0.03–27.50 g body weight and clove oil (**A**), and 27.50–399.73 g body weight and clove oil primary function relationship (**B**). The black dots represent the weight of the crabs involved in the experiment.

3.4. Hemolymph Biochemical Parameters and Hepatopancreas Antioxidant Capabilities with Clove Oil Treatment

Under the effect of clove oil, the contents of hemolymph glucose, cortisol, and lactate of crabs were not significantly different at AG, RG, R3G, and R6G stages in each treatment group (Figure 6A–C, $p > 0.05$). There was no significant difference in serum ALT and AST activity at AG, RG, R3G, and R6G stages in each treatment (Figure 6D,E, $p > 0.05$). There were no significant changes in MDA, SOD, CAT, and GSH-Px in the hepatopancreas of mud crabs in each treatment group (Figure 7A–D, $p > 0.05$).

3.5. Changes in Metabolites in Crabs Treated with Clove Oil

The representative 600 MHz ^{1}H NMR spectra of the nervous system extracts obtained from mud crab (including the control and clove oil treatment group) were shown in Figure 8 and the ^{1}H-^{1}H correlation was provided by two-dimensional spectrum. A total of 24 metabolites were identified in the ^{1}H NMR spectra of the nervous system and were classified in Table 2. The OPLS-DA model further distinguished the metabolic changes in mud crabs treated with clove oil. The results showed that glycine content in the AG group was significantly increased compared with the CG group, and the contents of lactate and 2-pyridinemethanol decreased significantly (Figure 9A, $p < 0.05$). However, no significant changes in metabolites were observed in the RG group compared to the CG group ($p > 0.05$), thus precluding the establishment of an OPLS-DA model for both groups. The contents of the metabolites such as glutamate, alanine, aspartic acid, betaine, taurine, and 2-pyridinemethanol in the RG group were significantly higher than in the AG group (Figure 9B, $p < 0.05$). The correlation coefficients of the significantly altered metabolites are summarized in Table 3.

Table 2. NMR data of metabolites detected in the nervous system extracts.

Key	Metabolites	Moieties	δ^1 H (ppm) and Multiplicity [a]
1	Leucine	aβCH_2, γCH, δCH_3, $\delta' CH_3$	1.73(m), 1.66(m), 0.98(d), 0.96(d)
2	Lactate	αCH, βCH_3, COOH	4.13(q), 1.33(d)
3	Alanine	αCH, βCH_3, COOH	3.77(q), 1.48(d)
4	Acetate	CH_3, COOH	1.92(s)
5	Methionine	βCH_2, γCH_2, S-CH_3	2.15(m), 2.65(t), 2.14(s)

Table 2. *Cont.*

Key	Metabolites	Moieties	δ^1 H (ppm) and Multiplicity [a]
6	Glutamate	δCO, αCH, βCH_2, γCH_2, COOH	3.78(m), 2.12(m), 2.05(m), 2.35(dt)
7	Glutamine	αCH, βCH_2, γCH_2	3.77(t), 2.14 (m), 2.46(m)
8	Aspartate	αCH, βCH_2, γCOOH	3.90(dd), 2.83(dd), 2.68(dd)
9	Sarcosine	CH_2, N-CH_3, COOH	3.61(s), 2.74(s)
10	Lysine	βCH_2, γCH_2, δCH_2, εCH_2	1.92(m), 1.48(d), 1.73(m), 3.03(t)
11	Arginine	γCH_2, δCH_2, εC	1.72(m), 3.27(t)
12	Betaine	CH_3, CH_2, COO-	3.26(s), 3.90(s)
13	Taurine	CH_2SO_3, CH_2NH_2	3.26(t), 3.43(t)
14	Glycine	αCH_2, COOH	3.56(s)
15	β-Glucose	C_1H, C_2H, C_3H	4.65(d), 3.22(dd), 3.45(m)
16	α-Glucose	C_1H, C_2H, C_3H	5.24(d), 3.54(dd), 3.72(m)
17	Tyrosine	Ring$C_{2,6}H$, Ring$C_{3,5}H$	7.21(d), 6.90 (d)
18	Tryptophan	RingC_2H, RingC_3, RingC_4H, RingC_5H, RingC_6H, RingC_7H, RingC_8, RingC_9	7.33(s), 7.55(d), 7.31(m), 7.20 (m), 7.74(d)
19	Trimethylamine-N-oxide	CH_3	3.26(s)
20	Uridine	C_2H, C_3H, $C_{1'}H$	7.99(d), 5.98(d), 5.97(m)
21	Fumarate	CH, COOH	6.52(s)
22	2-Pyridinemethanol	C_2, C_3H, C_4H, C_5H, C_6H, C_7H	8.71(dd), 7.97(td), 8.55(td), 8.04(dd), 4.37(s)
23	Uracil	C_4H, C_5H, C_1, C_2	7.54(d), 5.81(d)
24	Phenylalanine	Ring $C_{2,6}H$, Ring$C_{3,5}H$, RingC_4, RingC_1	7.33(q), 7.43(t), 7.38(m)

[a] Multiplicity: s, singlet; d, doublet; t, triplet; q, quartet; m, multiplet; dd, doublet of doublets; dt, doublet of triples; td, triples of doublet.

Table 3. The correlation coefficients (r) of significantly altered metabolites in the aqueous extracts of the nervous system of mud crabs after clove oil treatment.

Metabolites	Correlation Coefficients (r)	
	AG/CG	RG/AG
Lactate	−0.70	-
Alanine	-	0.56
Glutamate	-	0.67
Aspartate	-	0.71
Betaine	-	0.64
Taurine	-	0.60
Glycine	0.55	-
2-Pyridinemethanol	−0.66	0.65

The positive and negative signs of the correlation coefficients indicate the positive and negative correlation of the content, respectively. $|r| = 0.522$ was used as the corresponding cutoff value of the correlation coefficients for the statistical significance based on the discriminant significance. "-" means the correlation coefficient $|r|$ is less than the cutoff value. CG, control group; AG, anesthesia group; RG, resuscitation group.

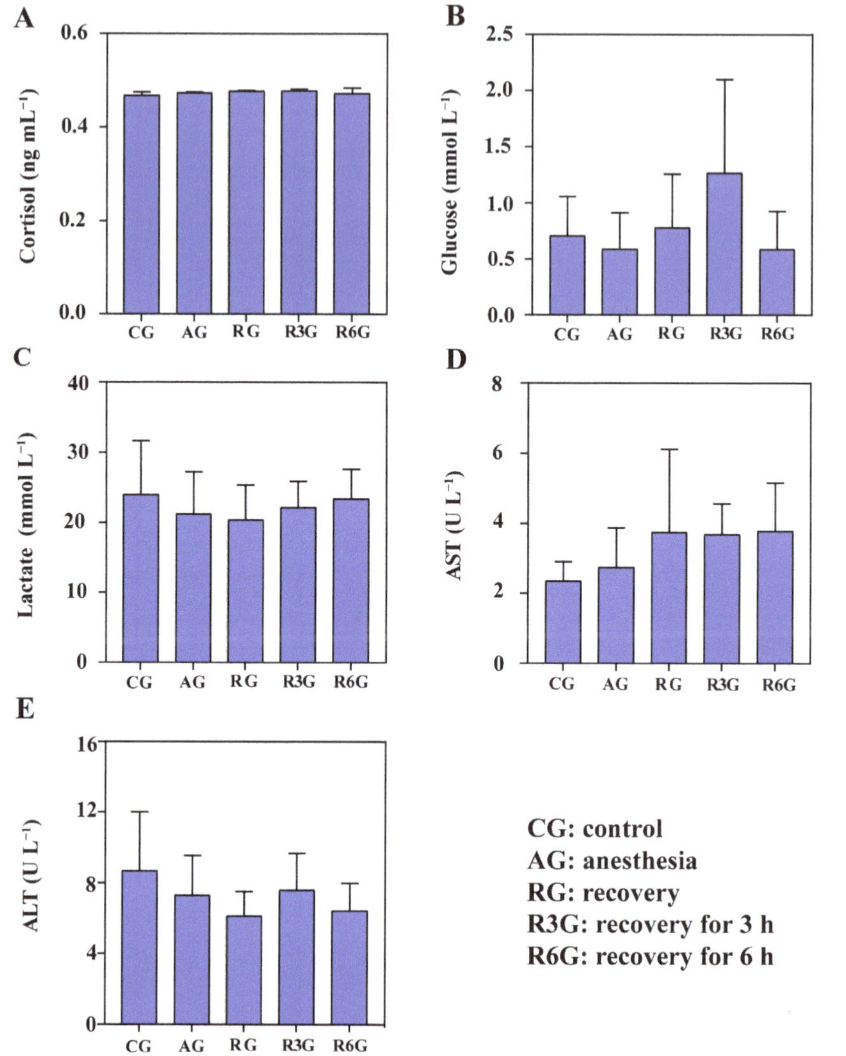

Figure 6. The changes in hemolymph cortisol (**A**), glucose (**B**), lactate (**C**), ALT activity (**D**), and AST activity (**E**) in mud crab during anesthesia (AG) and recovery (RG, R3G, R6G) with clove oil. There are statistical differences in bar charts with different superscripts ($p < 0.05$, $n = 6$). Same below.

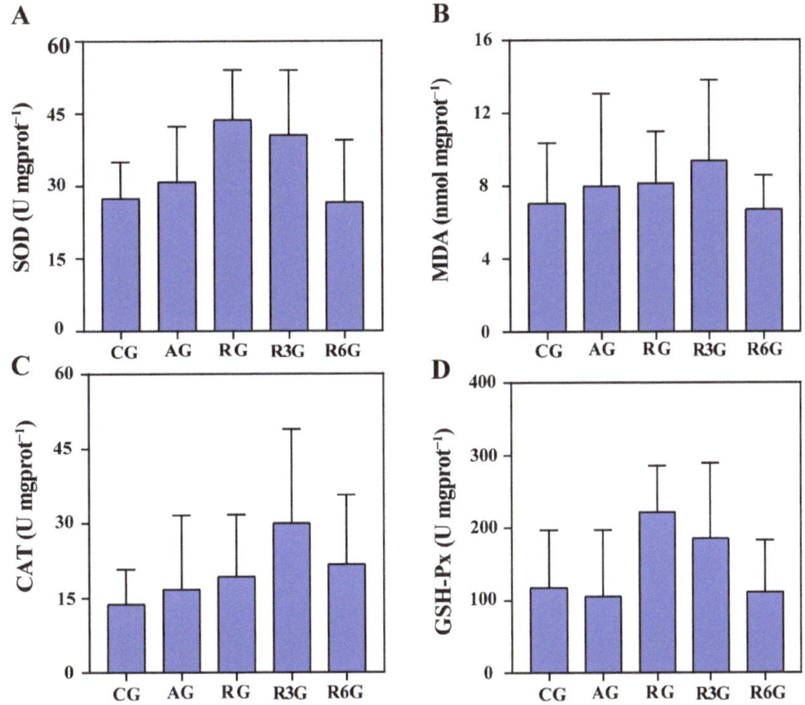

Figure 7. Hepatopancreas SOD (**A**), MDA (**B**), CAT (**C**), and GSH-Px (**D**) levels of mud crab anesthetized with clove oil.

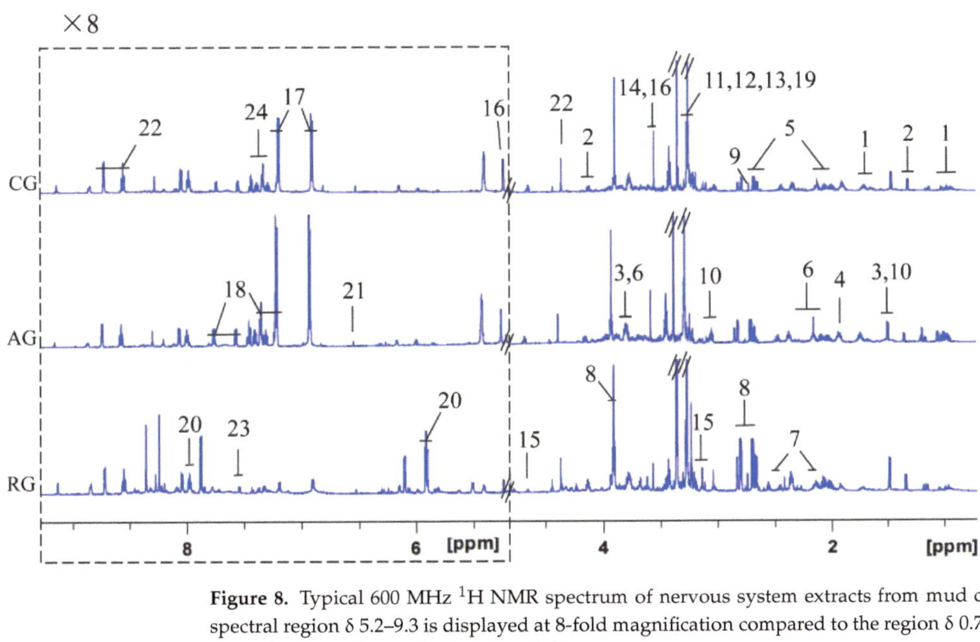

Figure 8. Typical 600 MHz ^{1}H NMR spectrum of nervous system extracts from mud crab. The spectral region δ 5.2–9.3 is displayed at 8-fold magnification compared to the region δ 0.7–4.8. CG: control group; AG: anesthesia group; RG: recovery group.

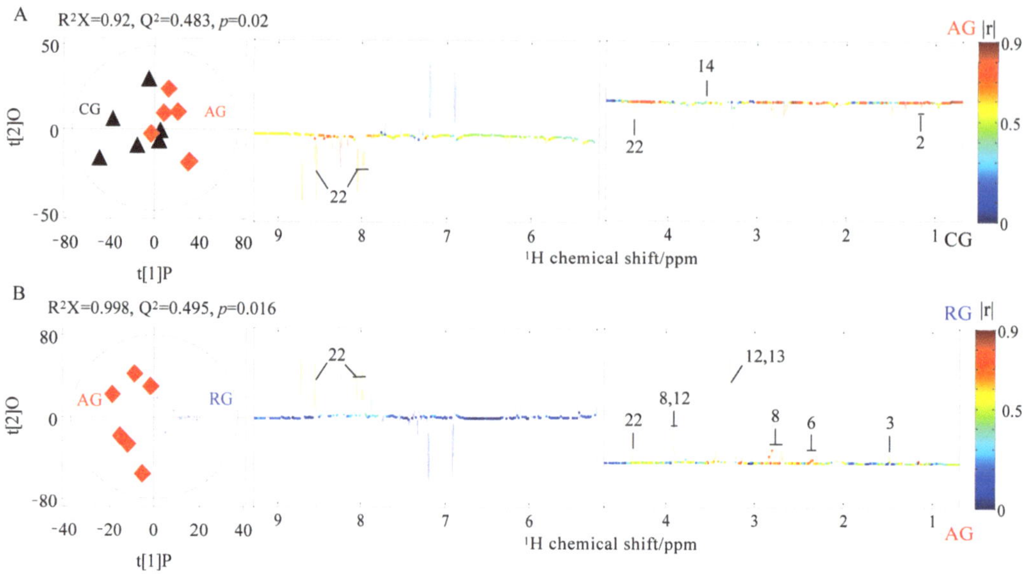

Figure 9. OPLS-DA score plot (**left**) and corresponding color-coded correlation coefficient plot (**right**) obtained by comparisons between nervous system extract spectra of mud crabs in control group CG (black triangle) and anesthesia treatment group (red oblique square) (**A**), and anesthesia treatment group and recovery group (blue snowflake) (**B**).

4. Discussion

Anesthetics are crucial in inducing sedation in aquatic organisms and reducing operational stress. However, to date, no studies have investigated the effectiveness of anesthetics in mud crabs. Currently, clove oil, MS-222, ethanol, and magnesium chloride have been employed in crustaceans such as blue crabs, crayfish, gammarids, pacific white shrimp, and European lobsters [3,8,9,25]. Nevertheless, lower survival rates were detected in crabs exposed to MS-222 and magnesium chloride compared with clove oil and ethanol in this study. Consequently, MS-222 and magnesium chloride anesthetics were excluded as effective options for mud crabs.

As a complement to physiology, behavior is a valuable tool for evaluating the adaptation of animals to an aquaculture environment [33,34]. Behavioral parameters, such as moving distance and speed, are effective indicators for evaluating stress levels. Previous studies have found that exposure to blue light induces stress in abalone, leading to increased moving speed and distance, likely as an attempt to find a hiding place [35]. For fish, a high concentration of anesthetics increases moving distance and speed due to stimulation of the skin, gills, and taste-smell receptors [36]. In the present study, we found that crabs in ethanol anesthesia had higher moving distance and speed and manic and active behavior frequency than in clove oil, accompanied by limb amputation, indicating a higher stress level. Therefore, it could be preliminarily judged that clove oil is an optimal anesthetic for mud crabs.

Mud crabs, a species that are highly tolerant to air exposure, and mud crabs with different weights could be found in the market; thus, in the present study, we further assessed the dose–response relationship between clove oil and mud crabs of different body weights. The correlation between the optimal anesthetic concentration of clove oil depends on the weight of the crab. Our results show that there is a cubic function relationship for crabs weighing less than 27.50 g and a primary function relationship for crabs weighing between 27.50 g and 399.73 g. These findings indicated that the optimal anesthetic concentration of clove oil is highly dependent on body weight.

Anesthetics are often stressors, stimulating cortisol release in aquatic animals [37,38]. Fish, during anesthesia, may experience short-term hypoxia due to inadequate ventilation, leading to increased cortisol levels [39]. Cortisol, in turn, activates gluconeogenesis and glycogenolysis, increasing blood glucose production to meet elevated energy demands. Additionally, hypoxia results in anaerobic glucose metabolism and lactate formation [40]. As a result, blood cortisol, glucose, and lactate levels are commonly used as valuable parameters in aquatic biological anesthesia studies. The findings of this study indicate that the optimal concentration of clove oil had no significant effect on hemolymph cortisol, glucose, and lactate contents in mud crabs during both the anesthesia and recovery stages. In contrast, a separate study found that the glucose of freshwater prawns (*Macrobrachium rosenbergii*) was higher when exposed to 30 min with eugenol [41]. Another study found that clove oil significantly increased the cortisol, glucose, lactate, hemoglobin, and hematocrit levels of the spotted knifejaw, but these parameters returned to near-normal levels after 6 h [42].

Alanine transaminase (ALT) and aspartate transaminase (AST) are ubiquitous transaminases released into the circulation after tissue damage and dysfunction, reflecting the degree of liver damage in aquatic animals [43]. The ALT activity of freshwater prawns was higher when exposed to eugenol for 30 min [41]. In the present study, the optimal concentration of clove oil had no significant effect on hemolymph ALT and AST activity in mud crabs during both the anesthesia and recovery stages. Therefore, the results from the present study suggest that there is no stress response in mud crabs (7.04 ± 1.54 g) under the optimal concentration of clove oil, and using clove oil as an anesthetic for mud crabs is highly safe.

The hepatopancreas, a crucial metabolic and immune organ, exhibits a high antioxidant defense capacity, and its stress resistance is intricately regulated by enzymes, hormones, and neuronal signals [42]. Superoxide dismutase (SOD), catalase (CAT), and glutathione peroxidase (GSH-Px) are key members of antioxidant enzymes, playing essential roles in balancing oxidative stress products, while malondialdehyde (MDA) is a byproduct of lipid peroxidation reactions [27]. High-concentration anesthetics, as stress sources, can induce oxidative stress in the hepatopancreas and alter the antioxidant state of aquatic organisms [44]. For example, the CAT activity in the white shrimp treated with eugenol did not differ from that of the other treatments [45]. Another study found that eugenol anesthesia in Nile tilapia can lead to significant and transient increases in GSH and CAT levels during anesthesia, which are entirely reversed during recovery. However, MDA and SOD do not show significant changes at all stages on average [38]. In our study, no significant changes were found in the activities of antioxidant enzymes and the content of MDA in the hepatopancreas of mud crabs before and after anesthesia with clove oil. Therefore, clove oil does not induce oxidative damage in mud crabs.

The mechanism of anesthetics could be determined through neurotransmitter changes [18]. Neurotransmitters, chemical messengers released by neurons, mediate intercellular communication, regulating central and peripheral physiological functions in the nervous system [46]. The central nervous system of the mud crab is composed of a brain and ventral nerve cord, which are similar to other decapods [47]. In our study, non-targeted NMR-based metabolomic analysis indicated a significant increase in inhibitory neurotransmitter glycine in the anesthesia group compared to the recovery group. Glycine binds to receptors, stimulating the transmembrane flow of chloride ions, preventing depolarization and neuronal discharge induced by excitatory neurotransmitters, and causing neuronal hyperpolarization, thereby reducing neuronal excitability [48]. This may explain why clove oil exerts its anesthetic effect on mud crabs. Additionally, glutamic acid and aspartic acid in the recovery group significantly increased [49,50]. These excitatory neurotransmitters in the central nervous system potentially contribute to the recovery of mud crabs after anesthesia. Importantly, excitatory and inhibitory neurotransmitters recovered to normal levels after anesthesia, suggesting that clove oil does not negatively affect the neurotransmitters of mud crabs.

5. Conclusions

In summary, our comprehensive research, encompassing behavioral, physiological, and metabolomic analyses, unequivocally establishes clove oil as the optimal anesthetic for mud crabs. The mechanism underlying its efficacy involves the modulation of excitatory and inhibitory neurotransmitter levels. Furthermore, our investigation reveals a clear correlation between the optimal anesthetic concentration of clove oil and the weight of mud crabs. These equations underscore the robust relationship between the anesthetic concentration and the weight of mud crabs. In conclusion, clove oil emerges as a safe and optimal anesthetic agent for mud crabs, offering valuable insights for research and practical aquaculture applications.

Supplementary Materials: The following supporting information can be downloaded at: https://www.mdpi.com/article/10.3390/antiox12122124/s1, Table S1: The definition of indicators involved in the experiment.

Author Contributions: Conceptualization, C.S.; methodology, S.Q.; data curation, Y.Y., C.W., C.M., R.L., Q.W., X.W. and Y.Z.; writing—original draft, L.Z.; writing—review and editing, S.C. and C.S.; project administration, C.S.; funding acquisition, C.S. All authors have read and agreed to the published version of the manuscript.

Funding: The research was sponsored by the Province Key Research and Development Program of Zhejiang (2021C02047), the Key Scientific and Technological Grant of Zhejiang for Breeding New Agricultural Varieties (2021C02069-6), the National Natural Science Foundation of China (Grant Nos. 32172994 and 31972783), 2025 Technological Innovation for Ningbo (2019B10010), the Agriculture Research System of China of MOF and MARA (Agriculture Research System of China of Ministry of Finance and Ministry of Agriculture and Rural Affairs), the K.C. Wong. Magna Fund of Ningbo University, and the Scientific Research Foundation of the Graduate School of Ningbo University (IF2022152).

Institutional Review Board Statement: This study was approved by the Ethics Committee of Ningbo University (no. 20190410-042) and conducted according to relevant national and international guidelines.

Informed Consent Statement: Not applicable.

Data Availability Statement: The data presented in this study are available in review.

Conflicts of Interest: The authors declare no conflict of interest.

References

1. Tarkhani, R.; Imani, A.; Jamali, H.; Moghanlou, K.S. Anaesthetic efficacy of eugenol on Flowerhorn (*Amphilophus labiatus* × *Amphilophus trimaculatus*). *Aquac. Res.* **2017**, *48*, 3207–3215. [CrossRef]
2. Roth, B.; Skra, T. Pre mortem capturing stress of Atlantic herring (*Clupea harengus*) in purse seine and subsequent effect on welfare and flesh quality. *Fish Res.* **2021**, *244*, 106124. [CrossRef]
3. Rotllant, G.; Llonch, P.; Garcia del Arco, J.A.; Chic, O.; Flecknell, P.; Sneddon, L.U. Methods to induce analgesia and anesthesia in crustaceans: A supportive decision tool. *Biology* **2023**, *12*, 387. [CrossRef] [PubMed]
4. Purbosari, N.; Warsiki, E.; Syamsu, K.; Santoso, J. Natural versus synthetic anesthetic for transport of live fish: A review. *Aquac. Fish.* **2019**, *4*, 129–133. [CrossRef]
5. Spandana, U.; Pallepati, L.; Gandrapu, P.; Sowjanya, S. Short and spicy evergreen tree: Clove. *World J. Pharm. Sci.* **2018**, *6*, 122–128.
6. Spoors, F.; James, M.A.; Mendo, T.; McKnight, J.C.; Bonnelycke, E.M.S.; Khan, N. Investigating clove oil and its derivatives as anaesthetic agents for decapod crustaceans to improve welfare commercially and at slaughter. *Front. Anim. Sci.* **2023**, *4*, 1180977. [CrossRef]
7. Topic Popovic, N.; Strunjak-Perovic, I.; Coz-Rakovac, R.; Barisic, J.; Jadan, M.; Persin Berakovic, A.; Sauerborn Klobucar, R. Tricaine methane-sulfonate (MS-222) application in fish anaesthesia. *J. Appl. Ichthyol.* **2012**, *28*, 553–564. [CrossRef]
8. Stanley, C.E.; Adams, R.; Nadolski, J.; Amrit, E.; Barrett, M.; Bohnett, C.; Cooper, R.L. The effects of tricaine mesylate on arthropods: Crayfish, crab and drosophila. *Invertebr. Neurosci.* **2020**, *20*, 10. [CrossRef]
9. Perrot-Minnot, M.J.; Balourdet, A.; Musset, O. Optimization of anesthetic procedure in crustaceans: Evidence for sedative and analgesic-like effect of MS-222 using a semi-automated device for exposure to noxious stimulus. *Aquat. Toxicol.* **2021**, *240*, 105981. [CrossRef]

10. Yang, H.; Qiu, X.; Zhao, Y.; Ye, Y.; Song, W.; Mu, C.; Li, R.; Wang, C. The influence of anesthetics on neurotransmitters in the nervous system of *Sepia pharaonis*. *Aquaculture* **2021**, *535*, 736368. [CrossRef]
11. Becker, A.J.; Ramos, P.B.; Monserrat, J.M.; Wasielesky, W., Jr.; Baldisserotto, B. Behavioural and biochemical responses in adult Pacific white shrimp, *Litopenaeus vannamei*, exposed to the essential oil of *Cymbopogon citratus*. *Aquac. Res.* **2021**, *52*, 6205–6217. [CrossRef]
12. Palomera, M.A.A.; Zaragoza, O.B.D.R.; Galvan, S.; Vega-Villasante, F. Evaluation of natural extracts with anesthetic properties in juveniles *Macrobrachium tenellum*. *Pan-Am. J. Aquat. Sci.* **2016**, *11*, 251–257.
13. Fregin, T.; Bickmeyer, U. Electrophysiological investigation of different methods of anesthesia in lobster and crayfish. *PLoS ONE* **2016**, *11*, e0162894. [CrossRef] [PubMed]
14. Roohi, Z.; Imanpoor, M.R. The efficacy of the oils of spearmint and methyl salicylate as new anesthetics and their effect on glucose levels in common carp (*Cyprinus carpio* L., 1758) juveniles. *Aquaculture* **2015**, *437*, 327–332. [CrossRef]
15. Arafa, S.; Sadok, S.; Abed, A.E. Assessment of magnesium chloride as an anaesthetic for adult sea urchins (*Paracentrotus lividus*): Incidence on mortality and spawning. *Aquac. Res.* **2007**, *38*, 1673–1678. [CrossRef]
16. Hajek, G.J.; Choczewski, M.; Dziaman, R.; Kłyszejko, B. Evaluation of immobilizing methods for the Chinese mitten crab, *Eriocheir sinensis* (Milne-Edwards). *Electr. J. Pol. Agric. Univ.* **2009**, *12*, 18.
17. Cowing, D.; Powell, A.; Johnson, M. Evaluation of different concentration doses of eugenol on the behaviour of *Nephrops norvegicus*. *Aquaculture* **2015**, *442*, 78–85. [CrossRef]
18. Yang, H.; Zhao, Y.; Song, W.; Ye, Y.; Wang, C.; Mu, C.; Li, R. Evaluation of the efficacy of potential anesthetic agents on cuttlefish (*Sepia pharaonis*) juveniles. *Aquacult. Rep.* **2020**, *18*, 100524. [CrossRef]
19. Gao, X.; Pang, G.; Luo, X.; You, W.; Ke, C. Effects of light cycle on motion behaviour and melatonin secretion in *Haliotis discus hannai*. *Aquaculture* **2021**, *532*, 735981. [CrossRef]
20. Tremaroli, V.; Workentine, M.L.; Weljie, A.M.; Vogel, H.J.; Ceri, H.; Viti, C.; Tatti, E.; Zhang, P.; Hynes, A.P.; Turner, R.J.; et al. Metabolomic investigation of the bacterial response to a metal challenge. *Appl. Environ. Microbiol.* **2009**, *75*, 719–728. [CrossRef]
21. Lin, W.; He, Y.; Li, R.; Mu, C.; Wang, C.; Shi, C.; Ye, Y. Adaptive changes of swimming crab (*Portunus trituberculatus*) associated bacteria helping host against dibutyl phthalate toxification. *Environ. Pollut.* **2023**, *324*, 121328. [CrossRef] [PubMed]
22. Tang, L.; Wang, H.; Wei, H.; Ye, C.; Chen, L.; Yao, H.; Shi, C.; Mu, C.; Wang, C. Overwintering behaviour affects distinctive flavour of mud crab *Scylla paramamosain* in commercial farms. *Aquac. Res.* **2020**, *51*, 29–40. [CrossRef]
23. Anderson, W.G.; Mckinley, R.S.; Colavecchia, M. The use of clove oil as an anesthetic for rainbow trout and its effects on swimming performance. *N. Am. J. Fish Manag.* **1997**, *17*, 301–307. [CrossRef]
24. Abbo, L.A.; Himebaugh, N.E.; DeMelo, L.M.; Hanlon, R.T.; Crook, R.J. Anesthetic efficacy of magnesium chloride and ethyl alcohol in temperate octopus and cuttlefish species. *J. Am. Assoc. Lab. Anim. Sci.* **2021**, *60*, 556–567. [CrossRef] [PubMed]
25. de Souza Valente, C. Anaesthesia of decapod crustaceans. *Vet. Anim. Sci.* **2022**, *16*, 100252. [CrossRef] [PubMed]
26. Marking, L.L.; Meyer, F.P. Are better anesthetics needed in fisheries? *Fisheries* **1985**, *10*, 2–5. [CrossRef]
27. Chen, S.; Wu, X.; Ren, Z.; Mu, C.; Song, W.; Li, R.; Liu, L.; Ye, Y.; Shi, C.; Wang, C.; et al. Effects of dietary supplementation recombined PtALF8 protein (rPtALF8) on the growth performance, antioxidant capacity and gut microbial composition in swimming crab, *Portunus trituberculatus*. *Aquaculture* **2021**, *537*, 736456. [CrossRef]
28. Ye, Y.; An, Y.; Li, R.; Mu, C.; Wang, C. Strategy of metabolic phenotype modulation in *Portunus trituberculatus* exposed to low salinity. *J. Agric. Food Chem.* **2014**, *62*, 3496–3503. [CrossRef]
29. Eriksson, L.; Trygg, J.; Wold, S. CV-ANOVA for significance testing of PLS and OPLS® models. *J. Chemometr.* **2008**, *22*, 594–600. [CrossRef]
30. Cloarec, O.; Dumas, M.E.; Trygg, J.; Craig, A.; Barton, R.H.; Lindon, J.C.; Nicholson, J.K.; Holmes, E. Evaluation of the orthogonal projection on latent structure model limitations caused by chemical shift variability and improved visualization of biomarker changes in ^{1}H NMR spectroscopic metabonomic studies. *Anal. Chem.* **2005**, *77*, 517–526. [CrossRef]
31. Shi, C.; Zeng, T.; Li, R.; Wang, C.; Ye, Y.; Mu, C. Dynamic metabolite alterations of *Portunus trituberculatus* during larval development. *J. Oceanol. Limnol.* **2019**, *37*, 361–372. [CrossRef]
32. Shi, C.; Ye, Y.; Pei, F.; Mu, C.; Wang, C. Survival and Metabolic Modulation of Swimming Crab *Portunus trituberculatus* During Live Transport. *Front. Mar. Sci.* **2021**, *8*, 724156. [CrossRef]
33. Claireaux, G.; McKenzie, D.J.; Genge, A.G.; Chatelier, A.; Aubin, J.; Farrell, A.P. Linking swimming performance, cardiac pumping ability and cardiac anatomy in rainbow trout. *J. Exp. Biol.* **2005**, *208*, 1775–1784. [CrossRef] [PubMed]
34. Millot, S.; Pean, S.; Leguay, D.; Vergnet, A.; Chatain, B.; Begout, M.L. Evaluation of behavioral changes induced by a first step of domestication or selection for growth in the European sea bass (*Dicentrarchus labrax*): A self-feeding approach under repeated acute stress. *Aquaculture* **2010**, *306*, 211–217. [CrossRef]
35. Zhang, M.; Gao, X.; Lyu, M.; Lin, S.; Luo, X.; Ke, C.; You, W. Behavioral and physiological responses of *Haliotis discus hannai* to different spectral compositions. *Aquaculture* **2022**, *555*, 738228. [CrossRef]
36. Aydin, B.; Orhan, N. Effects of thymol and carvacrol anesthesia on the electrocardiographic and behavioral responses of the doctor fish *Garra rufa*. *Aquaculture* **2021**, *533*, 736134. [CrossRef]
37. Boaventura, T.P.; Souza, C.F.; Ferreira, A.L.; Favero, G.C.; Baldissera, M.D.; Heinzmann, B.M.; Baldisserotto, B.; Luz, R.K. Essential oil of *Ocimum gratissimum* (Linnaeus, 1753) as anesthetic for *Lophiosilurus alexandri*: Induction, recovery, hematology, biochemistry and oxidative stress. *Aquaculture* **2020**, *529*, 735676. [CrossRef]

38. Zahran, E.; Risha, E.; Rizk, A. Comparison propofol and eugenol anesthetics efficacy and effects on general health in Nile Tilapia. *Aquaculture* **2021**, *534*, 736251. [CrossRef]
39. Sneddon, L.U. Clinical anesthesia and analgesia in fish. *J. Exot. Pet Med.* **2012**, *21*, 32–43. [CrossRef]
40. Van Ginneken, V.; Boot, R.; Murk, T.; van den Thillart, G.; Balm, P. Blood plasma substrates and muscle lactic-acid response after exhaustive exercise in common carp and trout: Indications for a limited lactate-shuttle. *Anim. Biol.* **2004**, *54*, 119–130. [CrossRef]
41. Saydmohammed, M.; Pal, A.K. Anesthetic effect of eugenol and menthol on handling stress in *Macrobrachium rosenbergii*. *Aquaculture* **2009**, *298*, 162–167. [CrossRef]
42. Jia, Y.; Xie, T.; Gao, Y.; Qin, H.; Guan, C. Anesthetics efficacy and physiological response of MS222 and clove oil in spotted knifejaw *Oplegnathus punctatus*. *Aquacult. Rep.* **2022**, *25*, 101201. [CrossRef]
43. Yousefi, M.; Farsani, M.N.; Ghafarifarsani, H.; Hoseinifar, S.H.; Van Doan, H. The effects of dietary supplementation of mistletoe (*Viscum album*) extract on the growth performance, antioxidant, and innate, immune responses of rainbow trout (*Oncorhynchus mykiss*). *Aquaculture* **2021**, *536*, 736385. [CrossRef]
44. Zahl, I.H.; Samuelsen, O.; Kiessling, A. Anaesthesia of farmed fish: Implications for welfare. *Fish Physiol. Biochem.* **2012**, *38*, 201–218. [CrossRef] [PubMed]
45. Parodi, T.V.; Cunha, M.A.; Heldwein, C.G.; de Souza, D.M.; Martins, A.C. The anesthetic efficacy of eugenol and the essential oils of *Lippia alba* and *Aloysia triphylla* in post-larvae and sub-adults of *Litopenaeus vannamei* (Crustacea, Penaeidae). *Comp. Biochem. Physiol. C Toxicol. Pharmacol.* **2012**, *155*, 462–468. [CrossRef] [PubMed]
46. Stepniewski, T.M.; Mancini, A.; Ågren, R.; Torrens-Fontanals, M.; Semache, M.; Bouvier, M.; Sahlholm, K.; Breton, B.; Selent, J. Mechanistic insights into dopaminergic and serotonergic neurotransmission–concerted interactions with helices 5 and 6 drive the functional outcome. *Chem. Sci.* **2021**, *12*, 10990–11003. [CrossRef]
47. Kornthong, N.; Tinikul, Y.; Khornchatri, K.; Saeton, J.; Magerd, S.; Suwansa-Ard, S.; Kruangkum, T.; Hanna, P.J.; Sobhon, P. Neuronal classification and distribution in the central nervous system of the female mud crab, *Scylla olivacea*. *Microsc. Res. Tech.* **2014**, *77*, 189–200. [CrossRef]
48. Hernandes, M.S.; Troncone, L.R. Glycine as a neurotransmitter in the forebrain: A short review. *J. Neural Transm.* **2009**, *116*, 1551–1560. [CrossRef]
49. Monaghan, D.T.; Bridges, R.J.; Cotman, C.W. The excitatory amino acid receptors: Their classes, pharmacology, and distinct properties in the function of the central nervous system. *Annu. Rev. Pharmacol. Toxicol.* **1989**, *29*, 365–402. [CrossRef]
50. Bradford, S.E.; Nadler, J.V. Aspartate release from rat hippocampal synaptosomes. *Neuroscience* **2004**, *128*, 751–765. [CrossRef]

MDPI AG
Grosspeteranlage 5
4052 Basel
Switzerland
Tel.: +41 61 683 77 34

Antioxidants Editorial Office
E-mail: antioxidants@mdpi.com
www.mdpi.com/journal/antioxidants

www.ingramcontent.com/pod-product-compliance
Lightning Source LLC
Chambersburg PA
CBHW041407210926
51850CB00013B/3202